The Essential Tension

Thomas S. Kuhn

The Essential Tension

Selected Studies in Scientific Tradition and Change

The University of Chicago Press
Chicago and London

The University of Chicago Press, Chicago 60637
The University of Chicago Press, Ltd., London

Printed in the United States of America

03 02 01 00 99 98 97 96 95 94 9 10 11 12 13

Library of Congress Cataloging in Publication Data

Kuhn, Thomas S.
The essential tension.

Includes bibliographical references and index.
1. Science—Philosophy—Collected works. 2. Science—History—Collected works. I. Title.
Q175.K954 501 77-78069
ISBN 0-226-45806-7 (paper)

∞ The paper used in this publication meets the minimum requirements of the American National Standard for Information Sciences—Permanence of Paper for Printed Library Materials, ANSI Z39.48-1984.

For K. M. K.,
still my favorite eschatologist

Contents

Preface

Though I had played for some years with the idea of publishing a volume of selected papers, the project might never have become actual if Suhrkamp Verlag of Frankfurt had not asked permission to assemble some essays of mine in a volume of German translations. I had reservations both about their initial list and about authorizing translations I could not altogether control. But my doubts vanished entirely when an attractive German visitor, who has since become a friend, agreed to take editorial responsibility for a redesigned German volume. He is Lorenz Krüger, professor of philosophy at the University of Bielefeld, and the two of us worked closely and harmoniously together on the selection and arrangement of the volume's contents. It was he, in addition, who persuaded me to prepare a special preface, indicating the relation between the essays and my better-known work, whether as preparation for it or as development and correction. Such a preface should, he urged, be designed to help readers better understand some central but apparently obscure aspects of my view of scientific development. Since the present book is very nearly a version in the original English of the German volume published under his supervision,[1] I owe him a very special debt.

1. *Die Entstehung des Neuen: Studien zur Struktur der Wissenschaftsgeschichte* (Frankfurt, 1977). That volume includes a Foreword by Professor Krüger. In the transition to the English edition, I have eliminated

Inevitably, the effort required by the sort of preface Krüger envisaged is autobiographical, and my exertions have sometimes induced the sense that my past intellectual life was passing before my eyes. Nevertheless, the contents of this volume do not, in one central respect, match the autobiographical *aperçus* that my return to them has stimulated. *The Structure of Scientific Revolutions* did not appear until late in 1962, but the conviction that some such book needed to be written had come to me fifteen years before, while I was a student of physics at work on my doctoral dissertation. Shortly afterward I abandoned science for its history, and my published research was then for some years straightforwardly historical, usually taking narrative form. Originally I had planned to reprint some of those early essays here, hoping thereby to supply the autobiographical ingredient now lacking—some indication of the decisive role of historical practice in the development of my views. But experimenting with alternative tables of contents gradually persuaded me that historical narratives would fail to make the points I had in mind and might even prove significantly misleading. Though experience as a historian can teach philosophy by example, the lessons vanish from finished historical writing. An account of the episode that first led me to history may suggest what is involved, simultaneously supplying a useful base from which to consider the essays that follow.

A finished historical narrative consists largely of facts about the past, most of them apparently indisputable. Many readers therefore assume that the historian's primary task is to examine texts, extract the relevant facts from them, and recount those facts with literary grace in approximate chronological order. During my years as a physicist, that was my view of the historian's discipline, which I did not then take very seriously. When I changed my mind (and very shortly my craft), the historical narratives I produced were, by their nature, likely sources of the same misunderstanding. In history, more than in any other discipline I know, the finished product of research disguises the nature of the work that produced it.

and replaced a few parts of the Preface directed to a German audience. In addition I have somewhat tightened and polished the previously unpublished essays, "The Relations between the History and the Philosophy of Science" and "Objectivity, Value Judgment, and Theory Choice." The former now also has a new conclusion, one I could probably not have prepared in this form before reading the book cited in note 7 below.

My own enlightenment began in 1947, when I was asked to interrupt my current physics project for a time in order to prepare a set of lectures on the origins of seventeenth-century mechanics. For that purpose, I needed first to discover what the predecessors of Galileo and Newton had known about the subject, and preliminary inquiries soon led me to the discussions of motion in Aristotle's *Physica* and to some later works descended from it. Like most earlier historians of science, I approached these texts knowing what Newtonian physics and mechanics were. Like them, too, I asked of my texts the questions: How much about mechanics was known within the Aristotelian tradition, and how much was left for seventeenth-century scientists to discover? Being posed in a Newtonian vocabulary, those questions demanded answers in the same terms, and the answers then were clear. Even at the apparently descriptive level, the Aristotelians had known little of mechanics; much of what they had had to say about it was simply wrong. No such tradition could have provided a foundation for the work of Galileo and his contemporaries. They necessarily rejected it and began the study of mechanics over again.

Generalizations of that sort were widely current and apparently inescapable. But they were also puzzling. When dealing with subjects other than physics, Aristotle had been an acute and naturalistic observer. In such fields as biology or political behavior, his interpretations of phenomena had often been, in addition, both penetrating and deep. How could his characteristic talents have failed him so when applied to motion? How could he have said about it so many apparently absurd things? And, above all, why had his views been taken so seriously for so long a time by so many of his successors? The more I read, the more puzzled I became. Aristotle could, of course, have been wrong—I had no doubt that he was—but was it conceivable that his errors had been so blatant?

One memorable (and very hot) summer day those perplexities suddenly vanished. I all at once perceived the connected rudiments of an alternate way of reading the texts with which I had been struggling. For the first time I gave due weight to the fact that Aristotle's subject was change-of-quality in general, including both the fall of a stone and the growth of a child to adulthood. In his physics, the subject that was to become mechanics was at best a still-not-quite-isolable special case. More consequential was my recognition that the permanent ingredients of Aristotle's universe, its ontologically primary and indestructible elements, were not material bodies but rather the qualities which, when imposed on some portion of omnipresent neutral matter, constituted an individual material body or substance. Position itself was, however, a quality in Aristotle's physics, and a body that changed its position therefore remained the same body only in the problematic sense that the child is the individual it becomes. In a universe where qualities were primary, motion was necessarily a change-of-state rather than a state.

Though drastically incomplete and far too baldly stated, those aspects of my new understanding of Aristotle's enterprise should indicate what I mean by the discovery of a new way to read a set of texts. After I achieved this one, strained metaphors often became naturalistic reports, and much apparent absurdity vanished. I did not become an Aristotelian physicist as a result, but I had to some extent learned to think like one. Thereafter I had few problems understanding why Aristotle had said what he did about motion or why his statements had been taken so seriously. I still recognized difficulties in his physics, but they were not blatant and few of them could properly be characterized as mere mistakes.

Since that decisive episode in the summer of 1947, the search for best, or best-accessible, readings has been central to my historical research (and has also been systematically eliminated from the narratives that report its results). Lessons learned while reading Aristotle have also informed my readings of men like Boyle and Newton, Lavoisier and Dalton, or Boltzmann and Planck. Briefly stated, those lessons are two. First, there are many ways to read a text, and the ones most accessible to a modern are often inappropriate when applied to the past. Second, that plasticity of texts does not place all ways of reading on a par, for some of them (ultimately, one hopes, only one) possess a plausibility and coherence absent from others. Trying to transmit such lessons to students, I offer them a maxim: When reading the works of an important thinker, look first for the apparent absurdities in the text and ask yourself how a sensible person could have written them. When you find an answer, I continue, when those passages make sense, then you may find that more central passages, ones you previously thought you understood, have changed their meaning.[2]

If this volume were addressed primarily to historians, that autobiographical fragment would not be worth recording. What I as a physicist had to discover for myself, most historians learn by example in the course of professional training. Consciously or not, they are all practitioners of the hermeneutic method. In my case, however, the discovery of hermeneutics did more than make history seem consequential. Its most immediate and decisive effect was instead on my view of science. That is the aspect of my encounter with Aristotle that has led to my recounting it here.

Men like Galileo and Descartes, who laid the foundation for seventeenth-century mechanics, were raised within the Aristotelian scientific tradition, and it made essential contributions to their achievement. Nevertheless, a key ingredient of that achievement was their creation of the way of reading texts that had initially so misled me, and they often participated in such misreadings themselves. Descartes, for example, early in *Le monde*, ridicules Aristotle by quoting his definition of motion in Latin, declining to translate on the ground that the definition makes equally little sense in French, and then proving his point by producing the missing translation. Aristotle's definition had, however, made sense for centuries before, probably at one time to Descartes himself. What my reading of Aristotle seemed therefore to disclose was a global sort of change in the way men viewed nature and applied language to it, one that could not properly be described as constituted by additions to knowledge or by the mere piecemeal correction of mistakes. That sort of change was shortly to be described by Herbert Butterfield as "putting on a different kind of thinking-cap,"[3] and puzzlement about it quickly led me to books on Gestalt psychology and related fields. While discovering history, I had discovered my first scientific revolution, and my subsequent search for best readings has often been a search for other episodes of the same sort. They are the ones that can be recognized and understood only by recapturing out-of-date ways of reading out-of-date texts.

2. More on this subject will be found in T. S. Kuhn, "Notes on Lakatos," *Boston Studies in Philosophy of Science* 8 (1971): 137–46.

3. Herbert Butterfield, *Origins of Modern Science, 1300–1800* (London, 1949), p. 1. Like my own understanding of the transformation of early modern science, Butterfield's was greatly influenced by the writings of Alexandre Koyré, especially his *Etudes galiléennes* (Paris, 1939).

Because one of its central concerns is the nature and the relevance to philosophy of the historian's craft, a lecture entitled "The Relations between the History and the Philosophy of Science" is the first of the essays reprinted below. Delivered in the spring of 1968, it has not previously appeared in print, for I had always intended first to extend its closing remarks on what philosophers might gain by taking history more seriously. For present purposes, however, that deficiency may be remedied by other essays in this volume, and the lecture itself can be read as an effort to deal in somewhat greater depth with the issues already introduced in this preface. Knowledgeable readers may think it dated, which in one sense it is. In the almost nine years since its presentation many more philosophers of science have conceded the relevance of history to their concerns. But, though the interest in history that has resulted is welcome, it has so far largely missed what I take to be the central philosophical point: the fundamental conceptual readjustment required of the historian to recapture the past or, conversely, of the past to develop toward the present.

Three of the five remaining essays in part 1 require no more than passing mention. The paper "Concepts of Cause in the Development of Physics" is clearly a by-product of the exposure to Aristotle described above. If that exposure had not taught me the integrity if his quadripartite analysis of causes, I might never have recognized the manner in which the seventeenth-century rejection of formal causes in favor of mechanical or efficient ones has constrained subsequent discussions of scientific explanation. The fourth essay, which deals with energy conservation, is the only one in part 1 written before my book on scientific revolutions, and my few remarks about it are interspersed below with those on other papers from the same period. The sixth, "The Relations between History and the History of Science," is in some sense a companion piece to the paper with which part 1 opens. A number of historians have thought it unfair, and it is doubtless both personal and polemical. But since its publication I have discovered that the frustrations it expresses are almost universally shared by those whose primary concern is with the development of scientific ideas.

Though written for other purposes, the essays "The History of Science" and "Mathematical versus Experimental Traditions" have a more direct relevance to themes developed in my *Structure of Scientific Revolutions*. The opening pages of the former may, for example, help to explain why the approach to history on which the book depends began to be applied to the sciences only after the first third of this century. Simultaneously, they may suggest a revealing oddity: the early models of the sort of history that has so influenced me and my *historical* colleagues is the product of a post-Kantian European tradition which I and my *philosophical* colleagues continue to find opaque. In my own case, for example, even the term "hermeneutic," to which I resorted briefly above, was no part of my vocabulary as recently as five years ago. Increasingly, I suspect that anyone who believes that history may have deep philosophical import will have to learn to bridge the longstanding divide between the Continental and English-language philosophical traditions.

In its penultimate section, "The History of Science" also provides the beginning of an answer to a line of criticism persistently directed to my book. Both general historians and historians of science have sometimes complained that my account of scientific development is too exclusively based on factors internal to the sciences themselves; that I fail to locate scientific communities in the society which supports them and from which their members are drawn; and that I therefore appear to believe that scientific development is immune to the influences of the social, economic, religious, and philosophical environment in which it occurs. Clearly my book has little to say about such external influences, but it ought not be read as denying their existence. On the contrary, it can be understood as an attempt to explain why the evolution of the more highly developed sciences is more fully, though by no means completely, insulated from its social milieu than that of such disciplines as engineering, medicine, law, and the arts (excepting, perhaps, music). Furthermore, if read in that way, the book may supply some preliminary tools to those who aim to explore the ways in which and the avenues through which external influences are made manifest.

Evidence for the existence of such influences will be found in other papers reprinted below, especially in "Energy Conservation" and "Mathematical versus Experimental Traditions." But the special relevance of the latter to my book on scientific revolutions is of another sort. It underscores the existence of a significant mistake in my earlier presentation and simultaneously suggests ways in which the error may ultimately be eliminated. Throughout *The*

Structure of Scientific Revolutions I identify and differentiate scientific communities by subject matter, implying, for example, that such terms as "physical optics," "electricity," and "heat" can serve to designate individual scientific communities just because they also designate subject matters for research. Once pointed out, the anachronism is obvious. I would now insist that scientific communities must be discovered by examining patterns of education and communication before asking which particular research problems engage each group. The effect of that approach on the concept of paradigms is indicated in the sixth of the essays in part 2 and is elaborated with respect to other aspects of my book in the extra chapter added to its second edition. The essay "Mathematical versus Experimental Traditions" exhibits the same approach applied to some longstanding historical controversies.

The relations between *Structure* and the essays reprinted in part 2 are too obvious to require discussion, and I shall therefore approach them differently, saying what I can about their role or about the stages they record in the development of my thoughts on scientific change. As a result, this preface will for a time again become explicitly autobiographical. After stumbling upon the concept of a scientific revolution in 1947, I first took time to finish my physics dissertation and then began to educate myself in the history of science.[4] The first opportunity to present my developing ideas was provided by an invitation to deliver a series of Lowell Lectures in the spring of 1951, but the primary result of that venture was to convince me that I did not yet know either enough history or enough about my ideas to proceed toward publication. For a period that I expected to be short but that lasted seven years, I set my more philosophical interests aside and worked straightforwardly at history. Only in the late 1950s, after finishing a book on the Copernican revolution[5] and receiving a tenured university appointment, did I consciously return to them.

The position my views had by then reached is indicated by the paper that opens part 2, "The Historical Structure of Scientific

4. The first portion of the time required for self-education was supplied by an appointment as a Junior Fellow of the Harvard Society of Fellows. Without it, I doubt that the transition could have been managed successfully.

5. *The Copernican Revolution: Planetary Astronomy in the Development of Western Thought* (Cambridge, Mass., 1957).

Discovery." Though not written until late in 1961 (by which time my book on revolutions was substantially complete), the ideas it presents and the main examples it employs were all, for me, old ones. Scientific development depends in part on a process of non-incremental or revolutionary change. Some revolutions are large, like those associated with the names of Copernicus, Newton, or Darwin, but most are much smaller, like the discovery of oxygen or the planet Uranus. The usual prelude to changes of this sort is, I believed, the awareness of anomaly, of an occurrence or set of occurrences that does not fit existing ways of ordering phenomena. The changes that result therefore require "putting on a different kind of thinking-cap," one that renders the anomalous lawlike but that, in the process, also transforms the order exhibited by some other phenomena, previously unproblematic. Though only implicit, that conception of the nature of revolutionary change also underlies the paper "Energy Conservation" reprinted in part 1, particularly its opening pages. It was written during the spring of 1957, and I am quite certain "The Historical Structure of Scientific Discovery" could have been written at that time, probably a good deal earlier.

A consequential advance in my understanding of my topic was closely associated with the preparation of the second paper in part 2, "The Function of Measurement," a subject I had not previously been inclined to consider at all. Its origin was an invitation to address the Social Science Colloqium at the University of California, Berkeley, in October 1956, and it was revised and extended to roughly its present form during the spring of 1958. The second section, Motives for Normal Measurement, was a product of those revisions, and its second paragraph contains the first description of what I had, in its title, come very close to calling "normal science." Rereading that paragraph now, I am struck by the sentence: "The bulk of scientific practice is thus a complex and consuming mopping-up operation that consolidates the ground made available by the most recent theoretical breakthrough and thus provides essential preparation for the breakthrough to follow." The transition from that way of putting the point to "Normal Science as Puzzle Solving," the title of chapter 4 of *Structure*, did not require many additional steps. Though I had recognized for some years that periods governed by one or another traditional mode of practice must necessarily intervene between revolutions, the special nature of that tradition-bound practice had in large part previously escaped me.

The next paper, "The Essential Tension," supplies the title for this volume. Prepared for a conference held in June 1959 and first published in that conference's proceedings, it displays a modest further development of the notion of normal science. From an autobiographical viewpoint, however, its primary importance is its introduction of the concept of paradigms. That concept had come to me only a few months before the paper was read, and by the time I employed it again in 1961 and 1962 its content had expanded to global proportions, disguising my original intent.[6] The closing paragraph of "Second Thoughts on Paradigms," also reprinted below, hints at how that expansion took place. This autobiographical preface may be an appropriate place to extend the hint.

I spent the year 1958/59 as a fellow at the Center for Advanced Study in the Behavioral Sciences at Stanford, California, intending to write a draft of the book on revolutions during my fellowship. Soon after arriving, I produced the first version of a chapter on revolutionary change, but attempts to prepare a companion chapter on the normal interlude between revolutions gave me great trouble. At that time I conceived normal science as the result of a consensus among the members of a scientific community. Difficulties arose, however, when I tried to specify that consensus by enumerating the elements about which the members of a given community supposedly agreed. In order to account for the way they did research and, especially, for the unanimity with which they ordinarily evaluated the research done by others, I had to attribute to them agreement about the defining characteristics of such quasi-theoretical terms as "force" and "mass," or "mixture" and "compound." But experience, both as a scientist and as a historian, suggested that

6. Immediately after completing a first draft of *Structure* in the beginning of 1961, I wrote what for some years I took to be a revised version of "The Essential Tension" for a conference held at Oxford the following July. That paper was published in A. C. Crombie, ed., *Scientific Change* (London and New York, 1963), pp. 347–69, under the title "The Function of Dogma in Scientific Research." Comparing it with "The Essential Tension" (conveniently available in C. W. Taylor and F. Barron, eds., *Scientific Creativity: Its Recognition and Development* [New York, 1963], pp. 341–54) highlights both the speed and the extent of the expansion of my notion of paradigm. Because of that expansion the two papers seem to be making different points, something I had by no means intended.

such definitions were seldom taught and that occasional attempts to produce them often evoked pronounced disagreement. Apparently, the consensus I had been seeking did not exist, but I could find no way to write the chapter on normal science without it.

What I finally realized early in 1959 was that no consensus of quite that kind was required. If scientists were not taught definitions, they were taught standard ways to solve selected problems in which terms like "force" or "compound" figured. If they accepted a sufficient set of these standard examples, they could model their own subsequent research on them without needing to agree about which set of characteristics of these examples made them standard, justified their acceptance. That procedure seemed very close to the one by which students of language learn to conjugate verbs and to decline nouns and adjectives. They learn, for example, to recite, *amo, amas, amat, amamus, amatis, amant*, and they then use that standard form to produce the present active tense of other first conjugation Latin verbs. The usual English word for the standard examples employed in language teaching is "paradigms," and my extension of that term to standard scientific problems like the inclined plane and conical pendulum did it no apparent violence. It is in that form that "paradigm" enters "The Essential Tension," an essay prepared within a month or so of my recognition of its utility. ("[Textbooks] exhibit concrete problem solutions that the profession has come to accept as paradigms, and they then ask the student . . . to solve for himself problems very closely related in both method and substance to those through which the textbook or the accompanying lecture has led him.") Though the text of the essay elsewhere suggests what was to occur during the next two years, "consensus" rather than "paradigm" remains the primary term there used when discussing normal science.

The concept of paradigms proved to be the missing element I required in order to write the book, and a first full draft was prepared between the summer of 1959 and the end of 1960. Unfortunately, in that process, paradigms took on a life of their own, largely displacing the previous talk of consensus. Having begun simply as exemplary problem solutions, they expanded their empire to include, first, the classic books in which these accepted examples initially appeared and, finally, the entire global set of commitments shared by the members of a particular scientific community. That more global use of the term is the only one most readers of the book have recognized, and the inevitable result has been confusion: many of the things there said about paradigms apply only to the original sense of the term. Though both senses seem to me important, they do need to be distinguished, and the word "paradigm" is appropriate only to the first. Clearly, I have made unnecessary difficulties for many readers.[7]

The remaining five papers in this volume require little individual discussion. Only "A Function for Thought Experiments" was written before my book, on the shape of which it had little influence; "Second Thoughts on Paradigms" is the first written, though last published, of three attempts to recover the original sense of paradigms;[8] and "Objectivity, Value Judgment, and Theory Choice" is a previously unpublished lecture that aims to answer the charge that I make theory choice entirely subjective. These papers may speak for themselves, together with the two I have not yet mentioned. Rather than take them up one at a time, I shall close this preface by isolating two aspects of a single theme that binds all five together.

Traditional discussions of scientific method have sought a set of rules that would permit any *individual* who followed them to produce sound knowledge. I have tried to insist, instead, that, though science is practiced by individuals, scientific knowledge is intrinsically a *group* product and that neither its peculiar efficacy nor the manner in which it develops will be understood without reference to the special nature of the groups that produce it. In this sense my work has been deeply sociological, but not in a way that permits that subject to be separated from epistemology.

Convictions like these are implicit throughout the essay "Logic of Discovery or Psychology of Research?" in which I compare my views with those of Sir Karl Popper. (The hypotheses of individ-

7. Wolfgang Stegmüller has been especially successful in finding his way through these difficulties. In the section "What Is a Paradigm?" in his *Structure and Dynamics of Theories*, trans. W. Wohlhueter (Berlin, Heidelberg, and New York, 1976), pp. 170–80, he discusses three senses of the term, and the second, his "Class II," captures precisely my original intent.

8. "Second Thoughts" was prepared for a conference held in March 1969. After completing it, I retraced some of the same ground in "Reflections on My Critics," the closing chapter of I. Lakatos and A. Musgrave, eds., *Criticism and the Growth of Knowledge* (Cambridge, 1970). Finally, still in 1969, I prepared the extra chapter for the second edition of *Structure*.

uals are tested, the commitments shared by his group being presupposed; group commitments, on the other hand, are not tested, and the process by which they are displaced differs drastically from that involved in the evaluation of hypotheses; terms like "mistake" function unproblematically in the first context but may be functionless in the second; and so on.) They become explicitly sociological at the end of that paper and throughout the lecture on theory choice, where I attempt to explain how shared values, though impotent to dictate an individual's decisions, may nevertheless determine the choice of the group which shares them. Very differently expressed, the same concerns underlie the final essay in this volume, in which I exploit the license permitted a commentator to explore the ways in which differences in shared values (and in audience) may decisively affect the developmental patterns characteristic of science and art. Additional, but more knowledgeable and systematic, comparisons of the value systems that govern the practitioners of varied disciplines seem to me urgently needed at this time. Probably they should begin with more closely related groups, for example physicists and engineers or biologists and physicians. The epilogue to "The Essential Tension" is relevant in this connection.

In the literature of sociology of science, the value system of science has been especially discussed by R. K. Merton and his followers. Recently that group has been repeatedly and sometimes stridently criticized by sociologists who, drawing on my work and sometimes informally describing themselves as "Kuhnians," emphasize that values vary from community to community and from time to time. In addition, these critics point out that, whatever the values of a given community may be, one or another of them is repeatedly violated by its members. Under these circumstances, they think it absurd to conceive the analysis of values as a significant means of illuminating scientific behavior.[9]

The preceding remarks and the papers they introduce should, however, indicate how seriously misdirected I take that line of criticism to be. My own work has been little concerned with the specification of scientific values, but it has from the start presup-

9. The *locus classicus* for this sort of criticism is S. B. Barnes and R. G. A. Dolby, "The Scientific Ethos: A Deviant Viewpoint," *Archives Européennes de sociologie* 11 (1970): 3–25. It has surfaced frequently since, especially in the journal *Social Studies of Science* (formerly *Science Studies*).

posed their existence and role.[10] That role does not require that values be identical in all scientific communities or, in any given community, at all periods of time. Nor does it demand that a value system be so precisely specified and so free from internal conflict that it could, even in abstract principle, unequivocally determine the choices that individual scientists must make. For that matter, the significance of values as guides to action would not be reduced if values were, as some claim, mere rationalizations that have evolved to protect special interests. Unless bound by a conspiracy theory of history or sociology, it is hard not to recognize that rationalizations usually affect those who propound them even more than those to whom they are addressed.

The later parts of "Second Thoughts on Paradigms" and the whole of "A Function for Thought Experiments" explore another central problem raised by considering scientific knowledge as the product of special groups. One thing that binds the members of any scientific community together and simultaneously differentiates them from the members of other apparently similar groups is their possession of a common language or special dialect. These essays suggest that in learning such a language, as they must to participate in their community's work, new members acquire a set of cognitive commitments that are not, in principle, fully analyzable within that language itself. Such commitments are a consequence of the ways in which the terms, phrases, and sentences of the language are applied to nature, and it is its relevance to the language-nature link that makes the original narrower sense of "paradigm" so important.

When writing the book on revolutions, I described them as episodes in which the meanings of certain scientific terms changed, and I suggested that the result was an incommensurability of viewpoints and a partial breakdown of communication between the proponents of different theories. I have since recognized that "meaning change" names a problem rather than an isolable phenomenon, and I am now persuaded, largely by the work of Quine, that the problems of incommensurability and partial communication should be treated in another way. Proponents of different theories (or different paradigms, in the broader sense of the term) speak different

10. For an early expression see *The Structure of Scientific Revolutions*, 2d ed. (Chicago, 1970), pp. 152–56, 167–70. These passages were transcribed unchanged from the first edition of 1962.

languages—languages expressing different cognitive commitments, suitable for different worlds. Their abilities to grasp each other's viewpoints are therefore inevitably limited by the imperfections of the processes of translation and of reference determination. Those issues are currently the ones that concern me most, and I hope before long to have more to say about them.

I

Historiographic Studies

1 The Relations between the History and the Philosophy of Science

Previously unpublished Isenberg Lecture, delivered at Michigan State University, 1 March 1968; revised October 1976.

The subject on which I have been asked to speak today is the relations between the history and the philosophy of science. For me, more than for most, it has deep personal as well as intellectual significance. I stand before you as a practicing historian of science. Most of my students mean to be historians, not philosophers. I am a member of the American Historical, not the American Philosophical, Association. But for almost ten years after I first encountered philosophy as a college freshman, it was my primary avocational interest, and I often considered making it my vocation, displacing theoretical physics, the only field in which I can claim to have been properly trained. Throughout those years, which lasted until around 1948, it never occurred to me that history or history of science could hold the slightest interest. To me then, as to most scientists and philosophers still, the historian was a man who collects and verifies facts about the past and who later arranges them in chronological order. Clearly the production of chronicles could have little appeal to someone whose fundamental concerns were with deductive inference and fundamental theory.

I shall later ask why the image of the historian as chronicler has such special appeal to both philosophers and scientists. Its continued and selective attraction is not due either to coincidence or to the nature of history, and it may therefore prove especially revealing. But my present point is still autobiographical. What drew me belatedly from physics and philosophy to history was the discovery that science, when encountered in historical source materials, seemed a very different enterprise from the one implicit in science pedagogy and explicit in standard philosophical accounts of scientific method. History might, I realized with astonishment, be relevant to the philosopher of science and perhaps also to the epistemologist in ways that transcended its classic role as a source of examples for previously occupied positions. It might, that is, prove to be a particularly consequential source of problems and of insights. Therefore, though I became a historian, my deepest interests remained philosophical, and in recent years those interests have become increasingly explicit in my published work. To an extent, then, I do both history and philosophy of science. Of course I therefore think about the relation between them, but I also live it, which is not the same thing. That duality of my involvement will inevitably be reflected in the way I approach today's topic. From this point my talk will divide into two quite different, though closely related parts. The first is a report, often quite personal, of the difficulties to be encountered in any attempt to draw the two fields closer together. The second, which deals with problems more explicitly intellectual, argues that the *rapprochement* is fully worth the quite special effort it requires.

Few members of this audience will need to be told that, at least in the United States, the history and the philosophy of science are separate and distinct disciplines. Let me, from the very start, develop reasons for insisting that they be kept that way. Though a new sort of dialogue between these fields is badly needed, it must be inter- not intra-disciplinary. Those of you aware of my involvement with Princeton University's Program in History and Philosophy of Science may find odd my insistence that there is no such field. At Princeton, however, the historians and the philosophers of science pursue different, though overlapping, courses of study, take different general examinations, and receive their degrees from different departments, either history or philosophy. What is particularly admirable in that design is that it provides an institutional basis for a dialogue between fields without subverting the disciplinary basis of either.

Subversion is not, I think, too strong a term for the likely result of an attempt to make the two fields into one. They differ in a number of their central constitutive characteristics, of which the most general and apparent is their goals. The final product of most historical research is a narrative, a story, about particulars of the past. In part it is a description of what occurred (philosophers and scientists often say, a *mere* description). Its success, however, depends not only on accuracy but also on structure. The historical narrative must render plausible and comprehensible the events it describes. In a sense to which I shall later return, history is an explanatory enterprise; yet its explanatory functions are achieved with almost no recourse to explicit generalizations. (I may point out here, for later exploitation, that when philosophers discuss the role of covering laws in history, they characteristically draw their examples from the work of economists and sociologists, not of historians. In the writings of the latter, lawlike generalizations are extraordinarily hard to find.) The philosopher, on the other hand, aims principally at explicit generalizations and at those with universal scope. He is no teller of stories, true or false. His goal is to discover and state what is true at all times and places rather than to impart understanding of what occurred at a particular time and place.

Each of you will want to articulate and to qualify those crass generalizations, and some of you will recognize that they raise deep problems of discrimination. But few will feel that distinctions of this sort are entirely empty, and I therefore turn from them to their consequences. It is these that make the distinction of aims important. To say that history of science and philosophy of science have different goals is to suggest that no one can practice them both at the same time. But it does not suggest that there are also great difficulties about practicing them alternately, working from time to time on historical problems and attacking philosophical issues in between. Since I obviously aim at a pattern of that sort myself, I am committed to the belief that it can be achieved. But it is nonetheless important to recognize that each switch is a personal wrench, the abandonment of one discipline for another with which it is not quite compatible. To train a student simultaneously in both would risk depriving him of any discipline at all. Becoming a philosopher is, among other things, acquiring a particular mental set toward the evaluation both of problems and of the techniques relevant to their solution. Learning to be a historian is also to acquire a special mental set, but the outcome of the two learning experiences is not at all the same. Nor, I think, is a compromise possible, for it presents problems of the same sort as a compromise between the duck and the rabbit of the well-known Gestalt diagram. Though most people can readily see the duck and the rabbit alternately, no amount of ocular exercise and strain will educe a duck-rabbit.

That view of the relation between enterprises is not at all the one I had at the time of my conversion to history twenty years ago. Rather it derives from much subsequent experience, sometimes painful, as a teacher and writer. In the former role I have, for example, repeatedly taught graduate seminars in which prospective historians and philosophers read and discussed the same classic works of science and philosophy. Both groups were conscientious and both completed the assignments with care, yet it was often difficult to believe that both had been engaged with the same texts. Undoubtedly the two had looked at the same signs, but they had been trained (programmed, if you will) to process them differently. Inevitably, it was the processed signs—for example their reading notes or their memory of the text—rather than the signs themselves that provided the basis for their reports, paraphrases, and contributions to discussion.

Subtle analytic distinctions that had entirely escaped the historians would often be central when the philosophers reported on their reading. The resulting confrontations were invariably educational for the historians, but the fault was not always theirs. Sometimes the distinctions dwelt upon by the philosophers were not to be found at all in the original text. They were products of the subsequent development of science or philosophy, and their introduction during the philosophers' processing of signs altered the argument. Or again, listening to the historians' paraphrase of a position, the philosophers would often point out gaps and inconsistencies that the historians had failed to see. But the philosophers could then sometimes be shocked by the discovery that the paraphrase was accurate, that the gaps were there in the original. Without quite knowing they were doing so, the philosophers had improved the argument while reading it, knowing what its subsequent form must be. Even with the text open before them it was regularly difficult and sometimes impossible to persuade them that the gap was really there, that the author had not seen the logic of the argument quite as they did. But if the philosophers could be brought to see that much, they could usually see something more important as well—that what they took to be gaps had in fact been introduced by analytic distinctions they had themselves supplied, that the original argument, if no longer viable philosophy, was sound in its own terms. At this point the whole text might begin to look different to them. Both the extent of the transformation and the pedagogic difficulty in deliberately bringing it about are reminiscent of the Gestalt switch.

Equally impressive, as evidence of different processing, was the range of textual material noticed and reported by the two groups. The historians always ranged more widely. Important parts of their reconstructions might, for example, be built upon passages in which the author had introduced a metaphor designed, he said, "to aid the reader." Or again, having noticed an apparent error or inconsistency in the text, the historian might spend some time explaining how a brilliant man could have slipped in this way. What aspect of the author's thought, the historian would ask, can be discovered by noting that an inconsistency obvious to us was invisible to him and was perhaps no inconsistency at all? For the philosophers, trained to construct an argument, not to reconstruct historical thought, both metaphors and errors were irrelevant and were sometimes not noticed at all. Their concern, which they pursued with a subtlety, skill, and persistence seldom found among the historians, was the explicit philosophical generalization and the arguments that could be educed in its defense. As a result, the papers they submitted at the end of the term were regularly shorter and usually far more coherent than those produced by the historians. But the latter, though often analytically clumsy, usually came far closer to reproducing the major conceptual ingredients in the thought of the men the two groups had studied together. The Galileo or Descartes who appeared in the philosophers' papers was a better scientist or philosopher but a less plausible seventeenth-century figure than the figure presented by the historians.

I have no quarrel with either of these modes of reading and reporting. Both are essential components as well as central products of professional training. But the professions are different, and they quite properly put different first things first. For the philosophers in my seminars the priority tasks were, first, to isolate the central elements of a philosophical position and, then, to criticize and develop them. Those students were, if you will, honing their wits against the developed opinions of their greatest predecessors. Many of them would continue to do so in their later professional life. The historians, on the other hand, were concerned with the viable and the general only in the forms that had, in fact, guided the men they studied. Their first concern was to discover what each one had thought, how he had come to think it, and what the consequences had been for him, his contemporaries, and his successors. Both groups thought of themselves as attempting to grasp the essentials of a past philosophical position, but their ways of doing the job were conditioned by the primary values of their separate disciplines, and their results were often correspondingly distinct. Only if the philosophers were converted to history or the historians to philosophy did additional work produce significant convergence.

A quite different sort of evidence of a deep interdisciplinary divide depends upon testimony so personal that it may convince only its author. Nevertheless, because the experience from which it derives is comparatively rare, the testimony seems worth recording. I have myself, at various times, written articles in physics, in history, and in something resembling philosophy. In all three cases the process of writing proves disagreeable, but the experience is not in other respects the same. By the time one begins to write a physics paper, the research is finished. Everything one needs is ordinarily contained in one's notes. The remaining tasks are selection, condensation, and translation to clear English. Usually only the last presents difficulties, and they are not ordinarily severe.

The preparation of a historical paper is different, but there is one important parallel. A vast amount of research has to be done before one begins to write. Books, documents, and other records must be located and examined; notes must be taken, organized, and organized again. Months or years may go into work of this sort. But the end of such work is not, as it is in science, the end of the creative process. Selected and condensed notes cannot simply be strung together to make a historical narrative. Furthermore, though chronology and narrative structure usually permit the historian to write steadily from notes and an outline for a considerable period, there are almost always key points at which his pen or typewriter refuses to function and his undertaking comes to a dead stop. Hours, days, or weeks later he discovers why he has been unable to proceed. Though his outline tells him what comes next, and though his notes provide all requisite information about it, there is no viable transition to that next part of the narrative from the point at which he has already arrived. Elements essential to the connection have been omitted from an earlier part of his story because at that point the narrative structure did not demand them. The historian must therefore go back, sometimes to documents and notetaking, and rewrite a substantial part of his paper in order that the connection to what comes next may be made. Not until the last page is written can he be altogether sure that he will not have to start again, perhaps from the very beginning.

Only the last part of this description applies to the preparation of an article in philosophy, and there the periods of circling back are far more frequent and the concomitant frustrations far more intense. Only the man whose memory span permits him to compose a whole paper in his head can hope for long periods of uninterrupted composition. But if the actual writing of philosophy shows some parallels to history, what comes before is altogether distinct. Excepting in the history of philosophy and perhaps in logic, there is nothing like the historian's period of preparatory research; in the literal sense there is in most of philosophy no equivalent for research at all. One starts with a problem and a clue to its solution, both often encountered in the criticism of the work of some other philosopher. One worries it—on paper, in one's head, in discussions with colleagues—waiting for the point at which it will feel ready to be written down. More often than not that feeling proves mistaken, and the worrying process begins again, until finally the article is born. To me, at least, that is what it feels like, as though the article had come all at once, not seriatim like the pieces of historical narrative.

If, however, there is nothing quite like research in philosophy, there is something else that takes its place and that is virtually unknown in physics and in history. Considering it will take us back directly to the differences between the perceptions and behaviors of the two groups of students in my seminars. Philosophers regularly criticize each other's work and the work of their predecessors with care and skill. Much of their discussion and publication is in this sense Socratic: it is a juxtaposition of views forged from each other through critical confrontation and analysis. The critic who proclaimed that philosophers live by taking in each other's washing was unsympathetic, but he caught something essential about the enterprise. What he caught was, in fact, what the philosophers in my seminars were doing: forging their own positions by an analytic confrontation with, in this case, the past. In no other field, I think, does criticism play so central a role. Scientists sometimes correct bits of each other's work, but the man who makes a career of piecemeal criticism is ostracized by the profession. Historians, too, sometimes suggest corrections, and they also occasionally direct diatribes at competing schools whose approach to history they disdain. But careful analysis is, in those circumstances, rare, and an explicit attempt to capture and preserve the novel insights generated by the other school is almost unknown. Though influenced in extremely important ways by the work of his predecessors and his colleagues, the individual historian, like the physicist and unlike the philosopher, forges his work from primary source material, from data that he has engaged in his research. Criticism may take the place of research, but the two are not equivalent, and they produce disciplines of very different sorts.

These are only first steps in a quasi-sociological account of history and philosophy as knowledge-producing enterprises. They should, however, be sufficient to suggest why, admiring both, I suspect that an attempt to make them one would be subversive. Those whom I have convinced or those who, for one or another reason, have needed no convincing will, however, have a different question. Given the deep and consequential differences between the two enterprises, what can they have to say to each other? Why have I insisted that an increasingly active dialogue between them is an urgent desideratum? To that question, particularly to one part of it, the remainder of my remarks this evening are directed.

Any answer must divide into two far-from-symmetrical parts, of which the first here requires no more than cursory summary. Historians of science need philosophy for reasons that are, at once, apparent and well known. For them it is a basic tool, like knowledge of science. Until the end of the seventeenth century, much of science was philosophy. After the disciplines separated, they continued to interact in often consequential ways. A successful attack on many of the problems central to the history of science is impossible for the man who does not command the thought of the main philosophic schools of the periods and areas he studies. Furthermore, since it is utopian to expect that any student of the history of science will emerge from graduate school with a command of the entire history of philosophy, he must learn to work this sort of material up for himself as his research requires it. The same holds true for some of the science he will need, and to both areas he must first be initiated by professionals, the men who know the subtleties and the traps of their disciplines and who can inculcate standards of professional acumen, skill, and rigor. There is no reason of principle why the historians in my seminars should have been clumsy when dealing with philosophical ideas. Given adequate prior training, most of them would not have been. Nor would the effects of such training have been limited to their performance when dealing with philosophical sources as such. Scientists are not often philosophers, but they do deal in ideas, and the analysis of ideas has long been the philosopher's province. The men who did most to establish the flourishing contemporary tradition in the history of science—I think particularly of A. O. Lovejoy and, above all, Alexandre Koyré—were philosophers before they turned to the history of scientific ideas. From them my colleagues and I learned to recognize the structure and coherence of idea systems other than our own. That search for the integrity of a discarded mode of thought is not what philosophers generally do; many of them, in fact, reject it as the glorification of past error. But the job can be done, and the philosopher's sensitivity to conceptual nuances is prerequisite to it. I cannot think that historians have learned their last lessons from this source.

These are sufficient reasons to urge the revivification of a more vigorous interaction between philosophers and historians of science, but they are also question begging. My assignment was the relation of the history of science to philosophy of science rather than to the history of philosophy. Can the historian of science also profit from a deep immersion in the literature of that special philosophical field? I have to answer that I very much doubt it. There have been philosophers of science, usually those with a vaguely neo-Kantian cast, from whom historians can still learn a great deal. I do urge my students to read Emile Meyerson and sometimes Léon Brunschvicg. But I recommend these authors for what they saw in historical materials not for their philosophies, which I join most of my contemporaries in rejecting. The living movements in philosophy of science, on the other hand, particularly as the field is currently practiced in the English-speaking world, include little that seems to me relevant to the historian. On the contrary, these movements aim at goals and perceive materials in ways more likely to mislead than to illuminate historical research. Though there is much about them that I admire and value, that is because my own concerns are by no means exclusively historical. No one in recent years has done so much to clarify and deepen my consideration of philosophical problems as my Princeton colleague C. G. Hempel. But my discourse with him and my acquaintance with his work does nothing for me at all when I work on, say, the history of thermodynamics or of the quantum theory. I commend his courses to my history students, but I do not especially urge that they enroll.

Those remarks will suggest what I had in mind in saying that the problem of the relations between history and philosophy of science divides into two parts, which are far from symmetrical. Though I do not think current philosophy of science has much relevance for the historian of science, I deeply believe that much writing on philosophy of science would be improved if history played a larger background role in its preparation. Before attempting to justify that belief, I must, however, introduce a few badly needed limitations. When speaking here of the history of science, I refer to that central part of the field that is concerned with the evolution of scientific ideas, methods, and techniques, not the increasingly significant portion that emphasizes the social setting of science, particularly changing patterns of scientific education, institutionalization, and support, both moral and financial. The philosophical import of the latter sort of work seems to me far more problematic than that of the former, and its consideration would, in any case, require a separate lecture. By the same token, when speaking of the philosophy of science, I have in mind neither those portions that shade over into applied logic nor, at least not with much assurance, those parts that are addressed to the implications of particular current theories for such longstanding philosophical problems as causation or space and time. Rather I am thinking of that central area that concerns itself with the scientific in general, asking, for example, about the structure of scientific theories, the status of theoretical entities, or the conditions under which scientists may properly claim to have produced sound knowledge. It is to this part of the philosophy of science, and very possibly to it alone, that the history of scientific ideas and techniques may claim relevance.

To suggest how this could be so, let me first point out a respect in which philosophy of science is almost unique among recognized philosophical specialties: the distance separating it from its subject matter. In fields like logic and, increasingly, the philosophy of mathematics, the problems that concern the professional are generated by the field itself. The difficulties of reconciling material implication with the "if . . . then" relation of normal discourse may be a reason for seeking alternative systems of logic, but it does not reduce the importance or fascination of the problems generated by standard axiom systems. In other parts of philosophy, most notably ethics and aesthetics, practitioners address themselves to experiences which they share with vast portions of humanity and which, are not, in any case, the special preserves of clearly demarcated professional groups. Though only the philosopher may be an aesthetician, the aesthetic experience is every man's. The philosophies of science and law are alone in addressing themselves to areas about which the philosopher *qua* philosopher knows little. And philosophers of law are far more likely than philosophers of science to have received significant professional training in their subject-matter field and to concern themselves with the same documents as the men about whose field they speak. That, I take it, is one reason why judges and lawyers read philosophy of law with far more regularity than scientists read philosophy of science.

My first claim, then, is that history of science can help to bridge the quite special gap between philosophers of science and science itself, that it can be for them a source of problems and of data. I do not, however, suggest that it is the only discipline that can do so. Actual experience in the practice of a science would probably be a more effective bridge than the study of its history. Sociology of science, if it ever develops sufficiently to embrace the cognitive content of science together with its organizational structure, might do as well. The historian's concern with development over time and the additional perspective available when studying the past may give history special advantages, to the first of which I shall later return. But my present point is only that history provides the most practical and available among several possible methods by which the philosopher might more closely acquaint himself with science.

Against this suggestion there is available a considerable arsenal. Some will argue that the gap, if unfortunate, does no great harm. Many more will insist that history cannot possibly supply a corrective. The part of philosophy of science currently under discussion does not, after all, direct itself to any particular scientific theory, except occasionally as illustrative. Its objective is theory in general.

Unlike history, furthermore, it is comparatively little concerned with the temporal development of theory, emphasizing instead the theory as a static structure, an example of sound knowledge at some particular, though unspecified, time and place. Above all, in philosophy of science, there is no role for the multitude of particulars, the idiosyncratic details, which seem to be the stuff of history. Philosophy's business is with rational reconstruction, and it need preserve only those elements of its subject essential to science as sound knowledge. For that purpose, it is argued, the science contained in college textbooks is adequate if not ideal. Or at least it is adequate if supplemented by an examination of a few scientific classics, perhaps Galileo's *Two New Sciences* together with the "Introduction" and "General Scholium" from Newton's *Principia.*

Having previously insisted that history and philosophy of science have very different goals, I can have no quarrel with the thesis that they may appropriately work from different sources. The difficulty, however, with the sorts of sources just examined is that, working from them, the philosopher's reconstruction is generally unrecognizable as science to either historians of science or to scientists themselves (excepting perhaps social scientists, whose image of science is drawn from the same place as the philosopher's). The problem is not that the philosopher's account of theory is too abstract, too stripped of details, too general. Both historians and scientists can claim to discard as much detail as the philosopher, to be as concerned with essentials, to be engaged in rational reconstruction. Instead the difficulty is the identification of essentials. To the philosophically minded historian, the philosopher of science often seems to have mistaken a few selected elements for the whole and then forced them to serve functions for which they may be unsuited in principle and which they surely do not perform in practice, however abstractly that practice be described. Though both philosophers and historians seek the essentials, the results of their search are by no means the same.

This is not the place to enumerate missing ingredients. Many of them are, in any case, discussed in my earlier work. But I do want to suggest what it is about history that makes it a possible source for a rational reconstruction of science different from that now current. For that purpose, furthermore, I must first insist that history is not itself the enterprise much contemporary philosophy takes it to be. I must, that is, argue briefly the case for what Louis

Mink has perceptively called "the autonomy of historical understanding."

No one, I think, still believes that history is mere chronicle, a collection of facts arranged in the order of their occurrence. It is, most would concede, an explanatory enterprise, one that induces understanding, and it must thus display not only facts but also connections between them. No historian has, however, yet produced a plausible account of the nature of these connections, and philosophers have recently filled the resulting void with what is known as the "covering law model." My concern with it is as an articulated version of a widely diffused image of history, one that makes the discipline seem uninteresting to those who seek lawlike generalizations, philosophers, scientists, and social scientists in particular.

According to proponents of the covering law model, a historical narrative is explanatory to the extent that the events it describes are governed by laws of nature and society to which the historian has conscious or unconscious access. Given the conditions that obtained at the point in time when the narrative opens, and given also a knowledge of the covering laws, one should be able to predict, perhaps with the aid of additional boundary conditions inserted along the way, the future course of some central parts of the narrative. It is these parts, and only these, that the historian may be said to have explained. If the laws permit only rough predictions, one speaks of having provided an "explanation sketch" rather than an explanation. If they permit no prediction at all, the narrative has provided no explanation.

Clearly the covering law model has been drawn from a theory of explanation in the natural sciences and applied to history. I suggest that, whatever its merits in the fields for which it was first developed, it is an almost total misfit in this application. Very likely there are or will be laws of social behavior capable of application to history. As they come into being, historians sooner or later use them. But laws of that sort are primarily the business of the social sciences, and except in economics very few are yet in hand. I have already pointed out that philosophers turn generally to writings by social scientists for the laws they attribute to historians. I now add that, when they do draw examples from historical writing, the laws they educe are at once obvious and dubious: for example, "Hungry men tend to riot." Probably, if the words "tend to" are heavily underscored, the law is valid. But does it follow that an account of starvation in eighteenth-century France is less essential to a narrative dealing with the first decade of the century, when there were no riots, than to one dealing with the last, when riots did occur?

Surely the plausibility of a historical narrative does not depend upon the power of a few scattered and doubtful laws like this one. If it did, then history would explain virtually nothing at all. With few exceptions, the facts that fill the pages of its narratives would be mere window dressing, facts for the sake of facts, unconnected to each other or to any larger goal. Even the few facts actually connected by law would become uninteresting, for precisely to the extent that they were "covered," they would add nothing to what everyone already knew. I am not claiming, let me be clear, that the historian has access to no laws and generalizations, nor that he should make no use of them when they are at hand. But I do claim that, however much laws may add substance to an historical narrative, they are not essential to its explanatory force. That is carried, in the first instance, by the facts the historian presents and the manner in which he juxtaposes them.

During my days as a philosophically inclined physicist, my view of history resembled that of the covering law theorists, and the philosophers in my seminars usually begin by viewing it in a similar way. What changed my mind and often changes their's is the experience of putting together a historical narrative. That experience is vital, for the difference between learning history and doing it is far larger than that in most other creative fields, philosophy certainly included. From it I conclude, among other things, that an ability to predict the future is no part of the historian's arsenal. He is neither a social scientist nor a seer. It is no mere accident that he knows the end of his narrative as well as the start before he begins to write. History cannot be written without that information. Though I have no alternate philosophy of history or of historical explanation to offer here, I can at least outline a better image of the historian's task and suggest why its performance might produce a sort of understanding.

The historian at work is not, I think, unlike the child presented with one of those picture puzzles of which the pieces are square; but the historian is given many extra pieces in the box. He has or can get the data, not all of them (what would that be?) but a very considerable collection. His job is to select from them a set that can be juxtaposed to provide the elements of what, in the child's case, would be a picture of recognizable objects plausibly juxtaposed and of what, for the historian and his reader, is a plausible narrative involving recognizable motives and behaviors. Like the child with the puzzle, the historian at work is governed by rules that may not be violated. There may be no empty spaces in the middle either of the puzzle or of the narrative. Nor may there be any discontinuities. If the puzzle displays a pastoral scene, the legs of a man may not be joined to the body of a sheep. In the narrative a tyrannical monarch may not be transformed by sleep alone to a benevolent despot. For the historian there are additional rules that do not apply to the child. Nothing in the narrative may, for example, do violence to the facts the historian has elected to omit from his story. That story must, in addition, conform to any laws of nature and society the historian knows. Violation of rules like these is ground for rejecting either the assembled puzzle or the historian's narrative.

Such rules, however, only limit but do not determine the outcome of either the child's or the historian's task. In both cases the basic criterion for having done the job right is the primitive recognition that the pieces fit to form a familiar, if previously unseen, product. The child has seen pictures, the historian behavior patterns, similar to these before. That recognition of similarity is, I believe, prior to any answers to the question, similar with respect to what? Though it can be rationally understood and perhaps even modelled on a computer (I once attempted something of the sort myself), the similarity relation does not lend itself to lawlike reformulation. It is global, not reducible to a unique set of prior criteria more primitive than the similarity relation itself. One may not replace it with a statement of the form. "*A* is similar to *B*, if and only if the two share the characteristics *c*, *d*, *e*, and *f*." I have elsewhere argued that the cognitive content of the physical sciences is in part dependent on the same primitive similarity relation between concrete examples, or paradigms, of successful scientific work, that scientists model one problem solution on another without at all knowing what characteristics of the original must be preserved to legitimate the process. Here I am suggesting that in history that obscure global relationship carries virtually the entire burden of connecting fact. If history is explanatory, that is not because its narratives are covered by general laws. Rather it is because the reader who says, "Now I know what happened," is simultaneously saying, "Now it makes sense; now I understand; what was for me previously a mere list of facts has fallen into a recognizable pattern." I urge that the experience he reports be taken seriously.

What has just been said is, of course, the early stage of a program for philosophical contemplation and research, not yet the solution of a problem. If many of you differ with me about its likely outcome, that is not because you are more aware than I of its incompleteness and difficulty, but because you are less convinced that the occasion demands so radical a break with tradition. That point, however, I shall not argue here. The object of the digression from which I now return has been to identify my convictions, not to defend them. What has troubled me about the covering law model is that it makes of the historian a social scientist *manqué*, the gap being filled by assorted factual details. It makes it hard to recognize that he has another and a profound discipline of his own, that there is an autonomy (and integrity) of historical understanding. If that claim now seems even remotely plausible, it prepares the way for my principal conclusion. When the historian of science emerges from the contemplation of sources and the construction of narrative, he may have a right to claim acquaintance with essentials. If he then says, "I cannot construct a viable narrative without giving a central place to aspects of science that philosophers ignore, nor can I find a trace of elements they consider essential," then he deserves an audience. What he is claiming is that the enterprise reconstructed by the philosopher is not, as to certain of its essentials, science.

What sort of lessons might the philosopher learn by taking the historian's narrative constructions more seriously? I shall close this lecture with a single global example, referring you to my earlier work for other illustrations, many of them dependent on the examination of individual cases. The overwhelming majority of historical work is concerned with process, with development over time. In principle, development and change need not play a similar role in philosophy, but in practice, I now want to urge, the philosopher's view of even static science, and thus of such questions as theory structure and theory confirmation, would be fruitfully altered if they did.

Consider, for example, the relation between empirical laws and theories, both of which I shall, for purposes of this brief conclusion, construe quite broadly. Despite real difficulties, which I have elsewhere perhaps overemphasized, empirical laws fit the received tradition in philosophy of science relatively well. They can, of course, be confronted directly with observation or experiment. More to my present point, when they first emerge, they fill an apparent gap, supplying information that was previously lacking. As science develops, they may be refined, but the original versions remain approximations to their successors, and their force is therefore either obvious or readily recaptured. Laws, in short, to the extent that they are purely empirical, enter science as net additions to knowledge and are never thereafter entirely displaced. They may cease to be of interest and therefore remain uncited, but that is another matter. Important difficulties do, I repeat, confront the elaboration of this position, for it is no longer clear just what it would be for a law to be purely empirical. Nevertheless, as an admitted idealization, this standard account of empirical laws fits the historian's experience quite well.

With respect to theories the situation is different. The tradition introduces them as collections or sets of law. Though it concedes that individual members of a set can be confronted with experience only through the deductive consequences of the set as a whole, it thereafter assimilates theories to laws as closely as possible. That assimilation does not fit the historian's experience at all well. When he looks at a given period in the past he can find gaps in knowledge later to be filled by empirical laws. The ancients knew that air was compressible but were ignorant of the regularity that quantitatively relates its volume and pressure; if asked, they would presumably have conceded the lack. But the historian seldom or never finds similar gaps to be filled by later theory. In its day, Aristotelian physics covered the accessible and imaginable world as completely as Newtonian physics later would. To introduce the latter, the former had to be literally displaced. After that occurred, furthermore, efforts to recapture Aristotelian theory presented difficulties of a very different nature from those required to recapture an empirical law. Theories, as the historian knows them, cannot be decomposed into constituent elements for purposes of direct comparison either with nature or with each other. That is not to say that they cannot be analytically decomposed at all, but rather that the lawlike parts produced by analysis cannot, unlike empirical laws, function individually in such comparisons.

A central tenet of Aristotle's physics was, for example, the impossibility of a void. Suppose that a modern physicist had told him that an arbitrarily close approximation to a void could now be produced in the laboratory. Probably Aristotle would have responded that a container emptied of air and other gases was not in his sense a void. That response would suggest that the impossibility of a void was not, in his physics, a merely empirical matter. Suppose now instead that Aristotle had conceded the physicist's point and announced that a void could, after all, exist in nature. Then he would have required a whole new physics, for his concept of the finite cosmos, of place within it, and of natural motion stand or fall together with his concept of the void. In that sense, too, the lawlike statement "there are no voids in nature" did not function within Aristotelian physics quite as a law. It could not, that is, be eliminated and replaced by an improved version, leaving the rest of the structure standing.

For the historian, therefore, or at least for this one, theories are in certain essential respects holistic. So far as he can tell, they have always existed (though not always in forms one would comfortably describe as scientific), and they then always cover the entire range of conceivable natural phenomena (though often without much precision). In these respects they are clearly unlike laws, and there are inevitably corresponding differences in the ways they develop and are evaluated. About these latter processes we know very little, and we shall not learn more until we learn properly to reconstruct selected theories of the past. As of today, the people taught to do that job are historians, not philosophers. Doubtless the latter could learn, but in the process, as I have suggested, they would likely become historians too. I would of course welcome them, but would be saddened if they lost sight of their problems in the transition, a risk that I take to be real. To avoid it I urge that history and philosophy of science continue as separate disciplines. What is needed is less likely to be produced by marriage than by active discourse.

2 Concepts of Cause in the Development of Physics

By permission from *Etudes d'épistémologie génétique* 25 (1971): 7–18, where it appeared as "Les notions de causalité dans le developpement de la physique."

Why should a historian of science be invited to address an audience of child psychologists on the development of causal notions in physics? A first answer is well known to all who are acquainted with the researches of Jean Piaget. His perceptive investigations of such subjects as the child's conception of space, of time, of motion, or of the world itself have repeatedly disclosed striking parallels to the conceptions held by adult scientists of an earlier age. If there are similar parallels in the case of the notion of cause, their elucidations should be of interest both to the psychologist and to the historian.

There is, however, also a more personal answer, perhaps applicable only to this historian and this group of child psychologists. Almost twenty years ago I first discovered, very nearly at the same time, both the intellectual interest of the history of science and the psychological studies of Jean Piaget. Ever since that time the two have interacted closely in my mind and in my work. Part of what I know about how to ask questions of dead scientists has been learned by examining Piaget's interrogations of living children. I vividly remember how that influence figured in my first meeting with Alexandre Koyré, the man who, more than any other historian, has been my *maître*. I said to him that it was Piaget's children from whom I had learned to understand Aristotle's physics. His response—that it was Aristotle's physics that had taught him to understand Piaget's children—only confirmed my impression of the importance of what I had learned. Even in those areas, like causality, about which we may not now quite agree, I am proud to acknowledge the ineradicable traces of Piaget's influence.

If the historian of physics is to succeed in an analysis of the notion of cause, he must, I think, recognize two related respects in which that concept differs from most of those with which he is accustomed to deal. As in other conceptual analyses, he must start from the observed occurrence of words like "cause" and "because" in the conversation and publication of scientists. But these words, unlike those relating to such concepts as position, motion, weight, time, and so on, do not occur regularly in scientific discourse, and when they do, the discourse is of a quite special sort. One is tempted to say, following a remark made for different reasons by M. Grize, that the term "cause" functions primarily in the metascientific, not the scientific, vocabulary of physicists.

That observation ought not suggest that the concept of cause is less important than more typical technical concepts like position, force, or motion. But it does suggest that the available tools of analysis function somewhat differently in the two cases. In analyzing the notion of cause the historian or philosopher must be far more sensitive than usual to nuances of language and behavior. He must observe not only the occurrences of terms like 'cause' but also the special circumstances under which such terms are evoked. Conversely, he must base essential aspects of his analysis on his observation of contexts in which, though a cause has apparently been supplied, no terms occur to indicate which parts of the total communication make reference to causes. Before he is finished, the analyst who proceeds in this way is likely to conclude that, as compared with, say, position, the concept of cause has essential linguistic and group-psychological components.

That aspect of the analysis of causal notions relates closely to a second one on which M. Piaget has insisted from the beginning of this conference. We must, he has said, consider the concept of cause under two headings, the narrow and the broad. The narrow concept derives, I take it, from the initially egocentric notion of an active agent, one that pushes or pulls, exerts a force or manifests a power. It is very nearly Aristotle's concept of the efficient cause, a notion that first functioned significantly in technical physics during the seventeenth-century analyses of collision problems. The broad conception is, at least at first glance, very different. M. Piaget has described it as the general notion of explanation. To describe the cause or causes of an event is to explain why it occurred. Causes figure in physical explanations, and physical explanations are generally causal. Recognizing that much, however, is to confront again the intrinsic subjectivity of some of the criteria governing the notion of cause. Both the historian and the psychologist are well aware that a sequence of words that provided an explanation at one stage in the development of physics or of the child may lead only to further questions at another. Is it sufficient to say that the apple falls to earth because of gravitational attraction, or must attraction itself be explained before questioning will cease? A specified deductive structure may be a necessary condition for the adequacy of a causal explanation, but it is not a sufficient condition. When analyzing causation, one must therefore inquire about the particular responses, short of *force majeur*, that will bring a regress of causal questions to a close.

The coexistence of two senses of cause also intensifies another of the problems encountered briefly above. For reasons at least partly historical, the narrow notion is often taken to be fundamental, and the broader concept is made to conform to it, often with resulting violence. Explanations that are causal in the narrow sense always do provide an agent and a patient, a cause and a subsequent effect. But there are other explanations of natural phenomena—we shall examine a few below—from which no earlier event or phenomenon, nor any active agent, emerges as *the cause*. Nothing is gained (and much linguistic naturalness is lost) by declaring such explanations to be noncausal: they lack nothing that, once supplied, could be construed as the missing cause. Nor can the questions be declared noncausal: asked under other circumstances, they would have evoked a narrowly causal response. If any line at all can be drawn between causal and noncausal explanations of natural phenomena, it will depend upon subtleties that are irrelevant here. Nor is it useful to transform such explanations, verbally or mathematically, into a form that does permit the isolation of an earlier state of affairs as the cause. Presumably the transformation can always be managed (sometimes by one of the ingenious techniques illustrated in the presentation of my fellow guest, Bunge), but the result is often to deprive the transformed expression of explanatory force.

A schematic epitome of the four main stages in the evolution of causal notions in physics will both document and deepen what has already been said. Simultaneously it will prepare the way for a few more general conclusions. Until about 1600 the principal tradition in physics was Aristotelian, and Aristotle's analysis of cause was dominant too. The latter, however, continued to be of use long after the former had been discarded, and it therefore merits separate examination at the start. According to Aristotle, every change, including coming into being, had four causes: material, efficient, formal, and final. These four exhausted the types of answers that could be given to a request for an explanation of change. In the case of a statue, for example, the material cause of its existence is the marble; its efficient cause is the force exerted on the marble by the sculptor's tools; its formal cause is the idealized form of the finished object, present from the start in the sculptor's mind; and the final cause is an increase in the number of beautiful objects accessible to the members of Greek society.

In principle, every change possessed all four causes, one of each type, but in practice the sort of cause invoked for effective explanation varied greatly from field to field. When considering the science of physics, Aristotelians ordinarily made use of only two causes, formal and final, and these regularly merged into one. Violent changes, those that disrupted the natural order of the cosmos, were of course attributed to efficient causes, to pushes and pulls, but changes of this sort were not thought capable of further explanation and thus lay outside of physics. That subject dealt only with the restoration and maintenance of natural order, and these depended upon formal causes alone. Thus, stones fell to the center of the universe because their nature or form could be entirely realized only in that position; fire rose to the periphery for the same reason; and celestial matter realized its nature by turning regularly and eternally in place.

During the seventeenth century, explanations of this sort came to seem logically defective, mere verbal play, tautologies, and the evaluation has endured. Molière's doctor, ridiculed for explaining opium's ability to put people to sleep in terms of its "dormative potency," remains today a stock figure of fun. That ridicule has been effective, and in the seventeenth century there was occasion for it. Nevertheless, there is no logical flaw in explanations of this sort. So long as people were able to explain, as the Aristotelians were, a relatively wide range of natural phenomena in terms of a relatively small number of forms, explanations in terms of forms were entirely satisfactory. They came to seem tautologies only when each distinct phenomenon seemed to necessitate the invention of a distinct form. Explanations of an exactly parallel sort are still immediately apparent in most of the social sciences. If they prove less powerful than one could wish, the difficulty is not in their logic but in the particular forms deployed. I shall shortly suggest that formal explanation now functions with extraordinary effectiveness in physics.

In the seventeenth and eighteenth centuries, however, its role was minimal. After Galileo and Kepler, who often pointed to simple mathematical regularities as formal causes that required no further analysis, all explanation was required to be mechanical. The only admissible forms were the shapes and positions of the ultimate corpuscles of matter. All change, whether of position or of some quality like color or temperature, was to be understood as the result of the physical impact of one group of particles on another. Thus Descartes explained the weight of bodies as resulting from the impact on their upper surface of particles from the surrounding aether. Aristotle's efficient causes, pushes and pulls, now dominated the explanation of change. Even Newton's work, which was widely interpreted as licensing nonmechanical interactions between particles, did little to reduce the dominion of efficient cause. It did, of course, do away with strict mechanism, and Newton was widely attacked by those who saw the introduction of action at a distance as a regressive violation of existing standards of explanation. (They were right. Eighteenth-century scientists could have introduced a new force for each sort of phenomenon. A few began to do so.) But Newtonian forces were generally treated in analogy to contact forces, and explanation remained dominantly mechanical. Particularly in the newer parts of physics—electricity, magnetism, the study of heat—explanation was largely conducted, throughout the eighteenth century, in terms of efficient causes.

During the nineteenth century, however, a change, which had begun earlier in mechanics, spread gradually through the whole of physics. As that field became increasingly mathematical, explanation came increasingly to depend upon the exhibition of suitable forms and the derivation of their consequences. In structure, though not in substance, explanation was again that of Aristotelian physics. Asked to explain a particular natural phenomenon, the physicist would write down an appropriate differential equation and deduce from it, perhaps conjoined with specified boundary conditions, the phenomenon in question. He might, it is true, then be challenged to justify his choice of differential equations. But that challenge would be directed to the particular formulation, not to the type of explanation. Whether he had chosen the correct one or not, it was a differential equation, a form that provided the explanation of what occurred. And as an explanation the equation was not further divisible. Without grave distortion, no active agent, no isolated cause temporally prior to the effect could be retrieved from it.

Consider, for example, the question why Mars moves in an elliptical orbit. The answer exhibits Newton's laws applied to an isolated system of two massive bodies interacting with an inverse-square attraction. Each of these elements is essential to the explanation, but none is the cause of the phenomenon. Nor are they prior to, rather than simultaneous with or later than, the phenomenon to be explained. Or consider the more limited question why Mars is at a particular position in the sky at a particular time. The answer is obtained from the preceding by inserting into the solution of the equation the position and velocity of Mars at some earlier time. Those boundary conditions do describe an earlier event connected by deduction from laws to the one to be explained. But it misses the point to call that earlier event, for which an infinity of others could be substituted, the cause of Mars's position at the specified later time. If boundary conditions supply the cause, then causes cease to be explanatory.

These two examples are also illuminating in a second respect. They are answers to questions that would not be asked, at least not by one physicist of another. What are introduced as answers above would be more realistically described as solutions to problems the physicist might pose for himself or exhibit to students. If we call them explanations, it is because, once they have been presented and understood, there are no more questions to ask: everything that the physicist can provide as explanation has already been given. There are, however, other contexts in which very similar questions would be asked, and in these contexts the structure of the answer would be different. Suppose Mars's orbit were observed not to be elliptical or that its position at a particular time were not quite the one predicted by the solution to the Newtonian two-body problem with boundary conditions. Then the physicist does ask (or did before these phenomena were well understood) what has gone wrong, why experience departs from his expectations. And the answer, in this case, does isolate a specific cause—here the gravitational attraction of another planet. Unlike regularities, anomalies are explained in terms that are causal in the narrow sense. Once again the resemblance to Aristotelian physics is striking. Formal causes explain nature's order, efficient causes its departures from order. Now, however, irregularity as well as regularity is in the province of physics.

These examples from celestial mechanics could be duplicated from other parts of mechanics, and from acoustics, electricity, optics, or thermodynamics as these subjects developed in the late eighteenth and early nineteenth centuries. But the point should already be clear. What may still need emphasis, however, is that the resemblance to Aristotelian explanation displayed by explanations in these fields is only structural. The forms deployed in nineteenth-century physical explanation were not at all like Aristotle's but were rather mathematical versions of the Cartesian and Newtonian forms, which had been dominant in the seventeenth and eighteenth centuries. This restriction to mechanical forms lasted, however, only until the closing years of the nineteenth century. Then, with the acceptance of Maxwell's equations for the electromagnetic field and with the recognition that these equations could not be derived from the structure of a mechanical aether, the list of forms the physicist might employ in explanations began to increase.

What has resulted in the twentieth century is one more revolution in physical explanation, this time not in its structure but in its substance. My fellow guest Halbwachs has pointed to many of its details. Here I shall attempt only a few very broad generalizations about it. The electromagnetic field, as a fundamental nonmechanical physical entity with formal properties describable only in mathematical equations, was only the entry point of the field concept into physics. The contemporary physicist recognizes other fields as well, and the number is still growing. For the most part they are employed to explain phenomena that were not even recognized in the nineteenth century, but they have also, for example in electromagnetism, displaced forces in some areas formerly reserved to them. As in the seventeenth century, what was once an explanation is an explanation no longer. Nor is it only fields, a new sort or entity, that are involved in the change. Matter has also acquired mechanically unimaginable formal properties—spin, parity, strangeness, and so on—each of them describable only in mathematical terms. Finally, the entry of an apparently ineradicable probabilistic element into physics has produced one other radical shift in the canons of explanation. There are now well-formed questions about observable phenomena, for example, the time at which an alpha particle leaves a nucleus, which physicists declare to be in principle unanswerable by science. As individual events, alpha-particle emission and many similar phenomena are uncaused. Any theory that did explain them would overthrow, rather than simply add to, the quantum theory. Perhaps some later transformation of physical theory will change that view or else make the relevant questions impossible to ask. But at this time few physicists regard the causal gap as an imperfection. That fact, too, may teach us something about causal explanation.

What is to be concluded from this brief sketch? As a minimal summary I suggest the following. Though the narrow concept of cause was a vital part of the physics of the seventeenth and eighteenth centuries, its importance declined in the nineteenth and has almost vanished in the twentieth. The main exceptions are explanations of occurrences that appear to violate existing physical theory, but in fact do not. These are explained by isolating the particular cause of the anomaly, by finding, that is, an element omitted from consideration in the initial solution of the problem. Except in these cases, however, the structure of physical explanation closely resembles that which Aristotle developed in analyzing formal causes. Effects are deduced from a few specified innate properties of the entities with which the explanation is concerned. The logical status of those properties and of the explanations deduced from them is the same as that of Aristotle's forms. Cause in physics has again become cause in the broader sense, that is, explanation.

Yet if modern physics resembles Aristotelian in the causal structure of its arguments, the particular forms that figure in physical explanation are today radically different from those of physics in antiquity and the Middle Ages. Even in the brief exposition above we have observed two major transitions in the types of forms that could function satisfactorily in physical explanation: from qualitative forms (innate gravity or levity) to mechanical forms and then from mechanical to mathematical. A more detailed account would have disclosed numerous subtler transitions as well. Transitions of this sort raise, however, a series of questions that demand comment, even though it must be brief and dogmatic. What brings about such changes in explanatory canons? What is their importance? And what is the relation of the older explanatory mode to the new?

With respect to the first of these questions, I suggest that in physics new canons of explanation are born with new theories on which they are, to a considerable extent, parasitic. New physical theories have, like Newton's, repeatedly been rejected by men who, while admitting the ability of the new view to resolve previously intractable problems, have nevertheless insisted that it explained nothing at all. Later generations, brought up to use the new theory for its power, have generally found it explanatory as well. The pragmatic success of a scientific theory seems to guarantee the ultimate success of its associated explanatory mode. Explanatory force may, however, be a long time coming. The experience of many contemporaries with quantum mechanics and relativity suggests that one may believe a new theory with deep conviction and still lack the retraining and habituation to receive it as explanatory. That comes only with time, but to date it has always come.

Being parasitic on new theories does not make new modes of explanation unimportant. The physicists' drive to understand and explain nature is an essential condition of his work. Accepted canons of explanation are part of what tells him which problems are still to be resolved, which phenomena remain unexplained. Furthermore, whatever problems a scientist works on, current canons of explanation do much to condition the sorts of solutions at which he is able to arrive. One cannot understand the science of any period without having grasped the explanatory canons accepted by its practitioners.

Finally, having sketched four stages in the development of causal notions in physics, I ask whether any overall pattern can be observed in their succession. Is there some sense in which the explanatory canons of modern physics are more advanced than those of, say, the eighteenth century and in which those of the eighteenth century transcend those of antiquity and the Middle Ages? In one sense the answer is clearly yes. The physical theory of each of these periods was vastly more powerful and precise than that of its predecessors. Explanatory canons, being integrally associated with physical theory itself, must necessarily have participated in the advance: the development of science permits the explanation of ever more refined phenomena. It is, however, only the phenomena, not the explanations, that are more refined in any obvious sense. Once abstracted from the theory within which it functioned, gravity is only different from an innate tendency toward the center, the concept of a field is merely different from that of a force. Considered by themselves as explanatory devices, without reference to what the theories that invoke them can explain, the permissible starting points for physical explanation do not seem intrinsically more advanced in a later than in an earlier age. There is even one sense in which revolutions in explanatory modes may be regressive. Though the evidence is far from conclusive, it does suggest that, as a science develops, it employs in explanations an ever increasing number of irreducibly distinct forms. With respect to explanation the simplicity of science may have decreased over historic time. Examination of that thesis would require another essay, but even the possibility of considering it suggests a conclusion that will be sufficient here. Studied by themselves, ideas of explanation and cause provide no obvious evidence of that progress of the intellect that is so clearly displayed by the science from which they derive.

3

Mathematical versus Experimental Traditions in the Development of Physical Science

Reprinted by permission from *The Journal of Interdisciplinary History* 7 (1976): 1–31.

Anyone who studies the history of scientific development repeatedly encounters a question, one version of which would be, "Are the sciences one or many?" Ordinarily that question is evoked by concrete problems of narrative organization, and these become especially acute when the historian of science is asked to survey his subject in lectures or in a book of significant scope. Should he take up the sciences one by one, beginning, for example, with mathematics, proceeding to astronomy, then to physics, to chemistry, to anatomy, physiology, botany, and so on? Or should he reject the notion that his object is a composite account of individual fields and take it instead to be knowledge of nature *tout court*?

This essay is the revised and extended version of a George Sarton Memorial Lecture, delivered in Washington, D.C., in 1972, at a joint session of the American Association for the Advancement of Science and the History of Science Society. A preliminary version had been read at Cornell University during the preceding month. In the three years that have elapsed since, I have benefited from the comments of colleagues too numerous to mention. Some special debts will be acknowledged in footnotes which follow. Here I record only my thanks for the encouragement and aid to clarification provided, during the course of revision, by two historians whose concerns overlap my own: Theodore Rabb and Quentin Skinner. The version that resulted was published in French translation in *Annales* 30 (1975): 975–98. A number of additional changes, mostly minor, have been introduced into the English version.

In that case, he is bound, insofar as possible, to consider all scientific subject matters together, to examine what men knew about nature at each period of time, and to trace the manner in which changes in method, in philosophical climate, or in society at large have affected the body of scientific knowledge conceived as one.

Given a more nuanced description, both approaches can be recognized as long-traditional and generally noncommunicating historiographic modes.[1] The first, which treats science as at most a loose-linked congeries of separate sciences, is also characterized by its practitioners' insistence on examining closely the technical content, both experimental and theoretical, of past versions of the particular specialty being considered. That is a considerable merit, for the sciences are technical, and a history which neglects their content often deals with another enterprise entirely, sometimes fabricating it for the purpose. On the other hand, historians who have aimed to write the history of a technical specialty have ordinarily taken the bounds of their topic to be those prescribed by recent textbooks in the corresponding field. If, for example, their subject is electricity, then their definition of an electrical effect often closely resembles the one provided by modern physics. With it in hand, they may search ancient, medieval, and early modern sources for appropriate references, and an impressive record of gradually accumulating knowledge of nature sometimes results. But that record is drawn from scattered books and manuscripts

1. For a somewhat more extended discussion of these two approaches, see Kuhn, "History of Science" in the *International Encyclopedia of the Social Sciences*, vol. 14 (New York, 1968), pp. 74–83 (pp. 105–26 below). Note also the way in which distinguishing between them both deepens and obscures the now far better known distinction between internalist and externalist approaches to the history of science. Virtually all the authors now regarded as internalists address themselves to the evolution of a single science or of a closely related set of scientific ideas; the externalists fall almost invariably into the group that has treated the sciences as one. But the labels "internalist" and "externalist" then no longer quite fit. Those who have concentrated primarily on individual sciences, e.g., Alexandre Koyré, have not hesitated to attribute a significant role in scientific development to extrascientific *ideas*. What they have resisted primarily is attention to socioeconomic and institutional factors as treated by such writers as B. Hessen, G. N. Clark, and R. K. Merton. But these nonintellectual factors have not always been much valued by those who took the sciences to be one. The "internalist-externalist debate" is thus frequently about issues different from the ones its name suggests, and the resulting confusion is sometimes damaging.

ordinarily described as works of philosophy, literature, history, scripture, or mythology. Narratives in this genre thus characteristically obscure the fact that most items they group as "electrical" —for example, lightning, the amber effect, and the torpedo (electric eel)—were not, during the period from which their descriptions are drawn, ordinarily taken to be related. One may read them carefully without discovering that the phenomena now called "electrical" did not constitute a subject matter before the seventeenth century and without finding even scattered hints about what then brought the field into existence. If a historian must deal with enterprises that did exist in the periods that concern him, then traditional accounts of the development of individual sciences are often profoundly unhistorical.

No similar criticism may be directed at the other main historiographic tradition, the one that treats science as a single enterprise. Even if attention is restricted to a selected century or nation, the subject matter of that putative enterprise proves too vast, too dependent on technical detail, and, collectively, too diffuse to be illuminated by historical analysis. Despite ceremonial bows to classics like Newton's *Principia* or Darwin's *Origin*, historians who view science as one have therefore paid little attention to its evolving content, concentrating instead on the changing intellectual, ideological, and institutional matrix within which it developed. The technical content of modern textbooks is thus irrelevant to their subject, and the works they produce have, especially in recent decades, been fully historical and sometimes intensely illuminating. The development of scientific institutions, values, methods, and world views is clearly in itself a worthy subject for historical research. Experience suggests, however, that it is by no means so nearly coextensive with the study of scientific development as its practitioners have ordinarily supposed. The relationship between the metascientific environment, on the one hand, and the development of particular scientific theories and experiments, on the other, has proved to be indirect, obscure, and controversial.

To an understanding of that relationship, the tradition which takes science to be one can in principle contribute nothing, for it bars *by presupposition* access to phenomena upon which the development of such understanding must depend. Social and philosophical commitments that fostered the development of a particular field at one period of time have sometimes hampered it at another;

if the period of concern is specified, then conditions that promoted advance in one science often seem to have been inimical to others.[2] Under these circumstances, historians who wish to illuminate actual scientific development will need to occupy a difficult middle ground between the two traditional alternatives. They may not, that is, assume science to be one, for it clearly is not. But neither may they take for granted the subdivisions of subject matter embodied in contemporary science texts and in the organization of contemporary university departments.

Textbooks and institutional organization are useful indices of the natural divisions the historian must seek, but they should be those of the period he studies. Together with other materials, they can then provide at least a preliminary roster of the various fields of scientific practice at a given time. Assembling such a roster is, however, only the beginning of the historian's task, for he needs also to know something about the relations between the areas of activity it names, asking, for example, about the extent of interaction between them and the ease with which practitioners could pass from one to the next. Inquiries of that sort can gradually provide a map of the complex structure of the scientific enterprise of a selected period, and some such map is prerequisite to an examination of the complex effects of metascientific factors, whether intellectual or social, on the development of the sciences. But a structural map alone is not sufficient. To the extent that the effects to be examined vary from field to field, the historian who aims to understand them will also have to examine at least representative parts of the sometimes recondite technical activities within the field or fields that concern him. Whether in the history or the sociology of science, the list of topics that can usefully be studied without attention to the content of the relevant sciences is extremely short.

Historical research of the sort just demanded has barely begun. My conviction that its pursuit will be fruitful derives not from new work, my own or someone else's, but from repeated attempts as a teacher to synthesize the apparently incompatible products of the

2. On this point, in addition to the material below, see Kuhn, "Scientific Growth: Reflections on Ben-David's 'Scientific Role,'" *Minerva* 10 (1972): 166–78.

two noncommunicating traditions just described.[3] Inevitably, all results of that synthesis are tentative and partial, regularly straining and sometimes overstepping the limits of existing scholarship. Nevertheless, schematic presentation of one set of those results may serve both to illustrate what I have had in mind when speaking of the changing natural divisions between the sciences and also to suggest the gains which might be achieved by closer attention to them. One consequence of a more developed version of the position to be examined below could be a fundamental reformulation of an already overlong debate about the origins of modern science. Another would be the isolation of an important novelty which, during the nineteenth century, helped to produce the discipline of modern physics.

The Classical Physical Sciences

My main theme may be introduced by a question. Among the large number of topics now included in the physical sciences, which ones were already in antiquity foci for the continuing ac-

3. These problems of synthesis go back to the very beginning of my career, at which time they took two forms which initially seemed entirely distinct. The first, sketched in note 1, above, was how to make socioeconomic concerns relevant to narratives about the development of scientific ideas. The second, highlighted by the appearance of Herbert Butterfield's admirable and influential *Origins of Modern Science* (London, 1949), concerned the role of experimental method in the Scientific Revolution of the seventeenth century. Butterfield's first four chapters plausibly explained the main conceptual transformations of early modern science as "brought about, not by new observations or additional evidence in the first instance, but by transpositions that were taking place inside the minds of the scientists themselves . . . [by their] putting on a different kind of thinking-cap" (p. 1). The next two chapters, "The Experimental Method in the Seventeenth Century" and "Bacon and Descartes," provided more traditional accounts of their subjects. Although they seemed obviously relevant to scientific development, the chapters which dealt with them contained little material actually put to work elsewhere in the book. One reason they did not, I belatedly recognized, was that Butterfield attempted, especially in his chapter "The Postponed Scientific Revolution in Chemistry," to assimilate the conceptual transformations in eighteenth-century science to the same model (not new observations but a new thinking-cap) which had succeeded so brilliantly for the seventeenth.

tivity of specialists? The list is extremely short. Astronomy is its oldest and most developed component; during the Hellenistic period, as research in that field advanced to a previously unprecedented level, it was joined by an additional pair, geometrical optics and statics, including hydrostatics. These three subjects—astronomy, statics, and optics—are the only parts of physical science which, during antiquity, became the objects of research traditions characterized by vocabularies and techniques inaccessible to laymen and thus by bodies of literature directed exclusively to practitioners. Even today Archimedes' *Floating Bodies* and Ptolemy's *Almagest* can be read only by those with developed technical expertise. Other subjects, which, like heat and electricity, later came to be included in the physical sciences, remained throughout antiquity simply interesting classes of phenomena, subjects for passing mention or for philosophic speculation and debate. (Electrical effects, in particular, were parceled out among several such classes.) Being restricted to initiates does not, of course, guarantee scientific advance, but the three fields just mentioned did advance in ways that required the esoteric knowledge and technique responsible for their isolation. If, furthermore, the accumulation of concrete and apparently permanent problem solutions is a measure of scientific progress, these fields are the only parts of what were to become the physical sciences in which unequivocal progress was made during antiquity.

At that time, however, the three were not practiced alone but were instead intimately associated with two others—mathematics and harmonics[4]—no longer ordinarily regarded as physical sci-

4. Henry Guerlac first urged on me the necessity of including music theory in the cluster of classical sciences. That I should initially have omitted a field no longer conceived as science indicates how easy it is to miss the force of the methodological precept offered in my opening pages. Harmonics was not, however, quite the field we would now call music theory. Instead, it was a mathematical science that attributed numerical proportions to the numerous intervals of various Greek scales or modes. Since there were seven of these, each available in three genera and in fifteen *tonoi* or keys, the discipline was complex, specification of some intervals requiring four- and five-digit numbers. Since only the simplest intervals were empirically accessible as the ratios of the lengths of vibrating strings, harmonics was also a highly abstract subject. Its relation to musical practice was at best indirect, and it remains obscure. Historically, harmonics dates from the fifth century B.C. and was highly developed by the time of Plato

ences. Of this pair, mathematics was even older and more developed than astronomy. Dominated, from the fifth century B.C., by geometry, it was conceived as the science of real physical quantities, especially spatial, and it did much to determine the character of the four others clustered around it. Astronomy and harmonics dealt with positions and ratios, respectively, and they were thus literally mathematical. Statics and geometric optics drew concepts, diagrams, and technical vocabulary from geometry, and they shared with it also a generally logical deductive structure common to both presentation and research. Not surprisingly, under these circumstances, men like Euclid, Archimedes, and Ptolemy, who contributed to one of these subjects, almost always made significant contributions to others as well. More than developmental level thus made the five a natural cluster, setting them apart from other highly evolved ancient specialties such as anatomy and physiology. Practiced by a single group and participating in a shared mathematical tradition, astronomy, harmonics, mathematics, optics, and statics are therefore grouped together here as the classical physical sciences or, more simply, as the classical sciences.[5] Indeed, even listing them as distinct topics is to some extent anachronistic. Evidence to be encountered below will suggest that, from some significant points of view, they might better be described as a single field, mathematics.

To the unity of the classical sciences one other shared characteristic was also prerequisite, and it will play an especially important role in the balance of this paper. Though all five fields, including

and Aristotle. Euclid is among the numerous figures who wrote treatises about it and whose work was largely superseded by Ptolemy's, a phenomenon familiar also in other fields. For these descriptive remarks and those in note 8, below, I am largely indebted to several illuminating conversations with Noel Swerdlow. Before they occurred, I had felt incapable of following Guerlac's advice.

5. The abbreviation "classical sciences" is a possible source of confusion, for anatomy and physiology were also highly developed sciences in classical antiquity, and they share a few, but by no means all, of the developmental characteristics here attributed to the classical physical sciences. These biomedical sciences were, however, parts of a second classical cluster, practiced by a distinct group of people, most of them closely associated with medicine and medical institutions. Because of these and other differences, the two clusters may not be treated together, and I restrict myself here to the physical sciences, partly on grounds of competence and partly to avoid excessive complexity. See, however, notes 6 and 9, below.

ancient mathematics, were empirical rather than *a priori*, their considerable ancient development required little refined observation and even less experiment. For a person schooled to find geometry in nature, a few relatively accessible and mostly qualitative observations of shadows, mirrors, levers, and the motions of stars and planets provided an empirical basis sufficient for the elaboration of often powerful theories. Apparent exceptions to this broad generalization (systematic astronomical observation in antiquity as well as experiments and observations on refraction and prismatic colors then and in the Middle Ages) will only reinforce its central point when examined in the next section. Although the classical sciences (including, in important respects, mathematics) were empirical, the data their development required were of a sort which everyday observation, sometimes modestly refined and systematized, could provide.[6] That is among the reasons why this cluster of fields could advance so rapidly under circumstances that did not significantly promote the evolution of a second natural group, the one to which my title refers as the products of an experimental tradition.

Before examining that second cluster, consider briefly the way in which the first developed after its origin in antiquity. All five of the classical sciences were actively pursued in Islam from the ninth century, often at a level of technical proficiency comparable with that of antiquity. Optics advanced notably, and the focus of mathematics was in some places shifted by the intrusion of algebraic techniques and concerns not ordinarily valued within the dominantly geometric Hellenistic tradition. In the Latin West, from the thirteenth century, further technical elaboration of these gen-

6. Elaborate or refined data generally become available only when their collection fulfills some perceived social function. That anatomy and physiology, which require such data, were highly developed in antiquity must be a consequence of their apparent relevance to medicine. That even that relevance was often hotly disputed (by the Empirics!) should help to account for the relative paucity, except in Aristotle and Theophrastus, of ancient data applicable to the more general taxonomic, comparative, and developmental concerns basic to the life sciences from the sixteenth century. Of the classical physical sciences, only astronomy required data of apparent social use (calendars and, from the second century B.C., horoscopy). If the others had depended upon the availability of refined data, they would probably have advanced no further than the study of topics such as heat.

erally mathematical fields was subordinated to a dominantly philosophical-theological tradition, important novelty being restricted primarily to optics and statics. Significant portions of the corpus of ancient and Islamic mathematics and astronomy were, however, preserved and occasionally studied for their own sake until they became again the objects of continuing erudite European research during the Renaissance.[7] The cluster of mathematical sciences then reconstituted closely resembled its Hellenistic progenitor. As these fields were practiced during the sixteenth century, however, research on a sixth topic was increasingly associated with them. Partly as a result of fourteenth-century scholastic analysis, the subject of local motion was separated from the traditional philosophic problem of general qualitative change, becoming a subject of study in its own right. Already highly developed within the ancient and medieval philosophical tradition, the problem of motion was a product of everyday observation, formulated in generally mathematical terms. It therefore fitted naturally into the cluster of mathematical sciences with which its development was thereafter firmly associated.

Thus enlarged, the classical sciences continued from the Renaissance onward to constitute a closely knit set. Copernicus specified the audience competent to judge his astronomical classic with the words, "Mathematics is written for mathematicians." Galileo, Kepler, Descartes, and Newton are only a few of the many seventeenth-century figures who moved easily and often consequentially from mathematics to astronomy, to harmonics, to statics, to optics, and to the study of motion. With the partial exception of harmonics, furthermore, the close ties between these relatively mathematical fields endured with little change into the early nineteenth century, long after the classical sciences had ceased to be the only parts of physical science subject to continuing intense scrutiny. The scientific subjects to which an Euler, Laplace, or Gauss principally contributed are almost identical with those illuminated earlier by Newton and Kepler. Very nearly the same list would encompass the

7. This paragraph has considerably benefited from discussions with John Murdoch, who emphasizes the historiographic problems encountered if the classical sciences are conceived of as continuing research traditions in the Latin Middle Ages. On this topic see his "Philosophy and the Enterprise of Science in the Later Middle Ages," in Y. Elkana, ed., *The Interaction between Science and Philosophy* (New York, 1974), pp. 51–74.

work of Euclid, Archimedes, and Ptolemy as well. Like their ancient predecessors, furthermore, the men who practiced these classical sciences in the seventeenth and eighteenth centuries had, with a few notable exceptions, little of consequence to do with experimentation and refined observation even though, after about 1650, such methods were for the first time intensively employed to study another set of topics later firmly associated with parts of the classical cluster.

One last remark about the classical sciences will prepare the way for consideration of the movement that promoted new experimental methods. All but harmonics[8] were radically reconstructed during the sixteenth and seventeenth centuries, and in physical science such transformations occurred nowhere else.[9] Mathematics made

8. Although harmonics was not transformed, its status declined greatly from the late fifteenth to the early eighteenth century. More and more it was relegated to the first section of treatises directed primarily to more practical subjects: composition, temperament, and instrument construction. As these subjects became more and more central to even quite theoretical treatises, music was increasingly divorced from the classical sciences. But the separation came late and was never complete. Kepler, Mersenne, and Descartes, all wrote on harmonics; Galileo, Huyghens, and Newton displayed interest in it; Euler's *Tentamen novae theoriae musicae* is in a longstanding tradition. After its publication in 1739, harmonics ceased to figure for its own sake in the research of major scientists, but an initially related field had already taken its place: the study, both theoretical and experimental, of vibrating strings, oscillating air columns, and acoustics in general. The career of Joseph Sauveur (1653–1716) clearly illustrates the transition from harmonics as music to harmonics as acoustics.

9. They did, of course, occur in the classical life sciences, anatomy and physiology. Also, these were the only parts of the biomedical sciences transformed during the Scientific Revolution. But the life sciences had always depended on refined observation and occasionally on experiment as well; they had drawn their authority from ancient sources (e.g., Galen) sometimes distinct from those important to the physical sciences; and their development was intimately involved with that of the medical profession and corresponding institutions. It follows that the factors to be discussed when accounting either for the conceptual transformation or for the newly enlarged range of the life sciences in the sixteenth and seventeenth centuries are by no means always the same as those most relevant to the corresponding changes in the physical sciences. Nevertheless, recurrent conversations with my colleague Gerald Geison reinforce my longstanding impression that they too can fruitfully be examined from a viewpoint like the one developed here. For that purpose the distinction between experimental and

the transition from geometry and "the art of the coss" to algebra, analytic geometry, and calculus; astronomy acquired noncircular orbits based on the newly central sun; the study of motion was transformed by new, fully quantitative laws; and optics gained a new theory of vision, the first acceptable solution to the classical problem of refraction, and a drastically altered theory of colors. Statics, conceived as the theory of machines, is an apparent exception. But as hydrostatics, the theory of fluids, it was extended during the seventeenth century to pneumatics, the "sea of air," and it can therefore be included in the list of reconstructed fields. These conceptual transformations of the classical sciences are the events through which the physical sciences participated in a more general revolution of Western thought. If, therefore, one thinks of the Scientific Revolution as a revolution of *ideas*, it is the changes in these traditional, quasi-mathematical fields which one must seek to understand. Although other vitally important things also happened to the sciences during the sixteenth and seventeenth centuries (the Scientific Revolution was not merely a revolution in thought), they prove to be of a different and to some extent independent sort.

The Emergence of Baconian Sciences

Turning now to the emergence of another cluster of research fields, I again begin with a question, this time with one about which there is much confusion and disagreement in the standard historical literature. What, if anything, was new about the experimental movement of the seventeenth century? Some historians have maintained that the very idea of basing science upon information acquired through the senses was novel. Aristotle, according to this view, believed that scientific conclusions could be deduced from axiomatic first principles; not until the end of the Renaissance did men escape his authority sufficiently to study nature rather than books. These residues of seventeenth-century rhetoric are, however, absurd. Aristotle's methodological writings contain many passages which are just as insistent upon the need for close observation as the writings of Francis Bacon. Randall and Crombie have isolated and studied an important medieval methodological tradition which,

mathematical traditions would be of little use, but a division between the medical and nonmedical life sciences might be crucial.

from the thirteenth century into the early seventeenth, elaborated rules for drawing sound conclusions from observation and experiment.[10] Descartes' *Regulae* and Bacon's *New Organon* owe much to that tradition. An empirical *philosophy* of science was no novelty at the time of the Scientific Revolution.

Other historians point out that, whatever people may have believed about the need for observations and experiments, they made them far more frequently in the seventeenth century than they had before. That generalization is doubtless correct, but it misses the essential qualitative differences between the older forms of experiment and the new. The participants in the new experimental movement, often called Baconian after its principal publicist, did not simply expand and elaborate the empirical elements present in the tradition of classical physical science. Instead they created a different sort of empirical science, one that for a time existed side by side with, rather than supplanting, its predecessor. A brief characterization of the occasional role played in the classical sciences by experiment and systematic observation will help to isolate the qualitative differences that distinguish the older form of empirical practice from its seventeenth-century rival.

Within the ancient and medieval tradition, many experiments prove on examination to have been "thought experiments," the construction in mind of potential experimental situations the outcome of which could safely be foretold from previous everyday experience. Others were performed, especially in optics, but it is often extremely difficult for the historian to decide whether a particular experiment discovered in the literature was mental or real. Sometimes the results reported are not what they would be now; on other occasions the apparatus required could not have been produced with existing materials and techniques. Real problems of historical decision result, and they also haunt students of Galileo. Surely he did experiments, but he is even more noteworthy as the man who brought the medieval thought-experimental tradition to its highest form. Unfortunately, it is not always clear in which guise he appears.[11]

0. A. C. Crombie, *Robert Grosseteste and the Origins of Experimental ence, 1100–1700* (Oxford, 1953); J. H. Randall, Jr., *The School of lua and the Emergence of Modern Science* (Padua, 1961).

1. For a useful and easily accessible example of medieval experimenta-ı, see Canto II of Dante's *Paradiso.* Passages located through the index

Finally, those experiments that clearly were performed seem invariably to have had one of two objects. Some were intended to demonstrate a conclusion known in advance by other means. Roger Bacon writes that, though one can in principle deduce the ability of flame to burn flesh, it is more conclusive, given the mind's propensity for error, to place one's hand in the fire. Other actual experiments, some of them consequential, were intended to provide concrete answers to questions posed by existing theory. Ptolemy's experiment on the refraction of light at the boundary between air and water is an important example. Others are the medieval optical experiments that generated colors by passing sunlight through globes filled with water. When Descartes and Newton investigated prismatic colors, they were extending this ancient and, more especially, medieval tradition. Astronomical observation displays a closely related characteristic. Before Tycho Brahe, astronomers did not systematically search the heavens or track the planets in their motions. Instead they recorded first risings, oppositions, and other standard planetary configurations of which the times and positions were needed to prepare ephemerides or to compute parameters called for by existing theory.

Contrast this empirical mode with the one for which Bacon became the most effective proponent. When its practitioners, men like Gilbert, Boyle, and Hooke, performed experiments, they seldom aimed to demonstrate what was already known or to determine a detail required for the extension of existing theory. Rather they wished to see how nature would behave under previously unobserved, often previously nonexistent, circumstances. Their typical products were the vast natural or experimental histories in which were amassed the miscellaneous data that many of them thought prerequisite to the construction of scientific theory. Closely examined, these histories often prove less random in choice and arrangement of experiments than their authors supposed. From 1650 at the latest, the men who produced them were usually guided by one or another form of the atomic or corpuscular philosophy. Their preference was thus for experiments likely to reveal the shape, arrangement, and motion of corpuscles; the analogies which underlie their juxtaposition of particular research reports often re-

entry "experiment, role of in Galileo's work" in Ernan McMullin, ed., *Galileo, Man of Science* (New York, 1965), will indicate how complex and controversial Galileo's relation to the medieval tradition remains.

veal the same set of metaphysical commitments.[12] But the gap between metaphysical theory on the one hand and particular experiments on the other was initially vast. The corpuscularism which underlies much seventeenth-century experimentation seldom demanded the performance or suggested the detailed outcome of any individual experiment. Under these circumstances, experiment was highly valued and theory often decried. The interaction that did occur between them was usually unconscious.

That attitude toward the role and status of experiment is only the first of the novelties which distinguish the new experimental movement from the old. A second is the major emphasis given to experiments which Bacon himself described as "twisting the lion's tail." These were the experiments that constrained nature, exhibiting it under conditions it could never have attained without the forceful intervention of man. The men who placed grain, fish, mice, and various chemicals seriatim in the artificial vacuum of a barometer or an air pump exhibit just this aspect of the new tradition.

Reference to the barometer and air pump highlights a third novelty of the Baconian movement, perhaps the most striking of all. Before 1590 the instrumental armory of the physical sciences consisted solely of devices for astronomical observation. The next hundred years witnessed the rapid introduction and exploitation of telescopes, microscopes, thermometers, barometers, air pumps, electric charge detectors, and numerous other new experimental devices. The same period was characterized by the rapid adoption by students of nature of an arsenal of chemical apparatus previously to be found only in the workshops of practical craftsmen and the retreats of alchemical adepts. In less than a century physical science became instrumental.

These marked changes were accompanied by several others, one of which merits special mention. The Baconian experimentalists scorned thought experiments and insisted upon both accurate and circumstantial reporting. Among the results of their insistence were sometimes amusing confrontations with the older experimental tradition. Robert Boyle, for example, pilloried Pascal for a book on hydrostatics in which, though the principles were found to be unexceptionable, the copious experimental illustrations had clearly

12. An extended example is provided by Kuhn, "Robert Boyle and Structural Chemistry in the Seventeenth Century," *Isis* 43 (1952): 12–36.

been mentally manufactured to fit. Monsieur Pascal does not tell us, Boyle complained, how a man is to sit at the bottom of a twenty-foot tub of water with a cupping glass held to his leg. Nor does he say where one is to find the superhuman craftsman able to construct the refined instruments upon which some of his other experiments depended.[13] Reading the literature of the tradition within which Boyle stands, the historian has no difficulty telling which experiments were performed. Boyle himself often names witnesses, sometimes supplying their patents of nobility.

Granting the qualitative novelty of the Baconian movement, how did its existence affect the development of science? To the conceptual transformations of the classical sciences, the contributions of Baconianism were very small. Some experiments did play a role, but they all have deep roots in the older tradition. The prism Newton purchased to examine "the celebrated phenomena of colors" descends from medieval experiments with water-filled globes. The inclined plane is borrowed from the classical study of simple machines. The pendulum, though literally a novelty, is first and foremost a new physical embodiment of a problem the medieval impetus theorists had considered in connection with the oscillatory motion of a vibrating string or of a body falling through the center of the earth and then back toward it. The barometer was first conceived and analyzed as a hydrostatic device, a water-filled pump-shaft without leaks designed to realize the thought experiment with which Galileo had "demonstrated" the limits to nature's abhorrence of a vacuum.[14] Only after an extended vacuum had been produced and the variation of column height with weather and altitude had been demonstrated did the barometer and its child the air pump join the cabinet of Baconian instruments.

Equally to the point, although the experiments just mentioned were of consequence, there are few like them, and all owe their special effectiveness to the closeness with which they could confront

13. "Hydrostatical Paradoxes, Made out by New Experiments" in A. Millar, ed., *The Works of the Honourable Robert Boyle* (London, 1744), 2:414–47, where the discussion of Pascal's book occurs on the first page.

14. For the medieval prelude to Galileo's approach to the pendulum, see Marshall Clagett, *The Science of Mechanics in the Middle Ages* (Madison, 1959), pp. 537–38, 570–71. For the road to Torricelli's barometer, see the too little known monograph by C. de Waard, *L'expérience barométrique, ses antécédents et ses explications* (Thouars [Deux-Sèvres], 1936).

the evolving theories of classical science, which had called them forth. The outcome of Torricelli's barometer experiment and of Galileo's with the inclined plane had been largely foreseen. Newton's prism experiment would have been no more effective than its traditional predecessors in transforming the theory of colors if Newton had not had access to the newly discovered law of refraction, a law sought within the classical tradition from Ptolemy to Kepler. For the same reason, the consequences of that experiment contrast markedly with those of the nontraditional experiments that during the seventeenth century revealed qualitatively novel optical effects like interference, diffraction, and polarization. The latter, because they were not products of classical science and could not be closely juxtaposed with its theories, had little bearing on the development of optics until the early nineteenth century. After all due qualification, some of it badly needed, Alexandre Koyré and Herbert Butterfield will prove to have been right. The transformation of the classical sciences during the Scientific Revolution is more accurately ascribed to new ways of looking at old phenomena than to a series of unanticipated experimental discoveries.[15]

Under those circumstances, numerous historians, Koyré included, have described the Baconian movement as a fraud, of no consequence to the development of science. That evaluation is, however, like the one it sometimes stridently opposed, a product of seeing the sciences as one. If Baconianism contributed little to the development of the classical sciences, it did give rise to a large number of new scientific fields, often with their roots in prior crafts. The study of magnetism, which derived its early data from prior experience with the mariner's compass, is a case in point. Electricity was spawned by efforts to find the relation of the magnet's attraction for iron to that of rubbed amber for chaff. Both these fields, furthermore, were dependent for their subsequent development upon the elaboration of new, more powerful, and more refined instruments. They are typical new Baconian sciences. Very nearly the same generalization applies to the study of heat. Long a topic for speculation within the philosophical and medical traditions, it was transformed into a subject for systematic investigation by the invention of the thermometer. Chemistry presents a case of

15. Alexandre Koyré, *Etudes galiléennes* (Paris, 1939); Butterfield, *Origins of Modern Science.*

a different and far more complex sort. Many of its main instruments, reagents, and techniques had been developed long before the Scientific Revolution. But until the late sixteenth century they were primarily the property of craftsmen, pharmacists, and alchemists. Only after a reevaluation of the crafts and of manipulative techniques were they regularly deployed in the experimental search for natural knowledge.

Since these fields and others like them were new foci for scientific activity in the seventeenth century, it is not surprising that their pursuit at first produced few transformations more striking than the repeated discovery of previously unknown experimental effects. If the possession of a body of consistent theory capable of producing refined predictions is the mark of a developed scientific field, the Baconian sciences remained underdeveloped throughout the seventeenth and much of the eighteenth centuries. Both their research literature and their patterns of growth were less like those of the contemporary classical sciences than like those discoverable in a number of the social sciences today. By the middle of the eighteenth century, however, experiment in these fields had become more systematic, increasingly clustering about selected sets of phenomena thought to be especially revealing. In chemistry, the study of displacement reactions and of saturation were among the newly prominent topics; in electricity, the study of conduction and of the Leyden jar; in thermometry and heat, the study of the temperature of mixtures. Simultaneously, corpuscular and other concepts were increasingly adapted to these particular areas of experimental research, the notions of chemical affinity or of electric fluids and their atmospheres providing particularly well-known examples.

The theories in which concepts like these functioned remained for some time predominantly qualitative and often correspondingly vague, but they could nonetheless be confronted by individual experiments with a precision unknown in the Baconian sciences when the eighteenth century began. Furthermore, as the refinements that permitted such confrontations continued into the last third of the century and became increasingly the center of the corresponding fields, the Baconian sciences rapidly achieved a state very like that of the classical sciences in antiquity. Electricity and magnetism became developed sciences with the work of Aepinus, Cavendish, and Coulomb; heat with that of Black, Wilcke, and Lavoisier; chemistry more gradually and equivocally, but not later than the time of Lavoisier's chemical revolution. At the beginning of the following century the qualitatively novel optical discoveries of the seventeenth century were for the first time assimilated to the older science of optics. With the occurrence of events like these, Baconian science had at last come of age, vindicating the faith, though not always the methodology, of its seventeenth-century founders.

How, during the almost two centuries of maturation, did the cluster of Baconian sciences relate to the cluster here called "classical"? The question has been far too little studied, but the answer, I think, will prove to be: not a great deal and then often with considerable difficulty—intellectual, institutional, and sometimes political. Into the nineteenth century the two clusters, classical and Baconian, remained distinct. Crudely put, the classical sciences were grouped together as "mathematics"; the Baconian were generally viewed as "experimental philosophy" or, in France, as "*physique expérimentale*"; chemistry, with its continuing ties to pharmacy, medicine, and the various crafts, was in part a member of the latter group, and in part a congeries of more practical specialties.[16]

This separation between the classical and Baconian sciences can be traced from the origin of the latter. Bacon himself was distrustful, not only of mathematics, but of the entire quasi-deductive structure of classical science. Those critics who ridicule him for failing to recognize the best science of his day have missed the point. He did not reject Copernicanism because he preferred the Ptolemaic system. Rather, he rejected both because he thought that no system so complex, abstract, and mathematical could contribute to either the understanding or the control of nature. His followers in the experimental tradition, though they accepted Copernican cosmology, seldom even attempted to acquire the mathematical skill and sophistication required to understand or pursue the classical sciences. That situation endured through the eighteenth century: Franklin, Black, and Nollet display it as clearly as Boyle and Hooke.

Its converse is far more equivocal. Whatever the causes of the Baconian movement, they impinged also on the previously estab-

16. For an early stage in the development of chemistry as a subject of intellectual concern, see Marie Boas, *Robert Boyle and Seventeenth-Century Chemistry* (Cambridge, 1958). For a vitally important later stage see Henry Guerlac, "Some French Antecedents of the Chemical Revolution," *Chymia* 5 (1959): 73–112.

lished classical sciences. New instruments entered those fields, too, especially astronomy. Standards for reporting and evaluating data changed as well. By the last decade of the seventeenth century confrontations like that between Boyle and Pascal are no longer imaginable. But, as previously indicated, the effect of these developments was a gradual refinement rather than a substantial change in the nature of the classical sciences. Astronomy had been instrumental and optics experimental before; the relative merits of quantitative telescopic and naked-eye observation were in doubt throughout the seventeenth century; excepting the pendulum, the instruments of mechanics were predominantly tools for pedagogic demonstration rather than for research. Under these circumstances, though the ideological gap between the Baconian and classical sciences narrowed, it by no means disappeared. Through the eighteenth century, the main practitioners of the established mathematical sciences performed few experiments and made fewer substantive contributions to the development of the new experimental fields.

Galileo and Newton are apparent exceptions. But only the latter is a real one, and both illuminate the nature of the classical-Baconian split. A proud member of the Lincei, Galileo was also a developer of the telescope, the pendulum escapement, an early form of thermometer, and other new instruments besides. Clearly he participated significantly in aspects of the movement here called Baconian. But, as Leonardo's career also indicates, instrumental and engineering concerns do not make a man an experimentalist, and Galileo's dominant attitude toward that aspect of science remained within the classical mode. On some occasions he proclaimed that the power of his mind made it unnecessary for him to perform the experiments he described. On others, for example when considering the limitations of water pumps, he resorted without comment to apparatus that transcended the capacity of existing technology. Boyle's critique of Pascal applies to Galileo without change. It isolates a figure who could and did make epochal contributions to the classical sciences but not, except through instrumental design, to the Baconian.

Educated during the years when British Baconianism was at its height, Newton did participate unequivocally in both traditions. But, as I. B. Cohen emphasized two decades ago, what results are two distinct lines of Newtonian influence, one traceable to Newton's *Principia*, the other to his *Opticks*.[17] That insight gains special significance if one notes that, though the *Principia* lies squarely within the tradition of the classical sciences, the *Opticks* is by no means unequivocally in the Baconian. Because his subject was optics, a previously well-developed field, Newton was able constantly to juxtapose selected experiments with theory, and it is from those juxtapositions that his achievements result. Boyle, whose *Experimental History of Colours* includes several of the experiments on which Newton founded his theory, made no such attempt, contenting himself with the remark that his results suggested speculations that might be worth pursuing.[18] Hooke, who discovered "Newton's rings," the first subject of the *Opticks*, book 2, accumulated data in much the same way. Newton, instead, selected and utilized them to elaborate theory, very much as his predecessors in the classical tradition had used the less recondite information usually provided by everyday experience. Even when he turned, as in the "Queries" to his *Opticks*, to such new Baconian topics as chemistry, electricity, and heat, Newton chose from the growing experimental literature those particular observations and experiments that could illuminate theoretical issues. Though no achievements so profound as those in optics could have been forthcoming in these still emerging fields, concepts like chemical affinity, scattered through the "Queries," proved a rich source for the more systematic and selective Baconian practitioners of the eighteenth century, and they therefore returned to them again and again. What they found in the *Opticks* and its "Queries" was a non-Baconian use of Baconian experiment, a product of Newton's deep and simultaneous immersion in the classical scientific tradition.

With the partial exception, however, of his continental contemporaries, Huyghens and Mariotte, Newton's example is unique. During the eighteenth century, by the beginning of which his scientific work was complete, no one else participated significantly in both traditions, a situation reflected also by the development of scientific institutions and career lines, at least into the nineteenth century. Although much additional research in this area is needed, the following remarks will suggest the gross pattern which research

17. I. B. Cohen, *Franklin and Newton* (Philadelphia, 1956).
18. Boyle, *Works*, 2:42–43.

is likely to refine. At least at the elementary level, the classical sciences had established themselves in the standard curriculum of the medieval university. During the seventeenth and eighteenth centuries the number of chairs devoted to them increased. The men who held them, together with those appointed to positions in the newly founded national scientific academies of France, Prussia, and Russia, were the principal contributors to the developing classical sciences. None of them is properly described as an amateur, though the term has often been applied indiscriminately to the practitioners of seventeenth- and eighteenth-century science as a whole. Practitioners of Baconian science were, however, usually amateurs, excepting only chemists, who found careers in pharmacy, industry, and some medical schools during the eighteenth century. For other experimental sciences the universities had no place before the last half of the nineteenth. Although some of their practitioners did receive positions in the various national scientific academies, they were there often second-class citizens. Only in England, where the classical sciences had begun to decline markedly before Newton's death, were they well represented, a contrast to be further developed below.

The example of the French Academy of Sciences is instructive in this respect, and its examination will simultaneously provide background for a point to be discussed in the next section. Guillaume Amontons (1663–1705), well known for his contributions to both the design and theory of such Baconian instruments as the thermometer and hygrometer, never rose in the academy beyond the status of *élève*, in which capacity he was attached to the astronomer Jean Le Fèvre. Pièrre Polinière (1671–1734), often cited as the man who introduced *physique expérimentale* to France, was never formally associated with the academy at all. Although the two main French contributors to eighteenth-century electrical sciences were academicians, the first, C. F. de C. Dufay (1698–1739), was placed in the chemistry section, while the second, the Abbé Nollet (1700–1770), was a member of the somewhat motley section officially reserved for practitioners of *arts mécaniques*. There, but only after his election to the Royal Society of London, Nollet rose through the ranks, succeeding among others, both the Comte de Buffon and Ferchauld de Réaumur. The famous instrument maker Abraham Bréguet, on the other hand, a man with the sorts of talent for which the mechanics section had been designed, found no place in the academy until, in 1816 at age sixty-nine, his name was inscribed on its rolls by royal ordinance.

What these isolated cases suggest is indicated also by the academy's formal organization. A section for *physique expérimentale* was not introduced until 1785, and it was then grouped in the mathematical division (with geometry, astronomy, and mechanics) rather than in the division for the more manipulative *sciences physiques* (anatomy, chemistry and metallurgy, botany and agriculture, and natural history and mineralogy). After 1815, when the new section's name was changed to *physique générale*, the experimentalists among its members were for some time very few. Looking at the eighteenth century as a whole, the contributions of academicians to the Baconian physical sciences were minor compared with those of doctors, pharmacists, industrialists, instrument makers, itinerant lecturers, and men of independent means. Again the exception is England, where the Royal Society was largely populated by such amateurs, rather than by men whose careers were first and foremost in the sciences.

The Origins of Modern Science

Return now briefly from the end of the eighteenth century to the middle of the seventeenth. The Baconian sciences were then in gestation, the classical being radically transformed. Together with concomitant changes in the life sciences, these two sets of events constitute what has come to be called the Scientific Revolution. Although no part of this essay purports to explain its extraordinarily complex causes, it is worth noting how different the question of causes becomes when the developments to be explained are subdivided.

That only the classical sciences were transformed during the Scientific Revolution is not surprising. Other fields of physical science scarcely existed until late in the period. To the extent that they did, furthermore, they lacked any significant body of unified technical doctrine to reconstruct. Conversely, one set of reasons for the transformation of the classical sciences lies within their own previous lines of development. Although historians differ greatly about the weight to be attached to them, few now doubt that some medieval reformulations of ancient doctrine, Islamic or Latin, were of major significance to figures like Copernicus, Galileo, and Kepler. No similar scholastic roots for the Baconian sciences are visible to me, despite the claims sometimes made for the methodological tradition that descends from Grosseteste.

Many of the other factors now frequently invoked to explain the Scientific Revolution did contribute to the evolution of both classical and Baconian sciences, but often in different ways and to different degrees. The effects of new intellectual ingredients—initially Hermetic and then corpuscular-mechanical—in the environment where early modern science was practiced provide a first example of such differences. Within the classical sciences, Hermetic movements sometimes promoted the status of mathematics, encouraged attempts to find mathematical regularities in nature, and occasionally licensed the simple mathematical forms thus discovered as formal causes, the terminus of the scientific causal chain.[19] Both Galileo and Kepler provide examples of this increasingly ontological role of mathematics, and the latter displays a second, more occult, Hermetic influence as well. From Kepler and Gilbert to Newton, though by then in an attenuated form, the natural sympathies and antipathies prominent in Hermetic thought helped to fill the void created by the collapse of the Aristotelian spheres that had previously kept the planets in their orbits.

After the first third of the seventeenth century, when Hermetic mysticism was increasingly rejected, its place, still in the classical sciences, was rapidly taken by one or another form of corpuscular philosophy derived from ancient atomism. Forces of attraction and repulsion between either gross or microscopic bodies were no longer favored, a source of much opposition to Newton. But within

19. The increased value ascribed to mathematics, as tool or as ontology, by many early-modern scientists has been recognized for almost half a century and was for many years described as a response to Renaissance Neoplatonism. Changing the label to "Hermeticism" does not improve the explanation of this aspect of scientific thought (though it has assisted in the recognition of other important novelties), and the change illustrates a decisive limitation of recent scholarship, one which I have not known how to avoid here. As currently used, "Hermeticism" refers to a variety of presumably related movements: Neoplatonism, Cabalism, Rosicrucianism, and what you will. They badly need to be distinguished: temporally, geographically, intellectually, and ideologically.

the infinite universe demanded by corpuscularism, there could be no preferred centers or directions. Natural enduring motions could only occur in straight lines and could only be disturbed by inter-corpuscular collisions. From Descartes on, that new perspective leads directly to Newton's first law of motion and—through the study of collisions, a new problem—to his second law as well. One factor in the transformation of the classical sciences was clearly the new intellectual climate, first Hermetic and then corpuscular, within which they were practiced after 1500.

The same new intellectual milieux affected the Baconian sciences, but often for other reasons and in different ways. Doubtless, the Hermetic emphasis on occult sympathies helps to account for the growing interest in magnetism and electricity after 1550; similar influences promoted the status of chemistry from the time of Paracelsus to that of van Helmont. But current research increasingly suggests that the major contribution of Hermeticism to the Baconian sciences and perhaps to the entire Scientific Revolution was the Faustian figure of the magus, concerned to manipulate and control nature, often with the aid of ingenious contrivances, instruments, and machines. Recognizing Francis Bacon as a transition figure between the magus Paracelsus and the experimental philosopher Robert Boyle has done more than anything else in recent years to transform historical understanding of the manner in which the new experimental sciences were born.[20]

For these Baconian fields, unlike their classical contemporaries, the effects of the transition to corpuscularism were equivocal, a first reason why Hermeticism endured longer in subjects like chemistry and magnetism than in, say, astronomy and mechanics. To declare that sugar is sweet because its round particles soothe the tongue is not obviously an advance on attributing to it a saccharine potency. Eighteenth-century experience was to demonstrate that the development of Baconian sciences often required guidance from concepts like affinity and phlogiston, not categorically unlike the natural sympathies and antipathies of the Hermetic movement. But

20. Frances A. Yates, "The Hermetic Tradition in Renaissance Science," in C. S. Singleton, ed., *Science and History in the Renaissance* (Baltimore, 1968), pp. 255–74; Paolo Rossi, *Francis Bacon: From Magic to Science*, trans. Sacha Rabinovitch (London, 1968).

corpuscularism did separate the experimental sciences from magic, thus promoting needed independence. Even more important, it provided a rationale for experiment, as no form of Aristotelianism or Platonism could have done. While the tradition governing scientific explanation demanded the specification of formal causes or essences, only data provided by the natural course of events could be relevant to it. To experiment or to constrain nature was to do it violence, thus hiding the role of the "natures" or forms which made things what they were. In a corpuscular universe, on the other hand, experimentation had an obvious relevance to the sciences. It could not change and might specially illuminate the mechanical conditions and laws from which natural phenomena followed. That was the lesson Bacon attached repeatedly to the fable of Cupid in chains.

A new intellectual milieu was not, of course, the sole cause of the Scientific Revolution, and the other factors most often invoked in its explanation also gain cogency when examined separately in classical and Baconian fields. During the Renaissance the medieval university's monopoly on learning was gradually broken. New sources of wealth, new ways of life, and new values combined to promote the status of a group formerly classified as artisans and craftsmen. The invention of printing and the recovery of additional ancient sources gave its members access to a scientific and technological heritage previously available, if at all, only within the clerical university setting. One result, epitomized in the careers of Brunelleschi and Leonardo, was the emergence from craft guilds during the fifteenth and sixteenth centuries of the artist-engineers whose expertise included painting, sculpture, architecture, fortification, water supply, and the design of engines of war and construction. Supported by an increasingly elaborate system of patronage, these men were at once employees and increasingly also ornaments of Renaissance courts and later sometimes of the city governments of northern Europe. Some of them were also informally associated with humanist circles, which introduced them to Hermetic and Neoplatonic sources. Those sources were not, however, what primarily legitimated their status as participants in a newly polite learning. Rather it was their ability to invoke and comment cogently upon such works as Vitruvius's *De architectura*, Euclid's *Geometry* and *Optics*, the pseudo-Aristotelian *Mechanical Prob-*

lems, and, after the mid-sixteenth century, both Archimedes' *Floating Bodies* and Hero's *Pneumatica*.[21]

The importance of this new group to the Scientific Revolution is indisputable. Galileo, in numerous respects, and Simon Stevin, in all, are among its products. What requires emphasis, however, is that the sources its members used and the fields they primarily influenced belong to the cluster here called classical. Whether as artists (perspective) or as engineers (construction and water supply), they mainly exploited works on mathematics, statics, and optics. Astronomy, too, occasionally entered their purview, though to a lesser extent. One of Vitruvius's concerns had been the design of precise sundials; the Renaissance artist-engineers sometimes extended it to the design of other astronomical instruments as well.

Although only here and there seminal, the concern of the artist-engineers with these classical fields was a significant factor in their reconstruction. It is probably the source of Brahe's new instruments and certainly of Galileo's concern with the strength of materials and the limited power of water pumps, the latter leading directly to Torricelli's barometer. Plausibly, but more controversially, engineering concerns, promoted especially by gunnery, helped to separate the problem of local motion from the larger philosophical problem of change, simultaneously making numbers rather than geometric proportions relevant to its further pursuit. These and related subjects are the ones that led to the inclusion of a section for *arts mécaniques* in the French academy and that caused that section to be grouped with the sections for geometry and astronomy. That it thereafter provided no natural home for the Baconian sciences finds its counterpart in the concerns of the Renaissance artist-engineers, which did not include the nonmechanical, nonmathematical aspects of such crafts as dyeing, weaving, glass-making, and navigation. These were, however, precisely the crafts that played so large a role in the genesis of the new experimental sciences. Bacon's programmatic statements called for natural histories of

21. P. Rossi, *Philosophy, Technology, and the Arts in the Early Modern Era*, trans. Salvator Attanasio (New York, 1970). Rossi and earlier students of the subject do not, however, discuss the possible importance of distinguishing between the crafts practiced by the artist-engineers and those later introduced to the learned world by figures like Vanoccio Biringuccio and Agricola. For some aspects of that distinction, introduced below, I am much indebted to conversation with my colleague, Michael S. Mahoney.

them all, and some of those histories of nonmechanical crafts were written.

Because the possible utility of even an analytic separation between the mechanical and nonmechanical crafts has not previously been suggested, what follows must be even more tentative than what precedes. As subjects for learned concern, however, the latter appear to have arrived later than the former. Presumably promoted at the start by Paracelsan attitudes, their establishment is demonstrated in such works as Biringuccio's *Pyrotechnia*, Agricola's *De re metallica*, Robert Norman's *Newe Attractive*, and Bernard Palissy's *Discours*, the earliest published in 1540. The status previously achieved by the mechanical arts doubtless helps to explain the appearance of books like these, but the movement which produced them is nevertheless distinct. Few practitioners of the nonmechanical crafts were supported by patronage or succeeded before the late seventeenth century in escaping the confines of craft guilds. None could appeal to a significant classical literary tradition, a fact that probably made the pseudo-classical Hermetic literature and the figure of the magus more important to them than to their contemporaries in the mathematical-mechanical fields.[22] Except in chemistry, among pharmacists and doctors, actual practice was seldom combined with learned discourse about it. Doctors do, however, figure in disproportionate numbers among those who wrote learned works not only on chemistry but also on the other nonmechanical crafts which provided data required for the development of the Baconian sciences. Agricola and Gilbert are only the earliest examples.

These differences between the two traditions rooted in prior crafts may help to explain still another. Although the Renaissance artist-engineers were socially useful, knew it, and sometimes based their claims upon it, the utilitarian elements in their writings are far less persistent and strident than those in the writings of men who drew upon the nonmechanical crafts. Remember how little

22. Although neither deals quite directly with this point, two recent articles suggest the way in which, first, Hermeticism and, then, corpuscularism could figure in seventeenth-century battles for intellectual-social status: P. M. Rattansi, "The Helmontian-Galenist Controversy in Restoration England," *Ambix* 12 (1964): 1–23; T. M. Brown, "The College of Physicians and the Acceptance of Iatromechanism in England, 1665–1695," *Bulletin of the History of Medicine*, 44 (1970): 12–30.

Leonardo cared whether or not the mechanical contrivances he invented could actually be built; or compare the writings of Galileo, Pascal, Descartes, and Newton with those of Bacon, Boyle, and Hooke. Present in both sets of writings, utilitarianism is central only to the second, a fact that may provide a clue to a last major difference between the classical and the Baconian sciences.

Excepting chemistry, which had found a variegated institutional base by the end of the seventeenth century, the Baconian and classical sciences flourished in different national settings from at least 1700. Practitioners of both can be found in most European countries, but the center for the Baconian sciences was clearly Britain, for the mathematical the Continent, especially France. Newton is the last British mathematician before the mid-nineteenth century who can compare with continental figures like the Bernoullis, Euler, Lagrange, Laplace, and Gauss. In the Baconian sciences, the contrast begins earlier and is less clear cut, but continental experimentalists with reputations to rival those of Boyle, Hooke, Hauksbee, Gray, Hales, Black, and Priestley are difficult to find before the 1780s. Furthermore, those who first come to mind tend to cluster in Holland and Switzerland, especially the former. Boerhaave, Musschenbroek, and de Saussure, all provide examples.[23] That geographical pattern needs more systematic investigation, but, if account is taken of relative populations and, especially, of relative productivity in the Baconian and classical sciences, it is likely to prove striking. Such investigation may also show that the national differences just sketched emerged only after the mid-seventeenth century, becoming slowly more striking during the generations that followed. Are not the differences between the eighteenth-century activities of the French Academy of Sciences and the Royal Society greater than those between the activities of the Accademia del Cimento, the Montmor Academy, and the British "Invisible College"?

Among the numerous competing explanations of the Scientific Revolution, only one provides a clue to this pattern of geographical divergences. It is the so-called Merton thesis, a redevelopment for the sciences of explanations for the emergence of capitalism pro-

23. Information relevant to this point is scattered throughout Pierre Brunet, *Les physiciens Hollandais et la méthode expérimentale en France au XVIIIe siècle* (Paris, 1926).

vided earlier by Weber, Troeltsch, and Tawney.[24] After their initial evangelical proselytizing phases, it is claimed, settled Puritan or protestant communities provided an "ethos" or "ethic" especially congenial to the development of science. Among its primary components were a strong utilitarian strain, a high valuation of work, including manual and manipulative work, and a distrust of system which encouraged each man to be his own interpreter first of Scripture and then of nature. Leaving aside, as others may not, the difficulties of identifying such an ethos and of determining whether it may be ascribed to all protestant or only to certain Puritan sects, the main drawbacks of this viewpoint have always been that it attempts to explain too much. If Bacon, Boyle, and Hooke seem to fit the Merton thesis, Galileo, Descartes, and Huyghens do not. It is in any case far from clear that postevangelical Puritan or protestant communities existed anywhere until the Scientific Revolution had been underway for some time. Not surprisingly the Merton thesis has been controversial.

Its appeal is, however, vastly larger if it is applied not to the Scientific Revolution as a whole, but rather to the movement which advanced the Baconian sciences. That movement's initial impetus toward power over nature through manipulative and instrumental techniques was doubtless supported by Hermeticism. But the corpuscular philosophies, which in the sciences increasingly replaced Hermeticism after the 1630s, carried no similar values, and Baconianism continued to flourish. That it did so especially in non-Catholic countries suggests that it may yet be worth discovering what, with respect to the sciences, a "Puritan" and an "ethos" are. Two isolated bits of biographical information may make that problem especially intriguing. Denis Papin, who built Boyle's second air pump and invented the pressure cooker, was a Huguenot driven from France by the mid-seventeenth-century persecutions. Abraham Bréguet, the instrument maker forced on the French Academy of Sciences in 1816, was an immigrant from Neuchâtel, to which his family had fled after the revocation of the Edict of Nantes.

24. R. K. Merton, *Science, Technology and Society in Seventeenth-Century England* (New York, 1970). This new edition of a work first published in 1938 includes a "Selected Bibliography: 1970," which provides useful guidance to the controversy that has continued since its initial appearance.

The Genesis of Modern Physics

My final topic must be presented as an epilogue, a tentative sketch of a position to be developed and modified by further research. But, having traced the generally separate development of the classical and Baconian sciences into the late eighteenth century, I must at least ask what happened next. Anyone acquainted with the contemporary scientific scene will recognize that the physical sciences no longer fit the pattern sketched above, a fact that has made the pattern itself difficult to see. When and how did the change occur? What was its nature?

Part of the answer is that the physical sciences during the nineteenth century participated in the rapid growth and transformation experienced by all learned professions. Older fields like medicine and law gained new institutional forms, more rigid and with intellectual standards more exclusive than any they had known before. In the sciences, from the late eighteenth century, the number of journals and societies increased rapidly, and many of them, unlike the traditional national academies and their publications, were restricted to individual scientific fields. Longstanding disciplines like mathematics and astronomy became for the first time professions with their own institutional forms.[25] Similar phenomena occurred only slightly more slowly in the newer Baconian fields, and one result was a loosening of ties which had previously bound them together. Chemistry, in particular, had by mid-century at the latest become a separate intellectual profession, still with ties to industry and to other experimental fields but with an identity now distinct from either. Partly for these institutional reasons and partly because of the effect on chemical research, first, of Dalton's atomic theory and, then, of increased attention to organic compounds, chemical concepts rapidly diverged from those used elsewhere in the physical sciences. As this occurred, topics like heat and electricity were increasingly barred from chemistry and left to experimental philosophy or to a new field, physics, that was increasingly taking its place.

A second important source of change during the nineteenth cen-

25. Everett Mendelsohn, "The Emergence of Science as a Profession in Nineteenth-Century Europe," in Karl Hill, ed., *The Management of Scientists* (Boston, 1964).

tury was a gradual shift in the perceived identity of mathematics. Until perhaps the middle of the century such topics as celestial mechanics, hydrodynamics, elasticity, and the vibrations of continuous and discontinuous media were at the center of professional mathematical research. Seventy-five years later, they had become "applied mathematics," a concern separate from and usually of lower status than the more abstract questions of "pure mathematics" which had become central to the discipline. Though courses in topics like celestial mechanics or even electromagnetic theory were sometimes still taught by members of mathematics faculties, they had become service courses, their subjects no longer on the frontier of mathematical thought.[26] The resulting separation between research in mathematics and in the physical sciences urgently needs more study, both for itself and for its effect on the development of the latter. That is doubly the case because it occurred in different ways and at different rates in different countries, a factor in the development of the additional national differences to be discussed below.

A third variety of change, especially relevant to the topics considered in this essay, was the remarkably rapid and full mathematization of a number of Baconian fields during the first quarter of the nineteenth century. Among the topics that now constitute the subject matter of physics, only mechanics and hydrodynamics had demanded advanced mathematical skills before 1800. Elsewhere the elements of geometry, trigonometry, and algebra were entirely sufficient. Twenty years later, the work of Laplace, Fourier, and Sadi Carnot had made higher mathematics essential to the study of heat; Poisson and Ampère had done the same for electricity and magnetism; and Jean Fresnel, with his immediate followers, had had a similar effect on the field of optics. Only as their new mathematical theories were accepted as models did a profession with an identity like that of modern physics become one of the sciences. Its

26. Relevant recollections about the relation of mathematics and mathematical physics in England, France, and the United States during the 1920s are contained in the interviews with Leon Brillouin, E. C. Kemble, and N. F. Mott on deposit in the various Archives for History of Quantum Physics. For information about these depositories, see T. S. Kuhn, J. L. Heilbron, P. F. Forman, and Lini Allen, *Sources for History of Quantum Physics: An Inventory and Report* (Philadelphia, 1967).

emergence demanded a lowering of the barriers, both conceptual and institutional, that had previously separated the classical and Baconian fields.

Why those barriers were lowered when and as they were is a problem demanding much additional research. But a major part of the answer will doubtless lie in the internal development of the relevant fields during the eighteenth century. The qualitative theories so rapidly mathematized after 1800 had come into existence only during and after the 1780s. Fourier's theory demanded the concept of specific heat and the consequent systematic separation of notions of heat and temperature. The contributions of Laplace and Carnot to thermal theory required in addition the recognition at the end of the century of adiabatic heating. Poisson's pioneering mathematization of static electrical and magnetic theory was made possible by the prior work of Coulomb, most of which appeared only in the 1790s.[27] Ampère's mathematization of the interaction between electric currents was supplied almost simultaneously with his discovery of the effects that his theory treated. Especially for the mathematization of electrical and thermal theory, recent developments in mathematical technique also played a role. Except perhaps in optics, the papers which between 1800 and 1825 made previously experimental fields fully mathematical could not have been written two decades before the burst of mathematization began.

Internal development, primarily of Baconian fields, will not, however, explain the manner in which mathematics was introduced after 1800. As the names of the authors of the new theories will already have suggested, the first mathematizers were uniformly French. Excepting in some initially little known papers by George Green and Gauss, nothing of the same sort occurred elsewhere before the 1840s, when the British and Germans began belatedly to adopt and adapt the example set by the French a generation before. Probably institutional and individual factors will prove primarily responsible for that early French leadership. Beginning very slowly in the 1760s, with the appointments of Nollet and then of Monge to teach *physique expérimentale* at the *Ecole du génie* at Mézières,

27. Aspects of the problem of mathematizing physics are considered in Kuhn, "The Function of Measurement in Modern Physical Science," *Isis* 52 (1961): 161–93 (pp. 178–224 below), where the distinction between classical and Baconian sciences was first introduced in print. Others are to be found in Robert Fox, *The Caloric Theory of Gases from Lavoisier to Regnault* (Oxford, 1971).

Baconian subjects increasingly penetrated the education of French military engineers.[28] That movement culminated in the establishment during the 1790s of the *Ecole polytechnique*, a new sort of educational institution at which students were exposed not only to the classical subjects relevant to the *arts mécaniques* but also to chemistry, the study of heat, and other related subjects. It can be no accident that all of those who produced mathematical theories of previously experimental fields were either teachers or students at the *Ecole polytechnique*. Also of great importance to the direction taken by their work was the magistral leadership of Laplace in extending Newtonian mathematical physics to nonmathematical subjects.[29]

For reasons that are currently both obscure and controversial, the practice of the new mathematical physics declined rapidly in France after about 1830. In part it participated in a general decline in the vitality of French science, but an even more important role was probably played by a reassertion of the traditional primacy of mathematics, itself after mid-century moving further away from the concrete concerns of physics. As physics after 1850 became mathematical in all its parts, remaining nonetheless dependent on refined experiment, French contributions for a century declined to a level unmatched in such previously comparable fields as chemistry and mathematics.[30] Physics required, as other sciences did not, the establishment of a firm bridge across the classical-Baconian divide.

28. Relevant information will be found in René Taton, "L'école royale du génie de Mézières," in R. Taton, ed., *Enseignement et diffusion des sciences en France au XVIIIe siècle* (Paris, 1964), pp. 559–615.

29. R. Fox, "The Rise and Fall of Laplacian Physics," *Historical Studies in the Physical Sciences* 4 (1976): 89–136; R. H. Silliman, "Fresnel and the Emergence of Physics as a Discipline," ibid., pp. 137–62.

30. Relevant information as well as guidance to the still sparse literature on this topic will be found in R. Fox, "Scientific Enterprise and the Patronage of Research in France, 1800–70," *Minerva* 11 (1973): 442–73; H. W. Paul, "La science française de la seconde partie du XIXe siècle vue par les auteurs anglais et américains," *Revue d'histoire des sciences* 27 (1974): 147–63. Note, however, that both are concerned primarily with the alleged decline in French science as a whole, an effect surely less pronounced and perhaps quite distinct from the decline of French physics. Conversations with Fox have reinforced my convictions and helped me to organize my remarks on these points.

What had begun in France during the first quarter of the nineteenth century had, therefore, later to be recreated elsewhere, initially in Germany and Britain after the mid-1840s. In both countries, as might by now be expected, existing institutional forms at first inhibited the cultivation of a field dependent upon easy communication between practitioners skilled in experiment on the one hand and mathematics on the other. Part of Germany's quite special success—attested by the preponderant role of Germans in the twentieth-century conceptual transformations of physics—must be due to the rapid growth and consequent plasticity of German educational institutions during the years when men like Neumann, Weber, Helmholtz, and Kirchhoff were creating a new discipline in which both experimentalists and mathematical theorists would be associated as practitioners of physics.[31]

During the first decades of this century that German model increasingly spread to the rest of the world. As it did so, the long-standing division between the mathematical and the experimental physical sciences was more and more obscured and may even seem to have disappeared. But, from another viewpoint, it is perhaps more accurately described as having been displaced—from a position between separate fields to the interior of physics itself, a location from which it continues to provide a source of both individual and professional tensions. It is only, I suggest, because physical theory is now everywhere mathematical that theoretical and experimental physics appear as enterprises so different that almost no one can hope to achieve eminence in both. No such dichotomy between experiment and theory has characterized fields like chemistry or biology in which theory is less intrinsically mathematical. Perhaps, therefore, the cleavage between mathematical and experimental science still remains, rooted in the nature of the human mind.[32]

31. Russel McCormmach, "Editor's Foreword," *Historical Studies in the Physical Sciences* 3 (1971): ix–xxiv.

32. Other frequently remarked but still little investigated phenomena also hint at a psychological basis for this cleavage. Many mathematicians and *theoretical* physicists have been passionately interested in and involved with music, some having had great difficulty choosing between a scientific and a musical career. No comparably widespread involvement is visible in the experimental sciences including experimental physics (nor, I think, in other disciplines without an apparent relationship to music). But music, or

part of it, was once a member of the cluster of mathematical sciences, never of the experimental. Also likely to be revealing is further study of a subtle distinction often remarked by physicists: that between a "mathematical" and a "theoretical" physicist. Both use much mathematics, often on the same problems. But the first tends to take the physics problem as conceptually fixed and to develop powerful mathematical techniques for application to it; the second thinks more physically, adapting the conception of his problem to the often more limited mathematical tools at his disposal. Lewis Pyenson, to whom I am indebted for helpful comments on my earliest draft, is developing interesting ideas on the evolution of the distinction.

4 Energy Conservation as an Example of Simultaneous Discovery

Reprinted by permission from Marshall Clagett, ed., *Critical Problems in the History of Science* (Madison: University of Wisconsin Press, 1959), pp. 321–56.

Between 1842 and 1847, the hypothesis of energy conservation was publicly announced by four widely scattered European scientists—Mayer, Joule, Colding, and Helmholtz—all but the last working in complete ignorance of the others.[1] The coincidence is conspicuous,

1. J. R. Mayer, "Bemerkungen über die Kräfte der unbelebten Natur," *Ann. d. Chem. u. Pharm.*, vol. 42 (1842). I have used the reprint in J. J. Weyrauch's excellent collection, *Die Mechanik der Wärme in gesammelten Schriften von Robert Mayer* (Stuttgart, 1893), pp. 23–30. This volume is cited below as Weyrauch, I. The same author's companion volume, *Kleinere Schriften und Briefe von Robert Mayer* (Stuttgart, 1893), is cited as Weyrauch, II.

James P. Joule, "On the Calorific Effects of Magneto-Electricity, and on the Mechanical Value of Heat," *Phil. Mag.*, vol. 23 (1843). I have used the version in *The Scientific Papers of James Prescott Joule* (London, 1884), pp. 123–59. This volume is cited below as Joule, *Papers.*

L. A. Colding, "Undersögelse on de almindelige Naturkraefter og deres gjensidige Afhaengighed og isaerdeleshed om den ved visse faste Legemers Gnidning udviklede Varme," *Dansk. Vid. Selsk.* 2 (1851): 121–46. I am indebted to Miss Kirsten Emilie Hedebol for preparing a translation of this paper. It is, of course, far fuller than the unpublished original, which Colding read to the Royal Society of Denmark in 1843, but it includes much information about that original. See also, L. A. Colding, "On the History of the Principle of the Conservation of Energy," *Phil. Mag.* 27 (1864): 56–64.

yet these four announcements are unique only in combining generality of formulation with concrete quantitative applications. Sadi Carnot, before 1832, Marc Séguin in 1839, Karl Holtzmann in 1845, and G. A. Hirn in 1854, all recorded their independent convictions that heat and work are quantitatively interchangeable, and all computed a value for the conversion coefficient or an equivalent.[2] The convertibility of heat and work is, of course, only a

H. von Helmholtz, *Ueber die Erhaltung der Kraft. Eine physikalische Abhandlung* (Berlin, 1847). I have used the annotated reprint in *Wissenschaftliche Abhandlungen von Hermann Helmholtz* (Leipzig, 1882), 1: 12–75. This set is cited below as Helmholtz, *Abhandlungen.*

2. Carnot's version of the conservation hypothesis is scattered through a notebook written between the publication of his memoir in 1824 and his death in 1832. The most authoritative version of the notes is E. Picard, *Sadi Carnot, biographie et manuscript* (Paris 1927); a more convenient source is the appendix to the recent reprint of Carnot's *Réflexions sur la puissance motrice du feu* (Paris, 1953). Notice that Carnot considered the material in these notes quite incompatible with the main thesis of his famous *Réflexions.* In fact, the essentials of his thesis proved to be salvageable, but a change in both its statement and its derivation was required.

Marc Séguin, *De l'influence des chemins de fer et de l'art de les construire* (Paris, 1839), pp. xvi, 380–96.

Karl Holtzmann, *Über die Wärme und Elasticität der Gase und Dämpfe* (Mannheim, 1845). I have used the translation by W. Francis in *Taylor's Scientific Memoirs,* 4 (1846):189–217. Since Holtzmann believed in the caloric theory of heat and used it in his monograph, he is a strange candidate for a list of discoverers of energy conservation. He also believed, however, that the same amount of work spent in compressing a gas isothermally must always produce the same increment of heat in the gas. As a result he made one of the early computations of Joule's coefficient and his work is therefore repeatedly cited by the early writers on thermodynamics as containing an important ingredient of their theory. Holtzmann can scarcely be said to have caught any part of energy conservation as we define that theory today. But for this investigation of simultaneous discovery the judgment of his contemporaries is more relevant than our own. To several of them Holtzmann seemed an active participant in the evolution of the conservation theory.

G. A. Hirn, "Etudes sur les principaux phénomènes que présentent les frottements médiats, et sur les diverses manières de déterminer la valeur mécanique des matières employées au graissage des machines," *Bulletin de la societé industrielle de Mulhouse* 26 (1854): 188–237; and "Notice sur les lois de la production du calorique par les frottements médiats," ibid., pp. 238–77. It is hard to believe that Hirn was completely ignorant of the special case of energy conservation, but the generality lacking in this second group of announcements occurs elsewhere in the literature of the period. Between 1837 and 1844, C. F. Mohr, William Grove, Faraday, and Liebig all described the world of phenomena as manifesting but a single "force," one which could appear in electrical, thermal, dynamical, and many other forms, but which could never, in all its transformations, be created or destroyed.[3] That so-called force is the one known to later scientists as energy.

work of Mayer, Joule, Helmholtz, Clausius, and Kelvin when he wrote the "Études" in 1854. But after reading his paper, I find his claim to independent discovery (presented in the "Notice") entirely convincing. Since none of the standard histories cite these articles or even recognizes the existence of Hirn's claim, it seems appropriate to sketch its basis here.

Hirn's investigation deals with the relative effectiveness of various engine lubricants as a function of pressure at the bearing, applied torque, etc. Quite unexpectedly, or so he says, his measurements led to the conclusion that: "The absolute quantity of caloric developed by mediated friction [e.g., friction between two surfaces separated by a lubricant] is directly and uniquely proportional to the mechanical work absorbed by this friction. And if we express the work in kilograms raised to the height of one meter and the quantity of caloric in calories, we find that the ratio of these two numbers is very nearly 0.0027 [corresponding to 370 kg.m./cal.], whatever the velocity and the temperature and whatever the lubricating material" (p. 202). Until almost 1860 Hirn had doubts about the validity of the law for impure lubricants or in the absence of lubrication (see particularly his *Récherches sur l'équivalent mécanique de la chaleur* [Paris, 1858], p. 83.) But despite these doubts, his work obviously displays one of the mid-nineteenth-century routes to an important part of energy conservation.

3. C. F. Mohr, "Ueber die Natur der Wärme," *Zeit. f. Phys.* 5 (1837): 419–45; and "Ansichten über die Natur der Wärme," *Ann. d. Chem. u. Pharm.* 24 (1837): 141–47.

William R. Grove, *On the Correlation of Physical Forces: Being the Substance of a Course of Lectures Delivered in the London Institution in the Year 1843* (London, 1846). Grove states that in this first edition he has introduced no new material since the lectures were delivered. The later and more accessible editions are greatly revised in the light of subsequent work.

Michael Faraday, *Experimental Researches in Electricity* (London, 1844), 2:101–4. The original "Seventeenth Series" of which this is a part was read to the Royal Society in March, 1840.

Justus Liebig, *Chemische Briefe* (Heidelberg, 1844), pp. 114–20. With this work, as with Grove's, one must beware of changes introduced in editions published after the conservation of energy was a recognized scientific law.

The history of science offers no more striking instance of the phenomenon known as simultaneous discovery.

Already we have named twelve men who, within a short period of time, grasped for themselves essential parts of the concept of energy and its conservation. Their number could be increased, but not fruitfully.[4] The present multiplicity sufficiently suggests that in

4. Since a few of my conclusions depend upon the particular list of names selected for study, a few words about the selection procedure seem essential. I have tried to include all the men who were thought by their contemporaries or immediate successors to have reached independently some significant part of energy conservation. To this group I have added Carnot and Hirn, whose work would surely have been so regarded if it had been known. Their lack of actual influence is irrelevant from the viewpoint of this investigation.

This procedure has yielded the present list of twelve names, and I am aware of only four others for whom a place might be claimed. They are von Haller, Roget, Kaufmann, and Rumford. Despite P. S. Epstein's impassioned defense (*Textbook of Thermodynamics* [New York, 1937], pp. 27–34), von Haller has no place on the list. The notion that fluid friction in the arteries and veins contributes to body heat implies no part of the notion of energy conservation. Any theory that accounts for frictional generation of heat can embrace von Haller's conception. A better case can be made for Roget, who did use the impossibility of perpetual motion to argue against the contact theory of galvanism (see note 27). I have omitted him only because he seems unaware of the possibility of extending the argument and because his own conceptions are duplicated in the work of Faraday, who did extend them.

Hermann von Kaufmann probably should be included. According to Georg Helm his work is identical with Holtzmann's (*Die Energetik nach ihrer geschichtlichen Entwickelung* [Leipzig, 1898], p. 64). But I have been unable to see Kaufmann's writings, and Holtzmann's case is already somewhat doubtful, so that it has seemed better not to overload the list. As to Rumford, whose case is the most difficult of all, I shall point out below that before 1825 the dynamical theory of heat did not lead its adherents to energy conservation. Until mid-century there was no necessary, or even likely, connection between the two sets of ideas. But Rumford was more than a dynamical theorist. He also said: "It would follow necessarily, from [the dynamical theory] . . . that the sum of the active forces in the universe must always remain constant" (*Complete Works* [London, 1876], 3:172), and this does sound like energy conservation. Perhaps it is. But if so, Rumford seems totally unaware of its significance. I cannot find the remark applied or even repeated elsewhere in his works. My inclination, therefore, is to regard the sentence as an easy echo, appropriate before a French audience, of the eighteenth-century theorem about the conservation of *vis viva*. Both Daniel Bernoulli and Lavoisier and Laplace had applied that theorem to the dynamical theory before (see note 95) without obtain-the two decades before 1850 the climate of European scientific thought included elements able to guide receptive scientists to a significant new view of nature. Isolating these elements within the works of the men affected by them may tell us something of the nature of simultaneous discovery. Conceivably, it may even give substance to those obvious yet totally unexpressive truisms: "A scientific discovery must fit the times," or "The time must be ripe." The problem is challenging. A preliminary identification of the sources of the phenomenon called simultaneous discovery is therefore the main objective of this paper.

Before proceeding toward that objective, however, we must briefly pause over the phrase "simultaneous discovery" itself. Does it sufficiently describe the phenomenon we are investigating? In the ideal case of simultaneous discovery two or more men would announce the same thing at the same time and in complete ignorance of each other's work, but nothing remotely like that happened during the development of energy conservation. The violations of simultaneity and mutual influence are secondary. But no two of our men even said the same thing. Until close to the end of the period of discovery, few of their papers have more than fragmentary resemblances retrievable in isolated sentences and paragraphs. Skillful excerpting is, for example, required to make Mohr's defense of the dynamical theory of heat resemble Liebig's discussion of the intrinsic limits of the electric motor. A diagram of the overlapping passages in the papers by the pioneers of energy conservation would resemble an unfinished crossword puzzle.

Fortunately no diagram is needed to grasp the most essential differences. Some pioneers, like Séguin and Carnot, discussed only a special case of energy conservation, and these two used very different approaches. Others, like Mohr and Grove, announced a universal conservation principle, but, as we shall see, their occasional attempts to quantify their imperishable "force" leave its concrete significance in doubt. Only in view of what happened later can we say that all these partial statements even deal with the same aspect of nature.[5] Nor is this problem of divergent discoveries restricted

ing anything like energy conservation. I know of no reason to suppose that Rumford saw further than they.

5. This may well explain why the pioneers seem to have profited so little from each other's work, even when they read it. Our twelve men were not, in fact, strictly independent. Grove and Helmholtz knew Joule's work and

to those scientists whose formulations were obviously incomplete. Mayer, Colding, Joule, and Helmholtz were not saying the same things at the dates usually given for their discoveries of energy conservation. More than *amour propre* underlies Joule's subsequent claim that the discovery he had announced in 1843 was different from the one published by Mayer in 1842.[6] In these years their papers have important areas of overlap, but not until Mayer's book of 1845 and Joule's publications of 1844 and 1847 do their theories become substantially coextensive.[7]

cited it in their papers of 1843 and 1847 (Grove, *Physical Forces*, pp. 39, 52; Helmholtz, *Abhandlungen*, 1:33, 35, 37, 55). Joule, in turn, knew and cited the work of Faraday (*Papers*, p. 189). Liebig, though he did not cite Mohr and Mayer, must have known their work, for it was published in his own journal. (See also G. W. A. Kahlbaum, *Liebig und Friedrich Mohr, Briefe, 1834–1870* [Braunschweig, 1897], for Liebig's knowledge of Mohr's theory.) Very possibly more precise biographical information would disclose other interdependencies as well.

But these interdependencies, at least the identifiable ones, seem unimportant. In 1847 Helmholtz seems to have been unaware both of the generality of Joule's conclusions and of their large-scale overlap with his own. He cites only Joule's experimental findings, and these very selectively and critically. Not until the priority controversies of the second half-century, does Helmholtz seem to have recognized the extent to which he had been anticipated. Much the same holds for the relation between Joule and Faraday. From the latter Joule took illustrations, but not inspiration. Liebig's case may prove even more revealing. He could have neglected to cite Mohr and Mayer simply because they provided no relevant illustration and did not even seem to be dealing with the same subject matter. Apparently the men whom we call early exponents of energy conservation could occasionally read each other's works without quite recognizing that they were talking about the same things. For that matter, the fact that so many of them wrote from different professional and intellectual backgrounds may account for the infrequency with which they even saw each other's writings.

6. J. P. Joule, "Sur l'équivalent mécanique du calorique," *Comptes rendus* 28 (1849): 132–35. I have used the reprint in Weyrauch, II, pp. 276–80. This is only the first salvo in the priority controversy, but it already shows what the controversy is going to be about. Which of two (and later more than two) different statements is to be equated with *the* conservation of energy?

7. J. R. Mayer, *Die organische Bewegung in ihrem Zusammenhange mit dem Stoffwechsel* (Heilbronn, 1845) in Weyrauch, I, pp. 45–128. Most of Joule's papers between 1843 and 1847 are relevant, but particularly: "On the Changes of Temperature Produced by the Rarefaction and Condensatio

In short, though the phrase "simultaneous discovery" points to the central problem of this paper, it does not, if taken at all literally, describe it. Even to the historian acquainted with the concepts of energy conservation, the pioneers do not all communicate the same thing. To each other, at the time, they often communicated nothing at all. What we see in their works is not really the simultaneous discovery of energy conservation. Rather it is the rapid and often disorderly emergence of the experimental and conceptual elements from which that theory was shortly to be compounded. It is these elements that concern us. We know why they were there: Energy *is* conserved; nature behaves that way. But we do not know why these elements suddenly became accessible and recognizable. That is the fundamental problem of this paper. Why, in the years 1830–50, did so many of the experiments and concepts required for a full statement of energy conservation lie so close to the surface of scientific consciousness?[8]

This question could easily be taken as a request for a list of all those almost innumerable factors that caused the individual pioneers to make the particular discoveries that they did. Interpreted in this way, it has no answer, at least none that the historian can give. But the historian can attempt another sort of response. A contemplative immersion in the works of the pioneers and their contemporaries may reveal a subgroup of factors which seem more significant than the others, because of their frequent recurrence, their specificity to the period, and their decisive effect upon indi-

of Air" (1845) and "On Matter, Living Force, and Heat" (1847) in *Papers*, pp. 172–89, 265–81.

8. This formulation has at least one considerable advantage over the usual version. It does not imply or even permit the question, "Who *really* discovered conservation of energy first?" As a century of fruitless controversy has demonstrated, a suitable extension or restriction in the definition of energy conservation will award the crown to almost any one of the pioneers, an additional indication that they cannot have discovered the same thing.

esent formulation also bars a second impossible question, "Did or Séguin, or Mohr, or any one of the other pioneers, at will) p the concept of energy conservation, even intuitively? Does he ng on the list of pioneers?" Those questions have no conceivable cept in terms of the respondent's taste. But whatever answer taste , Faraday (or Séguin, etc.) provides useful evidence about the led to the discovery of energy conservation.

vidual research.[9] The depth of my acquaintance with the literature permits, as yet, no definitive judgments. Nevertheless, I am already quite sure about two such factors, and I suspect the relevance of a third. Let me call them the "availability of conversion processes," the "concern with engines," and the "philosophy of nature." I shall consider them in order.

The availability of conversion processes resulted principally from the stream of discoveries that flowed from Volta's invention of the battery in 1800. According to the theory of galvanism most prevalent, at least, in France and England, the electric current was itself gained at the expense of forces of chemical affinity, and this conversion proved to be only the first step in a chain.[10] Electric cur-

9. These three criteria, particularly the second and third, determine the orientation of this study in a way that may not be immediately apparent. They direct attention away from the *prerequisites* to the discovery of energy conservation and toward what might be called the *trigger-factors* responsible for simultaneous discovery. For example, the following pages will show implicitly that all of the pioneers made significant use of the conceptual and experimental elements of calorimetry and that many of them also depended upon the new chemical conceptions derived from the work of Lavoisier and his contemporaries. These and many other developments within the sciences presumably had to occur before conservation of energy, as we know it, could be discovered. I have not, however, explicitly isolated elements like these below, because they do not seem to distinguish the pioneers from their predecessors. Since both calorimetry and the new chemistry had been the common property of all scientists for some years before the period of simultaneous discovery, they cannot have provided the immediate stimuli that triggered the work of the pioneers. As prerequisites for discovery, these elements have an interest and importance all their own. But their study is unlikely to illuminate very much the problem of simultaneous discovery to which this paper is directed. [This note has been added to the original manuscript in response to points raised during the discussion that followed the oral presentation.]

10. Faraday provides scarce and useful information about the progress of the significant controversy between the exponents of the chemical and contact theories of galvanism (*Experimental Researches*, 2: 18–20). According to his account, the chemical theory was dominant in France and England from at least 1825 on, but the contact theory was still dominant in Germany and Italy when Faraday wrote in 1840. Does the dominance of the contact theory in Germany account for the rather surprising way in which both Mayer and Helmholtz neglect the battery in their accounts of energy transformations?

rent invariably produced heat and, under appropriate conditions, light as well. Or, by electrolysis, the current could vanquish forces of chemical affinity, bringing the chain of transformations full circle. These were the first fruits of Volta's work; other more striking conversion discoveries followed during the decade and a half after 1820.[11] In that year Oersted demonstrated the magnetic effects of a current; magnetism, in turn, could produce motion, and motion had long been known to produce electricity through friction. Another chain of conversions was closed. Then, in 1822, Seebeck showed that heat applied to a bimetallic junction would produce a current directly. Twelve years later Peltier reversed this striking example of conversion, demonstrating that the current could, on occasions, absorb heat, producing cold. Induced currents, discovered by Faraday in 1831, were only another, if particularly striking, member of a class of phenomena already characteristic of nineteenth-century science. In the decade after 1827, the progress of photography added yet another example, and Melloni's identification of light with radiant heat confirmed a long-standing suspicion about the fundamental connection between two other apparently disparate aspects of nature.[12]

Some conversion processes had, of course, been available before 1800. Motion had already produced electrostatic charges, and the resulting attractions and repulsions had produced motion. Static generators had occasionally engendered chemical reactions, including dissociations, and chemical reactions produced both light and heat.[13] Harnessed by the steam engine, heat could produce motion, and motion, in turn, engendered heat through friction and percussion. Yet in the eighteenth century these were isolated phenomena; few seemed of central importance to scientific research; and

11. For the following discoveries see Sir Edmund Whittaker, *A History of the Theories of Aether and Electricity*, vol. 1, *The Classical Theories*, 2d ed. (London, 1951), pp. 81–84, 88–89, 170–71, 236–37. For Oersted's discovery see also, R. C. Stauffer, "Persistent Errors Regarding Oersted's Discovery of Electromagnetism," *Isis* 44 (1953): 307–10.

12. F. Cajori, *A History of Physics* (New York, 1922), pp. 158, 172–74. Grove makes a particular point of the early photographic processes (*Physical Forces*, pp. 27–32). Mohr gives great emphasis to Melloni's work (*Zeit. f. Phys.* 5 [1837]: 419).

13. For the chemical effects of static electricity see Whittaker, *Aether and Electricity*, 1:74, n. 2.

those few were studied by different groups. Only in the decade after 1830, when they were increasingly classified with the many other examples discovered in rapid succession by nineteenth-century scientists, did they begin to look like conversion processes at all.[14] By that time scientists were proceeding inevitably in the laboratory from a variety of chemical, thermal, electrical, magnetic, or dynamical phenomena to phenomena of any of the other types and to optical phenomena as well. Previously separate problems were gaining multiple interrelationships, and that is what Mary Sommerville had in mind when, in 1834, she gave her famous popularization of science the title, *On the Connexion of the Physical Sciences.* "The progress of modern science," she said in her preface, "especially within the last five years, has been remarkable for a tendency to . . . unite detached branches [of science, so that today] . . . there exists such a bond of union, that proficiency cannot be attained in any one branch without a knowledge of others."[15] Mrs. Sommerville's remark isolates the "new look" that physical science had acquired between 1800 and 1835. That new look, together with the discoveries that produced it, proved to be a major requisite for the emergence of energy conservation.

Yet, precisely because it produced a "look" rather than a single clearly defined laboratory phenomenon, the availability of conversion processes enters the development of energy conservation in an immense variety of ways. Faraday and Grove achieved an idea very close to conservation from a survey of the whole network of conversion processes taken together. For them conservation was quite literally a rationalization of the phenomenon Mrs. Sommerville described as the new "connexion." C. F. Mohr, on the other hand, took the idea of *conservation* from a quite different source, probably metaphysical.[16] But, as we shall see, it is only because he attempted to elucidate and defend this idea in terms of the new

14. The single exception is significant and is discussed at some length below. During the eighteenth century steam engines were occasionally regarded as conversion devices.

15. Mary Sommerville, *On the Connexion of the Physical Sciences* (London, 1834), unpaginated Preface.

16. Reasons for distinguishing Mohr's approach from that of Grove and Faraday will be examined below (note 83). The accompanying text will consider possible sources of Mohr's conviction about the conservation of "force."

conversion processes that Mohr's initial conception came to look like conservation *of energy*. Mayer and Helmholtz present still another approach. They began by applying their concepts of conservation to well-known older phenomena. But until they extended their theories to embrace the new discoveries, they were not developing the same theory as men like Mohr and Grove. Still another group, consisting of Carnot, Séguin, Holtzmann, and Hirn, ignored the new conversion processes entirely. But they would not be discoverers of energy conservation if men like Joule, Helmholtz, and Colding had not shown that the thermal phenomena with which these steam engineers dealt were integral parts of the new network of conversions.

There is, I think, excellent reason for the complexity and variety of these relationships. In an important sense, though one which will demand later qualification, the conservation of energy is nothing less than the theoretical counterpart of the laboratory conversion processes discovered during the first four decades of the nineteenth century. Each laboratory conversion corresponds in the theory to a transformation in the form of energy. That is why, as we shall see, Grove and Faraday could derive conservation from the network of laboratory conversions itself. But the very homomorphism between the theory, energy conservation, and the earlier network of laboratory conversion processes indicates that one did not have to start by grasping the network whole. Liebig and Joule, for example, started from a single conversion process and were led by the connection between the sciences through the entire network. Mohr and Colding started with a metaphysical idea and transformed it by application to the network. In short, just because the new nineteenth-century discoveries formed a network of connections between previously distinct parts of science, they could be grasped either individually or whole in a large variety of ways and still lead to the same ultimate result. That, I think, explains why they could enter the pioneers' research in so many different ways. More important, it explains why the researches of the pioneers, despite the variety of their starting points, ultimately converged to a common outcome. What Mrs. Sommerville had called the new connections between the sciences often proved to be the links that joined disparate approaches and enunciations into a single discovery.

The sequence of Joule's researches clearly illustrates the way in which the network of conversion processes actually marked out the experimental ground of energy conservation and thus provided the essential links between the various pioneers. When Joule first wrote in 1838, his exclusive concern with the design of improved electric motors effectively isolates him from all the other pioneers of energy conservation except Liebig. He was simply working on one of the many new problems born from nineteenth-century discovery. By 1840 his systematic evaluations of motors in terms of work and "duty" establishes a link to the researches of the steam engineers, Carnot, Séguin, Hirn, and Holtzmann.[17] But these connections vanished in 1841 and 1842, when Joule's discouragement with motor design forced him to seek instead a fundamental improvement in the batteries that drove them. Now he was concerned with new discoveries in chemistry, and he absorbed entirely Faraday's view of the essential role of chemical processes in galvanism. In addition, his research in these years was concentrated upon what turned out to have been two of the numerous conversion processes selected by Grove and Mohr to illustrate their vague metaphysical hypothesis.[18] The connections with the work of other pioneers are steadily increasing in number.

In 1843, prompted by the discovery of an error in his earlier work with batteries, Joule reintroduced the motor and the concept of mechanical work. The link to steam engineering was thus established, and simultaneously Joule's papers began, for the first time, to read like investigations of energy relations.[19] But even in 1843 the resemblance to energy conservation was incomplete. Only as

17. The first eleven items in Joule's *Papers* (pp. 1–53) are exclusively concerned with improving first motors and then electromagnets, and these items cover the period 1838–41. The systematic evaluations of motors in terms of the engineering concepts, work and "duty," occur on pp. 21–25, 48. For Joule's earliest published use of the concept work or its equivalent, see p. 4.

18. Joule's concern with batteries and more particularly with the electrical production of heat by batteries dominates the five major contributions in *Papers*, pp. 53–123. My remark that Joule was led to batteries by his discouragement with motor design is a conjecture, but it seems extremely probable.

19. See note 1. This is the paper in which Joule is usually said to have announced energy conservation.

Joule traced still other new connections during the years 1844–47 did his theory really encompass the views of such disparate figures as Faraday, Mayer, and Helmholtz.[20] Starting from an isolated problem, Joule had involuntarily traced much of the connective tissue between the new nineteenth-century discoveries. As he did so, his work was linked increasingly to that of the other pioneers, and only when many such links had appeared did his discovery resemble energy conservation.

Joule's work shows that energy conservation could be discovered by starting from a single conversion process and tracing the network. But, as we have already indicated, that is not the only way in which conversion processes could effect the discovery of energy conservation. C. F. Mohr, for example, probably drew his initial concept of conservation from a source independent of the new conversion processes, but then used the new discoveries to clarify and elaborate his ideas. In 1839, close to the end of a long and often incoherent defense of the dynamical theory of heat, Mohr suddenly burst out: "Besides the known 54 chemical elements, there is, in the nature of things, only one other agent, and that is called force; it can appear under various circumstances as motion, chemical affinity, cohesion, electricity, light, heat, and magnetism, and from any one of these types of phenomena all the others can be called forth."[21] A knowledge of energy conservation makes the import of these sentences clear. But in the absence of such knowledge, they would have been almost meaningless except that Mohr proceeded immediately to two systematic pages of experimental examples. The experiments were, of course, just the new and old conversion processes listed above, the new ones in the lead, and they are essential to Mohr's argument. They alone specify his subject and show its close similarity to Joule's.

Mohr and Joule illustrate two of the ways in which conversion processes could affect the discoverers of energy conservation. But, as my final example from the works of Faraday and Grove will indicate, these are not the only ways. Though Faraday and Grove reached conclusions much like Mohr's, their route to the conclusions includes none of the same sudden leaps. Unlike Mohr, they seem to have derived energy conservation directly from the ex-

20. See note 7.
21. *Zeit. f. Phys.* 5 (1837): 442.

perimental conversion processes that they had already studied so fully in their own researches. Because their route is continuous, the homomorphism of energy conservation with the new conversion processes appears most clearly of all in their work.

In 1834, Faraday concluded five lectures on the new discoveries in chemistry and galvanism with a sixth on the "Relations of Chemical Affinity, Electricity, Heat, Magnetism, and Other Powers of Matter." His notes supply the gist of this last lecture in the words: "We cannot say that any one [of these powers] is the cause of the others, but only that all are connected and due to one common cause." To illustrate the connection, Faraday then gave nine experimental demonstrations of "the production of any one [power] from another, or the conversion of one into another."[22] Grove's development seems parallel. In 1842 he included a remark almost identical with Faraday's in a lecture with the significant title, "On the Progress of Physical Science."[23] In the following year he expanded this isolated remark into his famous lecture series, *On the Correlation of Physical Forces.* "The position which I seek to establish in this Essay is," he said, "that [any one] of the various imponderable agencies . . . viz., Heat, Light, Electricity, Magnetism, Chemical Affinity, and Motion, . . . may, as a force, produce or be convertible into the other[s]; thus heat may mediately or immediately produce electricity, electricity may produce heat; and so of the rest."[24]

This is the concept of the universal convertibility of natural powers, and it is not, let us be clear, the same as the notion of conservation. But most of the remaining steps proved to be small and rather obvious.[25] All but one, to be discussed below, can be taken by applying to the concept of universal convertibility the perennially serviceable philosophic tags about the equality of cause and

22. Bence Jones, *The Life and Letters of Faraday* (London, 1870), 2:47.

23. *A Lecture on the Progress of Physical Science since the Opening of the London Institution* (London, 1842). Though the title page is dated 1842, the date is immediately followed by "[Not Published]." I do not know when the actual printing took place, but a prefatory remark of the author's indicates that the text itself was written very shortly after the lecture was delivered.

24. *Physical Forces*, p. 8.

25. Reasons for calling the remaining steps "obvious" are given in the closing paragraphs of this paper (see note 92).

effect or the impossibility of perpetual motion. Since any power can produce any other *and be produced by it*, the equality of cause and effect demands a uniform quantitative equivalence between each pair of powers. If there is no such equivalence, then a properly chosen series of conversions will result in the creation of power, that is, in perpetual motion.[26] In all its manifestations and conversions, power must be conserved. This realization came neither all at once, nor fully to all, nor with complete logical rigor. But it did come.

Though he had no general conception of conversion processes, Peter Mark Roget, in 1829, opposed Volta's contact theory of galvanism because it implied a creation of power from nothing.[27] Faraday independently reproduced the argument in 1840 and immediately applied it to conversions in general. "We have," he said, "many processes by which the form of the power may be so changed that an apparent *conversion* of one into another takes place. . . . But in no cases . . . is there a pure creation of force; a production of power without a corresponding exhaustion of something to supply it."[28] In 1842 Grove devised the argument once more in order to prove the impossibility of inducing an electric current from static magnetism, and in the following year he generalized still further.[29] If it were true, he wrote, "that motion [could] be subdivided or changed in character, so as to become heat, electricity, etc.; it ought to follow, that when we collect the dissipated and changed forces, and reconvert them, the initial motion, affecting the same amount of matter with the same velocity, should be reproduced, and so of the change of matter produced by the other forces."[30] In the context of Grove's exhaustive discussion of the known conversion processes, this quotation is a full statement of all but the quantitative components of energy conservation. Furthermore, Grove knew what was missing. "The great problem that remains to be solved, in regard to the correlation of physical

26. Strictly speaking, this derivation is valid only if all the transformations of energy are reversible, which they are not. But that logical shortcoming completely escaped the notice of the pioneers.

27. P. M. Roget, *Treatise on Galvanism* (London, 1829). I have seen only the excerpt quoted by Faraday, *Experimental Researches*, 2:103, n. 2.

28. *Experimental Researches*, 2:103.

29. *Progress of Physical Science*, p. 20.

30. *Physical Forces*, p. 47.

forces, is," he wrote, "the establishment of their equivalent of power, or their measurable relation to a given standard."[31] Conversion phenomena could carry scientists no further toward the enunciation of energy conservation.

Grove's case brings this discussion of conversion processes almost full circle. In his lectures energy conservation appears as the straightforward theoretical counterpart of nineteenth-century laboratory discoveries, and that was the suggestion from which I began. Only two of the pioneers, it is true, actually derived their versions of energy conservation from these new discoveries alone. But because such a derivation was possible, every one of the pioneers was decisively affected by the availability of conversion processes. Six of them dealt with the new discoveries from the start of their research. Without these discoveries, Joule, Mohr, Faraday, Grove, Liebig, and Colding would not be on our list at all.[32] The other six pioneers show the importance of conversion processes in a subtler but no less important way. Mayer and Helmholtz were late in turning to the new discoveries, but only when they did so, did they become candidates for the same list as the first six. Carnot, Séguin, Hirn, and Holtzmann are the most interesting of all. None of them even mentioned the new conversion processes. But their contributions, being uniformly obscure, would have vanished from history entirely if they had not been gathered into the larger network explored by the men we have already examined.[33] When con-

31. Ibid., p. 45.

32. I am not quite sure that this is true of Colding, particularly since I have not seen his unpublished paper of 1843. The early pages of his 1851 paper (note 1) contain many examples of conversion processes and are thus reminiscent of Mohr's approach. Also, Colding was a protegé of Oersted, whose chief renown derived from his discovery of electromagnetic conversions. On the other hand, most of the conversion processes cited explicitly by Colding date from the eighteenth century. In Colding's case, I suspect a prior tie between conversion processes and metaphysics (see note 83 and accompanying text). Very probably neither can be viewed as either logically or psychologically the more fundamental in the development of his thought.

33. Carnot's notes were not published until 1872 and then only because they contained anticipations of an important scientific law. Séguin had to call attention to the relevant passages in his book of 1839. Hirn did not bother to claim credit, but only attached a note denying plagiarism to his 1854 paper. That paper was published in an engineering journal that I

version processes did not govern an individual's work, they often governed that work's reception. If they had not been available, the problem of simultaneous discovery might not exist at all. Certainly it would look very different.

Nevertheless, the view which Grove and Faraday derived from conversion processes is not identical with what scientists now call the conservation of energy, and we must not underestimate the importance of the missing element. Grove's *Physical Forces* contains the layman's view of energy conservation. In an expanded and revised form it proved to be one of the most effective and sought-after popularizations of the new scientific law.[34] But this role was achieved only after the work of Joule, Mayer, Helmholtz, and their successors had provided a full quantitative substructure for the conception of force correlation. Anyone who has worked through a mathematical and numerical treatment of energy conservation may well wonder whether, in the absence of such substructure, Grove would have had anything to popularize. The "measurable relation to a given standard" of the various physical forces is an essential ingredient of energy conservation as we know it, and neither Grove, Faraday, Roget, nor Mohr was able even to approach it.

The quantification of energy conservation proved, in fact, insuperably difficult for those pioneers whose principal intellectual equipment consisted of concepts related to the new conversion processes. Grove thought he had found the clue to quantification in Dulong and Petit's law relating chemical affinity and heat.[35] Mohr believed he had produced the quantitative relationship when

have never seen cited by a scientist. Holtzmann's paper is the exception in that it was not obscure. But if other men had not discovered conservation of energy, Holtzmann's memoir would have continued to look like another one of the extensions of Carnot's memoir, for that is basically what it was (see note 2).

34. Between 1850 and 1875 Grove's book was reprinted at least six times in England, three times in America, twice in France, and once in Germany. The extensions were, of course, numerous, but I am aware of only two essential revisions. In the original discussion of heat (pp. 8–11), Grove suggested that macroscopic motion appears as heat only to the extent that it is *not* transformed to microscopic motion. In addition, of course, Grove's few attempts at quantification were quite off the track (see below).

35. *Physical Forces,* p. 46.

he equated the heat employed to raise the temperature of water 1° with the static force necessary to compress the same water to its original volume.[36] Mayer initially measured force by the momentum which it could produce.[37] These random leads were all totally unproductive, and of this group only Mayer succeeded in transcending them. To do so he had to use concepts belonging to a very different aspect of nineteenth-century science, an aspect to which I previously referred as the concern with engines, and whose existence I shall now take for granted as a well-known by-product of the Industrial Revolution. As we examine this aspect of science, we shall find the main source of the concepts—particularly of mechanical effect or work—required for the quantitative formulation of energy conservation. In addition, we shall find a multitude of experiments and of qualitative conceptions so closely related to energy conservation that they collectively provide something very like a second and independent route to it.

Let me begin by considering the concept of work. Its discussion will provide relevant background as well as opportunity for a few essential remarks on a more usual view about the sources of the quantitative concepts underlying energy conservation. Most histories or prehistories of the conservation of energy imply that the model for quantifying conversion processes was the dynamical theorem known almost from the beginning of the eighteenth century as the conservation of *vis viva*.[38] That theorem has a distinguished role in the history of dynamics, and it also turns out to have been a special case of energy conservation. It could have provided a model. Yet I think the prevalent impression that it did so is misleading. The conservation of *vis viva* was important to Helmholtz's derivation of energy conservation, and a special case (free fall) of the same dynamical theorem was ultimately of great assistance to Mayer. But these men also drew significant elements from a second generally separate tradition—that of water, wind,

36. *Zeit. f. Phys.* 5 (1837): 422–23.

37. Weyrauch, II, pp. 102–5. This is in his first paper, "Ueber die quantitative und qualitative Bestimmung der Kräfte," sent to Poggendorf in 1841 but not published until after Mayer's death. Before he wrote his second paper, the first to be published, Mayer had learned a bit more physics.

38. It would be more precise to say that most prehistories of energy conservation are principally lists of anticipations, and these occur particularly often in the early literature on *vis viva*.

and steam engineering—and that tradition is all important to the work of the other five pioneers who produced a quantitative version of energy conservation.

There is excellent reason why this should be so. *Vis viva* is mv^2, the product of mass by the square of velocity. But until a late date that quantity appears in the works of none of the pioneers except Carnot, Mayer, and Helmholtz. As a group the pioneers were scarcely interested in energy of motion, much less in using it as a basic quantitative measure. What they did use, at least those who were successful, was $f{\cdot}s$, the product of force times distance, a quantity known variously under the names mechanical effect, mechanical power, and work. That quantity does not, however, occur as an independent conceptual entity in the dynamical literature. More precisely it scarcely occurs there until 1820, when the French (and only the French) literature was suddenly enriched by a series of theoretical works on such subjects as the theory of machines and of industrial mechanics. These new books did make work a significant independent conceptual entity, and they did relate it explicitly to *vis viva*. But the concept was not invented for these books. On the contrary it was borrowed from a century of engineering practice where its use had usually been quite independent of both *vis viva* and its conservation. That source within the engineering tradition is all that the pioneers of energy conservation required and as much as most of them used.

Another paper will be needed to document this conclusion, but let me illustrate the considerations from which it derives. Until 1743 the general dynamical significance of the conservation of *vis viva* must be recaptured from its application to two special sorts of problems: elastic impact and constrained fall.[39] Force times dis-

39. The early eighteenth-century literature contains many general statements about the conservation of *vis viva* regarded as a metaphysical force. These formulations will be discussed briefly below. For the present notice only that none of them is suitable for application to the technical problems of dynamics, and it is with those formulations that we are here concerned. An excellent discussion of both the dynamical and metaphysical formulations is included in A. E. Haas, *Die Entwicklungsgeschichte des Satzes von der Erhaltung der Kraft* (Vienna, 1909), generally the fullest and most reliable prehistory of energy conservation. Other useful details can be found in Hans Schimank, "Die geschichtliche Entwicklung des Kraftbegriffs bis

tance has no relevance to the former, since elastic impact numerically conserves *vis viva.* In other applications, for example, the bachistochrone and isochronous pendulum, vertical displacement rather than force times distance appears in the conservation theorem. Huyghen's statement that the center of gravity of a system of masses can ascend no higher than its initial position of rest is typical.[40] Compare Daniel Bernoulli's famous formulation of 1738: Conservation of *vis viva* is "the equality of actual descent with potential ascent."[41]

The more general formulations, inaugurated by d'Alembert's *Traité* in 1743, suppress even vertical displacement, which might conceivably be called an embryonic conception of work. D'Alembert states that the forces acting on a system of interconnected bodies will increase its *vis viva* by the amount $\Sigma m_i u_i^2$, where the u_i are the velocities that the masses m_i would have acquired if moved freely over the same paths by the same forces.[42] Here, as in Daniel Bernoulli's subsequent version of the general theorem, force times distance enters only in certain particular applications to permit the computation of individual u_i's; it has neither general significance nor a name; *vis viva* is the conceptual parameter.[43] The same parameter dominates the later analytic formulations. Euler's *Mechanica*, Lagrange's *Mécanique analytique*, and Laplace's *Mé-*

zum Aufkommen der Energetik," in *Robert Mayer und das Energieprinzip, 1842–1942*, ed. H. Schimank and E. Pietsch (Berlin, 1942). I am indebted to Professor Erwin Hiebert for calling these two useful and little known works to my attention.

40. Christian Huyghens, *Horologium oscillatorium* (Paris, 1673). I have used the German edition, *Die Pendeluhr*, ed. A. Heckscher and A. V. Oettingen, Ostwald's Klassiker der Exakten Wissenschaften, no. 192 (Leipzig, 1913), p. 112.

41. D. Bernoulli, *Hydrodynamica, sive de viribus et motibus fluidorum, commentarii* (Basel, 1738), p. 12.

42. J. L. d'Alembert, *Traité de dynamique* (Paris, 1743). I have been able to see only the second edition (Paris, 1758), where the relevant material occurs on pp. 252–53. D'Alembert's discussion of the changes introduced since the first edition give no reason to suspect he has altered the original formulation at this point.

43. D. Bernoulli, "Remarques sur le principe de la conservation des forces vives pris dans un sense général," *Hist. Acad. de Berlin* (1748), pp. 356–64.

canique céleste give exclusive emphasis to central forces derivable from potential functions.[44] In these works the integral of force times differential path element occurs only in the derivation of the conservation law. The law itself equates *vis viva* with a function of position coordinates.

Not until 1782, in Lazare Carnot's *Essai sur les machines en général*, did force times distance begin to receive a name and a conceptual priority in dynamical theory.[45] Nor was this new dy-

44. L. Euler, *Mechanica sive motus scientia analytice exposita*, in *Opera omnia* (Leipzig and Berlin, 1911–), ser. 2, 2:74–77. The first edition was St. Petersburg, 1736.

J.-L. Lagrange, *Mécanique analytique* (Paris, 1788), pp. 206–9. I cite the first edition because the second, as reprinted in volumes 11 and 12 of Lagrange's *Oeuvres* (Paris, 1867–92), contains a very significant change. In the first edition, the conservation of *vis viva* is formulated only for time-independent constraints and for central or other integrable forces. It then takes the form $\Sigma m_i v_i^2 = 2H + 2\Sigma m_i \pi_i$, where H is a constant of integration and the π_i are functions of the position coordinates. In the second edition, Paris, 1811–15 (*Oeuvres*, 11:306–10), Lagrange repeats the above but restricts it to a particular class of elastic bodies in order to take account of Lazare Carnot's engineering treatise (note 45), which he cites. For a fuller account of the engineering problem treated by Carnot, he refers his readers to his own *Théorie des fonctions analytiques* (Paris, 1797), pp. 399–410, where his version of Carnot's engineering problem is formulated more explicitly. That formulation makes the impact of the engineering tradition quite apparent, for the concept work now begins to appear. Lagrange states that the increment of *vis viva* between two dynamical states of the system is $2(P) + 2(Q) + \ldots$, where (P)—Lagrange calls it an "aire"—is $\Sigma_i \int P_i \, dp_i$, and P_i is the force on the i'th body in the direction of the position coordinates p_i. These "aires" are, of course, just work.

P. S. Laplace, *Traité de mécanique céleste* (Paris, 1798–1825). The relevant passages are more readily found in *Oeuvres complètes* (Paris, 1878–1904), 1:57–61. Mathematically, this treatment of 1798 actually resembles Lagrange's 1797 formulation rather than the earlier 1788 form. But, as in the pre-engineering formulations, the conservation law which includes a work integral is rapidly passed over in favor of the more restricted statement employing a potential function.

45. L. N. M. Carnot, *Essai sur les machines en général* (Dijon, 1782). I have consulted this work in Carnot's *Oeuvres mathématiques* (Basel, 1797) but rely principally on the expanded and more influential second edition, *Principes fondementaux de l'équilibre et du mouvement* (Paris, 1803). Carnot introduces several terms for what we call work, the most important being, "force vive latent" and "moment d'activité" (ibid., pp. 38–43). Of

namical view of the concept work really worked out or propagated until the years 1819–39, when it received full expression in the works of Navier, Coriolis, Poncelet, and others.[46] All these works are concerned with the analysis of machines in motion. As a result, work—the integral force with respect to distance—is their fundamental conceptual parameter. Among other significant and typical results of this reformulation were the introduction of the term "work" and of units for its measure, the redefinition of *vis viva* as $\frac{1}{2}mv^2$ to preserve the conceptual priority of the measure work, and the explicit formulation of the conservation law in terms of the equality of work done and kinetic energy created.[47] Only when thus reformulated did the conservation of *vis viva* provide a

these he says, "The kind of quantity to which I have given the name *moment of activity* plays a very large role in the theory of machines in motion: for in general it is this quantity which one must economize as much as possible in order to derive from an agent [i.e., a source of power] all the [mechanical] effect of which it is capable" (ibid., p. 257).

46. A useful survey of the early history of this important movement is C. L. M. H. Navier, "Détails historiques sur l'emploi du principes des forces vives dans la théorie des machines et sur diverses roues hydrauliques," *Ann. Chim. Phys.* 9 (1818): 146–59. I suspect that Navier's edition of B. de F. Belidor's *Architecture hydraulique* (Paris, 1819) contains the first developed presentation of the new engineering physics, but I have not yet seen this work. The standard treatises are: G. Coriolis, *Du calcul de l'effet des machines, ou considérations sur l'emploi des moteurs et sur leur évaluation pour servir d'introduction à l'étude speciale des machines* (Paris, 1829); C. L. M. H. Navier, *Résumé des leçons données a l'école des ponts et chaussées sur l'application de la mécanique à l'établissement des constructions et des machines* (Paris, 1838), vol. 2; and J.-V. Poncelet, *Introduction à la mécanique industrielle*, ed. Kratz, 3d ed. (Paris, 1870). This work originally appeared in 1829 (part had appeared in lithoprint in 1827); the much enlarged and now standard edition from which the third is taken appeared in 1830–39.

47. The formal adoption of the term *work* (*travail*) is often credited to Poncelet (*Introduction*, p. 64), though many others had used it casually before; Poncelet also (pp. 74–75) gives a useful account of the units (*dynamique, dyname, dynamie*, etc.) commonly used to measure this quantity. Coriolis (*Du calcul de l'effet des machines*, p. iv) is the first to insist that *vis viva* be $\frac{1}{2}mv^2$, so that it will be numerically equal to the work it can produce; he also makes much use of the term *travail*, which Poncelet may have borrowed from him. The reformulation of the conservation law proceeds gradually from Lazare Carnot through all these later works.

convenient conceptual model for the quantification of conversion processes, and then almost none of the pioneers used it. Instead, they returned to the same older engineering tradition in which Lazare Carnot and his French successors had found the concepts needed for their new versions of the dynamical conservation theorem.

Sadi Carnot is the single exception. His manuscript notes proceed from the assertion that heat is motion to the conviction that it is molecular *vis viva* and that its increment must therefore be equal to work done. These steps imply an immediate command of the relation between work and *vis viva*. Mayer and Helmholtz might also have been exceptions, for both could have made good use of the French reformulation. But neither seems to have known it. Both began by taking work (or rather the product of weight times height) as the measure of "force," and each then rederived something very like the French reformulation for himself.[48] The other

48. As soon as he considers a quantitative problem in his first published paper, Mayer says: "A cause, which effects the raising of a weight, is a force; since this force brings about the fall of a body, we shall call it fall-force [Fallkraft]" (Weyrauch, I, p. 24). This is the engineering, not the theoretical dynamical, measure. By applying it to the problem of free fall, Mayer immediately derives $\frac{1}{2}mv^2$ (note the fraction) as the measure of energy of motion. The very crudeness of his derivation together with its lack of generality indicates his ignorance of the French engineering texts. The one French text he does mention in his writings (G. Lamé, *Cours de physique de l'école polytechnique*, 2d ed. [Paris, 1840]) does not deal with *vis viva* or its conservation at all.

Helmholtz uses the terms *Arbeitskraft, bewegende Kraft, mechanische Arbeit,* and *Arbeit* for his fundamental measurable force (Helmholtz, *Abhandlungen*, I, 12, 17–18). I have not as yet been able to trace these terms in the earlier German literature, but their parallels in the French and English engineering traditions are obvious. Also, the term *bewegende Kraft* is used by the translator of Clapeyron's version of Sadi Carnot's memoir as equivalent to the French *puissance motrice* (*Pogg. Ann.* 59 [1843]: 446), and Helmholtz cites this translation (p. 17, n. 1). To this extent the tie to the engineering tradition is explicit.

Helmholtz was not, however, aware of the French theoretical engineering tradition. Like Mayer, he derives the factor of ½ in the definition of energy of motion and is unaware of any precedent for it (p. 18). More significant, he fails completely to identify $\int Pdp$ as work or *Arbeitskraft*, and instead calls it the "sum of the tensions" (*Summe der Spannkräfte*) over the space dimension of the motion.

six pioneers who reached or came close to the quantification of conversion processes could not even have used the reformulation. Unlike Mayer and Helmholtz, they applied the concept work directly to a problem in which *vis viva* is constant from cycle to cycle and therefore does not enter. Joule and Liebig are typical. Both began by comparing the "duty" of the electric motor with that of the steam engine. How much weight, they both asked, can each of these engines raise through a fixed distance for a given expenditure of coal or zinc? That question is basic to their entire research programs as it is to the programs of Carnot, Séguin, Holtzmann, and Hirn. It is not, however, a question drawn from either the new or old dynamics.

But neither, except for its application to the electrical case, is it a novel question. The evaluation of engines in terms of the weight each could raise to a given height is implicit in Savery's engine descriptions of 1702 and explicit in Parent's discussion of water wheels in 1704.[49] Under a variety of names, particularly mechanical effect, weight times height provided the basic measure of engine achievement throughout the engineering works of Desagulier, Smeaton, and Watt.[50] Borda applied the same measure to hydraulic machines and Coulomb to wind and animal power.[51] These ex-

49. The unit implicit in Savery's work is really the horsepower, but this includes weight times height as a part. See H. W. Dickinson and Rhys Jenkins, *James Watt and the Steam Engine* (Oxford, 1927), pp. 353–54. Antoine Parent, "Sur le plus grande perfection possible des machines," *Hist. Acad. Roy.* (1704), pp. 323–38.

50. J. T. Desagulier, *A Course of Experimental Philosophy*, 3d ed., 2 vols. (London, 1763), particularly 1:132, and 2:412. This posthumous edition is practically a reprint of the second edition (London, 1749).

John Smeaton, "An Experimental Inquiry concerning the Natural Powers of Water and Wind to Turn Mills, and Other Machines, depending on a Circular Motion," *Phil. Trans.* 51 (1759): 51. Here the measure is weight times height per unit time. The time dependence is, however, dropped in his "An Experimental Examination of the Quantity and Proportion of Mechanic Power Necessary to be Employed in Giving Different Degrees of Velocity to Heavy Bodies," *Phil. Trans.* 66 (1776): 458.

For Watt see Dickinson and Jenkins, *James Watt*, pp. 353–56.

51. J. C. Borda, "Mémoires sur les roues hydrauliques," *Mem. l'Acad. Roy.* (1767), p. 272. Here the measure is weight times vertical speed. Height replaces speed in C. Coulomb, "Observation théorique et expérimentale sur l'effet des moulins à vent, et sur la figure de leurs ailes," ibid. (1781), p. 68, and "Resultat de plusieurs expériences destinée à determiner la quantité

amples, drawn from all parts of the eighteenth century, but increasing in density toward its close, could be multiplied almost indefinitely. Yet even these few should prepare the way for a little noted but virtually decisive statistic. Of the nine pioneers who succeeded, partially or completely, in quantifying conversion processes, all but Mayer and Helmholtz were either trained as engineers or were working directly on engines when they made their contributions to energy conservation. Of the six who computed independent values of the conversion coefficient, all but Mayer were concerned with engines either in fact or by training.[52] To make the computation they needed the concept work, and the source of that concept was principally the engineering tradition.[53]

The concept work is the most decisive contribution to energy conservation made by the nineteenth-century concern with engines. That is why I have devoted so much space to it. But the concern with engines contributed to the emergence of energy conservation in a number of other ways besides, and we must consider at least

d'action que les hommes peuvent fournir par leur travail journalier, suivant les differentes manières dont ils emploient leurs forces," *Mem. de l'Inst.* 2 (1799): 381.

52. Mayer states that he loved to build model water wheels as a boy and that he learned the impossibility of perpetual motion in studying them (Weyrauch, II, p. 390). He could have learned simultaneously the proper measure of the product of machines.

53. Professor Hiebert asks if the concept of mechanical work may not have emerged from elementary statics and particularly from the formulation that derives statics from the principle of virtual velocities. The point needs further research, but my present response must be at least equivocally negative. The elements of statics were an important item in the equipment of all eighteenth-century engineers and the principle of virtual velocities, or an equivalent, therefore recurs in eighteenth-century writings on engineering problems. Quite possibly the engineers could not have evolved the concept work without the aid of the pre-existing static principle. But, as the preceding discussion may indicate, if the eighteenth-century concept work did emerge from the far older principle of virtual velocities, it did so only when that principle was firmly embedded in the engineering tradition and only when that tradition turned its attention to the evaluation of power sources such as animals, falling water, wind, and steam. Therefore, reverting to the vocabulary of note 9, I suggest that the principle of virtual velocities may have been a prerequisite for the discovery of energy conservation but that it can scarcely have been a trigger. [This note added to original manuscript in response to points raised during discussion.]

a few of them. For example, long before the discovery of electrochemical conversion processes, men interested in steam and water engines had occasionally seen them as devices for transforming the force latent in fuel or falling water to the mechanical force that raises weight. "I am persuaded," said Daniel Bernoulli in 1738, "that if all the *vis viva* hidden in a cubic foot of coal were called forth and usefully applied to the motion of a machine, more could be achieved than by the daily labor of eight or ten men."[54] Apparently that remark, made at the height of the controversy over metaphysical *vis viva*, had no later influence. Yet the same perception of engines recurs again and again, most explicitly in the French engineering writers. Lazare Carnot, for example, says that "the problem of turning a mill stone, whether by the impact of water, or by wind, or by animal power . . . is that of consuming the maximum possible [portion] of the work delivered by these agents."[55] With Coriolis, water, wind, steam, and animals are all simply sources of work, and machines become devices for putting this in useful form and transmitting it to the load.[56] Here, engines by themselves lead to a conception of conversion processes very close to that produced by the new discoveries of the nineteenth century. That aspect of the engine problem may well explain why the steam engineers—Hirn, Holtzmann, Séguin, and Sadi Carnot—were led to the same aspect of nature as men like Grove and Faraday.

The fact that engines could and occasionally did look like conversion devices may also explain something more. Is this not the reason why engineering concepts proved so readily transferable to the more abstract problems of energy conservation? The concept work is only the most important example of such a transfer. Joule and Liebig reached energy conservation by asking an old engi-

54. *Hydrodynamica*, p. 231.

55. *De l'équilibre et du mouvement*, p. 258. Notice also that as soon as Lagrange turns to Carnot's problem (note 44), he speaks in the same way. In the *Fonctions analytiques*, he says that waterfalls, coal, gunpowder, animals, etc., all "contain a quantity of *vis viva*, which one can harness but which one cannot increase by any mechanical means. One may [therefore] always regard a machine as intended to destroy a given quantity of *vis viva* [in the load] by consuming some other given *vis viva* [from the source]" (*Oeuvres*, 9:410).

56. *Du calcul de l'effet des machines*, chap. 1. For Coriolis the conservation theorem applied to a perfect machine becomes the "Principle of the Transmission of Work."

neering question, "What is the 'duty'?" about the new conversion processes in the battery-driven electric motor. But that question—how much work for how much fuel?—embraces the notion of a conversion process. In retrospect, it even sounds like the request for a conversion coefficient. Joule, at least, finally answered the question by producing one. Or consider the following more surprising transfer of engineering concepts. Though its fundamental conceptions are incompatible with energy conservation, Sadi Carnot's *Réflexion sur la puissance motrice du feu* was cited by both Helmholtz and Colding as the outstanding application of the impossibility of perpetual motion to a nonmechanical conversion process.[57] Helmholtz may well have borrowed from Carnot's memoir the analytic concept of a cyclic process that played so large a role in his own classic paper.[58] Holtzmann derived his value of the conversion coefficient by a minor modification of Carnot's analytic procedures, and Carnot's own discussion of energy conservation repeatedly employs data and concepts from his earlier and fundamentally incompatible memoir. These examples may give at least a hint of the ease and frequency with which engineering concepts were applied in deriving the abstract scientific conservation law.

My final example of the productiveness of the nineteenth-century concern with engines is less directly tied to engines. Yet it underscores the multiplicity and variety of the relationships that make the engineering factor bulk so large in this account of simultaneous discovery. I have shown elsewhere that many of the pioneers shared an important interest in the phenomenon known as adiabatic com-

57. Helmholtz, *Abhandlungen*, 1:17. Colding, "Naturkraefter," *Dansk. Vid. Selsk.* 2 (1851): 123–24. Particularly interesting evidence about the apparent similarities between the theory of energy conservation and Carnot's incompatible theory of the heat engine is provided by Carlo Matteucci. His paper, "De la relation qui existe entre la quantité de l'action chimique et la quantité de chaleur, d'électricité et de lumière qu'elle produit," *Bibliothèque universelle de Genève, Supplement*, 4 (1847): 375–80, is an attack upon several of the early exponents of energy conservation. He describes his opponents as the group of physicists who "have tried to show that Carnot's celebrated principle about the motive force of heat can be applied to the other imponderable fluids."

58. Helmholtz, *Abhandlungen*, 1:18–19, gives Helmholtz's initial abstract formulation of the cyclic process.

pression.[59] Qualitatively, the phenomenon provided an ideal demonstration of the conversion of work to heat; quantitatively, adiabatic compression yielded the only means of computing a conversion coefficient with existing data. The discovery of adiabatic compression has, of course, little or nothing to do with the interest in engines, but the nineteenth-century experiments which the pioneers used so heavily often seem related to just this practical concern. Dalton, and Clément and Désormes, who did important early work on adiabatic compression, also contributed early fundamental measurements on steam, and these measurements were used by many of the engineers.[60] Poisson, who developed an early theory of adiabatic compression, applied it, in the same article, to the steam engine, and his example was immediately followed by Sadi Carnot, Coriolis, Navier, and Poncelet.[61] Séguin, though he uses a different sort of data, seems a member of the same group. Dulong, to whose classic memoir on adiabatic compression many of the pioneers referred, was a close collaborator of Petit, and during the period of their collaboration Petit produced a quantitative account of the steam engine that antedates Carnot's by eight years.[62] There is even a hint of government interest. The prize offered by the French *Institut national* and won in 1812 by the classic research on gases

59. T. S. Kuhn, "The Caloric Theory of Adiabatic Compression," *Isis* 49 (1958): 132–40.

60. John Dalton, "Experimental Essays on the Constitution of Mixed Gases; on the Force of Steam or Vapour from Water and Other Liquids in Different Temperatures, Both in a Torricellian Vacuum and in Air; on Evaporation; and on the Expansion of Gases by Heat," *Manch. Mem.* 5 (1802): 535–602. The second essay, though it grew out of Dalton's meteorological interests, was immediately exploited by both British and French engineers.

Clément and Désormes, "Mémoires sur la théorie des machines à feu," *Bulletin des sciences par la société philomatique* 6 (1819): 115–18; and "Tableau relatif à la théorie général de la puissance mécanique de la vapeur," ibid. 13 (1826): 50–53. The second paper appears in full in Crelle's *Journal für die Baukunst* 6 (1833): 143–64. For the contributions of these men to adiabatic compression, see my paper, cited in note 59.

61. S. D. Poisson, "Sur la chaleur des gaz et des vapeurs," *Ann.. Chim. Phys.* 23 (1823): 337–52. For Navier, Coriolis, and Poncelet, all of whom devote chapters to steam engine computations, see note 46.

62. A. T. Petit, "Sur l'emploi du principe des forces vives dans le calcul de l'effet des machines," *Ann. Chim. Phys.* 8 (1818): 287–305.

of Delaroche and Bérard may well have grown in part from government interest in engines.[63] Certainly Regnault's later work on the same topic did. His famous investigations of the thermal characteristics of gas and steam bear the imposing title, "Experiments, undertaken by order of the Minister of Public Works and at the instigation of the Central Commission for Steam Engines, to determine the principal laws and the numerical data which enter into steam engine calculations."[64] One suspects that without these ties to the recognized problems of steam engineering, the important data on adiabatic compression would not have been so accessible to the pioneers of energy conservation. In this instance the concern with engines may not have been essential to the work of the pioneers, but it certainly facilitated their discoveries.

Because the concern with engines and the nineteenth-century conversion discoveries embrace most of the new technical concepts and experiments common to more than a few of the discoverers of energy conservation, this study of simultaneous discovery might well end here. But a last look at the papers of the pioneers generates an uncomfortable feeling that something is still missing, something that is not perhaps a substantive element at all. This feeling would not exist if all the pioneers had, like Carnot and Joule, begun with a straightforward technical problem and proceeded by stages to the concept of energy conservation. But in the cases of Colding, Helmholtz, Liebig, Mayer, Mohr, and Séguin, the notion of an underlying imperishable metaphysical force seems prior to research and almost unrelated to it. Put bluntly, these pioneers seem to have held an idea capable of becoming conservation of energy for some time before they found evidence for it. The factors previously discussed in this paper may explain why they were ultimately able to clothe the idea and thus to make sense of it. But the discussion does not yet sufficiently account for the idea's ex-

63. F. Delaroche and J. Bérard, "Mémoire sur la determination de la chaleur specifique des differents gaz," *Ann. Chim. Phys.* 85 (1813): 72–110, 113–82. I know of no direct evidence relating the prize won by this memoir to the problems of steam engineering, but the Academy did offer a prize for improvement in steam engines as early as 1793. See H. Guerlac, "Some Aspects of Science during the French Revolution," *The Scientific Monthly* 80 (1955): 96.

64. In *Mém. de l'Acad.* 21 (1847): 1–767.

istence. One or two such cases among the twelve pioneers might not be troublesome. The sources of scientific inspiration are notoriously inscrutable. But the presence of major conceptual lacunae in six of our twelve cases is surprising. Though I cannot entirely resolve the problem it presents, I must at least touch upon it.

We have already noted a few of the lacunae. Mohr jumped without warning from a defense of the dynamical theory of heat to the statement that there is only one force in nature and that it is quantitatively unalterable.[65] Liebig made a similar leap from the duty of electric motors to the statement that the chemical equivalents of the elements determine the work retrievable from chemical processes by either electrical or thermal means.[66] Colding tells us that he got the idea of conservation in 1839, while still a student, but withheld announcement until 1843 so that he might gather evidence.[67] The biography of Helmholtz outlines a similar story.[68] Séguin confidently applied his concept of the convertibility of heat and motion to steam engine computations, even though his single attempt to confirm the idea had been totally fruitless.[69] Mayer's leap has repeatedly been noted, but its full size is not often remarked. From the light color of venous blood in the tropics, it is a

65. See note 21 and accompanying text.

66. *Chemische Briefe*, pp. 115–17.

67. Colding, "History of Conservation," *Phil. Mag.* 27 (1864): 57–58.

68. Leo Koenigsberger (*Hermann von Helmholtz*, tr. F. A. Welby [Oxford, 1906], pp. 25–26, 31–33) implies that Helmholtz's ideas about conservation were complete as early as 1843, and he states that by 1845 the attempt at experimental proof motivated all of Helmholtz's research. But Koenigsberger gives no evidence, and he cannot be quite correct. In two articles on physiological heat written during 1845 and 1846 (*Abhandlungen*, 1:8–11; 2:680–725), Helmholtz fails to notice that body heat may be expended in mechanical work (compare the discussion of Mayer, below). In the second of these papers he also gives the usual caloric explanation of adiabatic compression in terms of the change in heat capacity with pressure. In short, his ideas were by no means complete until 1847 or shortly before. But the papers of 1845 and 1846 do show that in these years Helmholtz was concerned to combat vitalism, which he thought implied the creation of force from nothing. Also they show that he already knew the work of Clapeyron and of Holtzmann, which he thought relevant. To this extent, at least, Koenigsberger must be right.

69. *Chemins de fer*, p. 383. Séguin had tried unsuccessfully to measure the difference in the quantities of heat abstracted from the boiler and delivered to the condenser of a steam engine.

small step to the conclusion that less internal oxidation is needed when the body loses less heat to the environment.[70] Crawford had drawn that conclusion from the same evidence in 1778.[71] Laplace and Lavoisier, in the 1780s, had balanced the same equation relating inspired oxygen to the body's heat losses.[72] A continuous line of research relates their work to the biochemical studies of respiration made by Liebig and Helmholtz in the early 1840s.[73] Though Mayer apparently did not know it, his observation of venous blood was simply a rediscovery of evidence for a well-known, though controversial, biochemical theory. But that theory was not the one to which Mayer leaped. Instead Mayer insisted that internal oxidation must be balanced against *both* the body's heat loss *and* the manual labor the body performs. To this formulation, the light color of tropical venous blood is largely irrelevant. Mayer's extension of the theory calls for the discovery that lazy men, rather than hot men, have light venous blood.

The persistent occurrence of mental jumps like these suggests that many of the discoverers of energy conservation were deeply predisposed to see a single indestructible force at the root of all natural phenomena. The predisposition has been noted before, and a number of historians have at least implied that it is a residue of a similar metaphysic generated by the eighteenth-century controversy over the conservation of *vis viva*. Leibniz, Jean and Daniel Bernoulli, Hermann, and du Châtelet, all said things like, "*Vis* [*viva*] never perishes; it may in truth appear lost, but one can always discover it again in its effects if one can see them."[74] There are a multitude of such statements, and their authors do attempt, however crudely, to trace *vis viva* into and out of nonmechanical phenomena. The parallel to men like Mohr and Colding is very

70. Weyrauch, I, pp. 12–14.

71. E. Farber, "The Color of Venous Blood," *Isis* 45 (1954): 3–9.

72. A. Lavoisier and P. S. Laplace, "Mémoire sur la chaleur," *Hist. de l'Acad.* (1780), pp. 355–408.

73. Helmholtz touches on much of this research in his paper of 1845, "Wärme, physiologisch," for the *Encyclopädische Wörterbuch der medicinischen Wissenschaften* (*Abhandlungen*, 2:680–725).

74. Haas, *Erhaltung*, p. 16, n. Quoted from *Institutions physiques de Madame la Marquise du Chastellet adressés à Mr. son Fils* (Amsterdam, 1742).

close. Yet eighteenth-century metaphysical sentiments of this sort seem an implausible source for the nineteenth-century predisposition we are examining. Though the technical *dynamical* conservation theorem has a continuous history from the early eighteenth century to the present, its metaphysical counterpart found few or no defenders after 1750.[75] To discover the *metaphysical* theorem, the pioneers of energy conservation would have had to return to books at least a century old. Neither their works nor their biographies suggest that they were significantly influenced by this particular bit of ancient intellectual history.[76]

Statements like those of both the eighteenth-century Leibnizians and the nineteenth-century pioneers of energy conservation can, however, be found repeatedly in the literature of a second philosophical movement, *Naturphilosophie*.[77] Positing organism as the fundamental metaphor of their universal science, the *Naturphilosophen* constantly sought a single unifying principle for all natural phenomena. Schelling, for example, maintained "that magnetic, electrical, chemical, and finally even organic phenomena would be interwoven into one great association . . . [which] extends over the

75. Haas, *Erhaltung*, p. 17.

76. None of the pioneers mention the eighteenth-century conservation literature in their original papers. Colding, however, says that he got his first glimpse of conservation while reading d'Alembert in 1839 (*Phil. Mag.* 27 [1864]: 58), and Koenigsberger says that Helmholtz had read d'Alembert and Daniel Bernoulli by 1842 (*Helmholtz*, p. 26). These two counterexamples do not, however, really modify my thesis. D'Alembert omitted all mention of the metaphysical conservation theorem from the first edition of his *Traité*, and in the second he explicitly disowned the view (Paris, 1758, beginning of the "Avertissement" and pp. xvii–xxiv). In fact, d'Alembert was among the first to insist on freeing dynamics from what he considered to be mere metaphysical speculations. To take his ideas from this source Colding would still have required a strong predisposition. Bernoulli's *Hydrodynamica* is a more appropriate source (see, for example, the text that accompanies note 54), but Koenigsberger makes the very plausible point that Helmholtz consulted Bernoulli in order to work out his preexisting conception of conservation.

77. The roots of *Naturphilosophie* can, of course, be traced back through Kant and Wolff to Leibniz, and Leibniz was the author of the metaphysical conservation theorem about which both Kant and Wolff wrote (Haas, *Erhaltung*, pp. 15–18). The two movements are not, therefore, entirely independent.

whole of nature."[78] Even before the discovery of the battery he insisted that "without doubt only a single force in its various guises is manifest in [the phenomena of] light, electricity, and so forth."[79] These quotations point to an aspect of Schelling's thought fully documented by Bréhier and more recently by Stauffer.[80] As a *Naturphilosoph*, Schelling constantly sought out conversion and transformation processes in the science of his day. At the beginning of his career chemistry seemed to him the basic physical science; from 1800 on he increasingly found in galvanism "the true border-phenomenon of both [organic and inorganic] natures."[81] Many of Schelling's followers, whose teaching dominated German and many neighboring universities during the first third of the nineteenth century, gave similar emphasis to the new conversion phenomena. Stauffer has shown that Oersted—a *Naturphilosoph* as well as a scientist—persisted in his long search for a relation between electricity and magnetism largely because of his prior philosophical conviction that one must exist. Once the interaction was discovered, electro-magnetism played a major role in Herbart's further elaboration of the scientific substructure of *Naturphilosophie*.[82] In short, many *Naturphilosophen* drew from their philosophy a view of physical processes very close to that which Faraday and Grove seem to have drawn from the new discoveries of the nineteenth century.[83]

78. Quoted by R. C. Stauffer, "Speculation and Experiment in the Background of Oersted's Discovery of Electromagnetism," *Isis* 48 (1957): 37, from Schelling's *Einleitung zu seinem Entwurf eines Systems der Naturphilosophie* (1799).

79. Quoted by Haas, *Erhaltung*, p. 45, n. 61, from Schelling's *Erster Entwurf eines Systems der Naturphilosophie* (1799).

80. Émile Bréhier, *Schelling* (Paris, 1912). This is the most helpful discussion I have found and should certainly be added to Stauffer's list of useful aids for studying the complex relations of science and *Naturphilosophie* (*Isis* 48 [1957]: 37, n. 21).

81. Stauffer, "Speculation and Experiment," p. 36, from Schelling's "Allgemeiner Deduktion des dynamischen Processes oder der Kategorien der Physik" (1800).

82. Haas, *Erhaltung*, p. 41.

[illegible] of course, impossible to distinguish sharply between the in[illegible] *Naturphilosophie* and that of conversion processes. Bréhier [illegible] pp. 23–24) and Windelband (*History of Philosophy*, trans. J. H. [illegible] ed. [New York, 1901], pp. 597–98) both emphasize that conver[illegible]esses were themselves a significant source of *Naturphilosophie*, so

Naturphilosophie could, therefore, have provided an appropriate philosophical background for the discovery of energy conservation. Furthermore, several of the pioneers were acquainted with at least its essentials. Colding was a protegé of Oersted's.[84] Liebig studied for two years with Schelling, and though he afterwards described these years as a waste, he never surrendered the vitalism he had then imbibed.[85] Hirn cited both Oken and Kant.[86] Mayer did not study *Naturphilosophie*, but he had close student friends who did.[87] Helmholtz's father, an intimate of the younger Fichte's and a minor *Naturphilosoph* in his own right, constantly exhorted his son to desert strict mechanism.[88] Though Helmholtz himself felt forced to excise all philosophical discussion from his classic mem-

that the two were often grasped together. This fact must qualify some of the dichotomies set up in the first part of this paper, for the distinction between the two sources of the conservation concept is often equally hard to apply to individual pioneers. I have already pointed out the difficulty in Colding's case (note 32). With Mohr and Liebig I am still inclined to give *Naturphilosophie* the psychological priority, because neither had dealt much with the new conversion processes in their own research and because both make such large leaps. Their cases appear in sharp contrast to those of Grove and Faraday, who seem to proceed by a continuous path from conversion processes to conservation. But this continuity may be deceptive. Grove (*Physical Forces*, pp. 25–27) mentions Coleridge, and Coleridge was the principal British exponent of *Naturphilosophie*. Since the problem presented by these examples seems to me both real and unresolved, I had better point out that it affects only the organization, not the main thesis, of this paper. Perhaps conversion processes and *Naturphilosophie* should be considered in the same section. Nevertheless, they would both have to be considered.

84. Povl Vinding, "Colding, Ludwig August," *Dansk Biografisk Leksikon* (Copenhagen, 1933–44), pp. 377–82. I am grateful to Roy and Ann Lawrence for providing me with a précis of this useful biographical sketch.

85. E. von Meyer, *A History of Chemistry*, trans. G. McGowan, 3d ed. (London, 1906), p. 274. J. T. Merz, *European Thought in the Nineteenth Century* (London, 1923–50), 1:178–218, particularly the last page.

86. G. A. Hirn, "Etudes sur les lois et sur les principes constituants de l'univers," *Revue d' Alsace* 1 (1850): 24–41, 127–42, 183–201; ibid., 2 (1851): 24–45. References to writings related to *Naturphilosophie* occur relatively often, though they are not very favorable. On the other hand, the very title of this piece suggests *Naturphilosophie*, and the title is appropriate to the contents.

87. B. Hell, "Robert Mayer," *Kantstudien* 19 (1914): 222–48.

88. Koenigsberger, *Helmholtz*, pp. 3–5, 30.

oir, he was able by 1881 to recognize important Kantian residues that had escaped his earlier censorship.[89]

Biographical fragments of this sort do not, of course, prove intellectual indebtedness. They may, however, justify strong suspicion, and they surely provide leads for further research. At the moment I shall only insist that this research should be done and that there are excellent reasons to suppose it will be fruitful. Most of those reasons are given above, but the strongest has not yet been noticed. Though Germany in the 1840s had not yet achieved the scientific eminence of either Britain or France, five of our twelve pioneers were Germans, a sixth, Colding, was a Danish disciple of Oersted's, and a seventh, Hirn, was a self-educated Alsatian who read the *Naturphilosophen*.[90] Unless the *Naturphilosophie* indigenous to the educational environment of these seven men had a productive role in the researches of some, it is hard to see why more than half of the pioneers should have been drawn from an area barely through its first generation of significant scientific productivity. Nor is this quite all. If proved, the influence of *Naturphilosophie* may also help to explain why this particular group of five Germans, a Dane, and an Alsatian includes five of the six pioneers in whose approaches to energy conservation we have previously noted such marked conceptual lacunae.[91]

89. Helmholtz, *Abhandlungen*, 1:68.

90. Much biographical and bibliographical material for the study of Hirn's life and work can be found in the *Bulletin de la société d'histoire naturelle de Colmar* 1 (1899): 183–335.

91. Séguin is the sixth, and the source of his idea remains a complete riddle. He attributes it (*Chemins de fer*, p. xvi) to his uncle Montgolfier about whom I have been able to get no relevant information.

The statistics above are not meant to imply that those exposed to *Naturphilosophie* were invariably affected by it; nor do I mean to argue that those whose work shows no conceptual lacunae were *ipso facto* not influenced by *Naturphilosophie* (see remarks on Grove in note 83). It is the *predominance* rather than the presence of pioneers from the area dominated by German intellectual traditions that constitutes the puzzle.

[The following paragraph was added to the original manuscript in response to points raised during the discussion.]

Professor Gillispie, in his paper, calls attention to a little-known movement in eighteenth-century France that shows striking parallels to *Naturphilosophie*. If this movement had still been prevalent in nineteenth-century France, my contrast between the German scientific tradition and that prevalent elsewhere in Europe would be questionable. But I find nothing re-

This preliminary discussion of simultaneous discovery must end here. Comparing it with the sources, primary and secondary, from which it derives, makes apparent its incompleteness. Almost nothing has been said, for example, about either the dynamical theory of heat or the conception of the impossibility of perpetual motion. Both bulk large in standard histories, and both would require discussion in a more extended treatment. But if I am right, these neglected factors and others like them would not enter a fuller discussion of simultaneous discovery with the urgency of the three discussed here. The impossibility of perpetual motion, for example, was an essential intellectual tool for most of the pioneers. The ways in which many of them arrived at the conservation of energy cannot be understood without it. Yet recognizing the intellectual tool scarcely contributes to an understanding of simultaneous discovery because the impossibility of perpetual motion had been endemic in scientific thought since antiquity.[92] Knowing the tool was there, our question has been: Why did it suddenly acquire a new significance and a new range of application? For us, that is the more significant question.

The same argument applies in part to my second example of neglected factors. Despite Rumford's deserved fame, the dynami-

sembling *Naturphilosophie* in any of the nineteenth-century French sources I have examined, and Professor Gillispie assures me that, to the best of his knowledge, the movement to which his paper draws attention had disappeared (except perhaps from parts of biology) by the turn of the century. Notice, in addition, that this eighteenth-century movement, which was particularly prevalent among craftsmen and inventors, may provide a clue to the puzzle of Montgolfier (see above).

92. E. Mach, *History and Root of the Principle of the Conservation of Energy*, trans. Philip E. B. Jourdain (Chicago, 1911), pp. 19–41; and Haas, *Erhaltung*, chap. 4. Remember also that in 1775 the French academy formally resolved to consider no more purported designs of perpetual motion machines. Almost all of our pioneers make use of the impossibility of perpetual motion, and none feels the slightest necessity of arguing about its validity. In contrast, they do find it necessary to argue at length about the validity of the concept of universal conversions. Grove, for example, opens his *Physical Forces* (pp. 1–3) with a plea for a fair hearing of a radical idea. The idea turns out to be the concept of universal conversions developed at great length in the text (pp. 4–44). The impossibility of perpetual motion is casually applied to this idea without argument in the last seven pages (pp. 45–52). Facts like these have led me to call the steps from universal conversions to an unquantified version of conservation "rather obvious."

cal theory of heat had been close to the surface of scientific consciousness almost since the days of Francis Bacon.[93] Even at the end of the eighteenth century, when temporarily eclipsed by the work of Black and Lavoisier, the dynamical theory was often described in scientific discussions of heat, if only for the sake of refutation.[94] To the extent that the conception of heat as motion figured in the work of the pioneers, we must principally understand why that conception gained a significance after 1830 that it had seldom possessed before.[95] Besides, the dynamical theory did not

93. For seventeenth-century theories of heat, see M. Boas, "The Establishment of the Mechanical Philosophy," *Osiris* 10 (1952): 412–541. Much information about eighteenth-century theories is scattered through: D. McKie and N. H. de V. Heathcote, *The Discovery of Specific and Latent Heat* (London, 1935), and H. Metzger, *Newton, Stahl, Boerhaave et la doctrine chimique* (Paris, 1930). Much other useful information will be found in G. Berthold, *Rumford und die Mechanische Wärmetheorie* (Heidelberg, 1875), though Berthold skips too rapidly from the seventeenth to the nineteenth century.

94. Since the caloric theory was scarcely presented in a developed form before the publication of Lavoisier's *Traité élémentaire de chimie* in 1789, it could hardly have eradicated the dynamical theory in the decade remaining before the publication of Rumford's work. For evidence that even the most pronounced caloricists continued to discuss it, see Armand Séguin, "Observations générales sur le calorique . . . reflexions sur la théorie de MM. Black, Crawford, Lavoisier, et Laplace," *Ann. de Chim.* 3 (1789): 148–242, and 5 (1790): 191–271, particularly, 3:182–90. The material theory of heat has, of course, roots far older than Lavoisier, but Rumford, Davy et al., are really opposing a new theory, not an old one. Their work, particularly Rumford's, may have kept the dynamical theory alive after 1800, but Rumford did not create the theory. It had not died.

95. It is too seldom recognized that until almost the mid-nineteenth century, brilliant scientists could apply the dynamical conservation of *vis viva* to the theory that heat is motion without at all recognizing that heat and work should then be convertible. Consider the following three examples. Daniel Bernoulli, in the often quoted paragraphs from Section X of his *Hydrodynamica* equates heat with particulate *vis viva* and derives the gas laws. Then, in paragraph 40, he applies this theory in computing the height from which a given weight must fall to compress a gas to a given fraction of its initial volume. His solution gives the energy of motion abstracted from the falling weight in order to compress the gas, but fails entirely to notice that this energy must be transferred to the gas particles and must therefore raise the gas's temperature. Lavoisier and Laplace, on pp. 357–59 of their classic memoir (note 72), apply the conservation of energy to the

figure very large. Only Carnot used it as an essential stepping stone. Mohr leaped from the dynamical theory to conservation, but his paper indicates that other stimuli might have served as well. Grove and Joule adhered to the theory but show substantially no dependence on it.[96] Holtzmann, Mayer, and Séguin opposed it—Mayer vehemently and to the end of his life.[97] The apparently close connections between energy conservation and the dynamical theory are largely retrospective.[98]

Compare these two neglected factors with the three we have discussed. The rash of conversion discoveries dates from 1800. Technical discussions of dynamical engines were scarcely a recurrent ingredient of scientific literature before 1760 and their density increased steadily from that date.[99] *Naturphilosophie* reached its

dynamical theory in order to show that for all experimental purposes the caloric and dynamical theories are precisely equivalent. J. B. Biot repeats the same argument, in his *Traité de physique expérimentale et mathématique* (Paris, 1816), 1:66–67, and elsewhere in the same chapter. Grove's mistake about heat (note 34) indicates that even the conception of conversion processes was sometimes insufficient to guide scientists away from this virtually universal mistake.

96. Grove, *Physical Forces*, pp. 7–8. Joule, *Papers*, pp. 121–23. Perhaps these two would not have developed their theories if they had not tended to regard heat as motion, but their published works indicate no such decisive connections.

97. Holtzmann's memoir is based on the caloric theory. For Mayer see Weyrauch, I, pp. 265–72, and II, p. 320, n. 2. For Séguin see *Chemins de fer*, p. xvi.

98. The ease and immediacy with which the dynamical theory was identified with energy conservation is indicated by the contemporary misinterpretations of Mayer quoted in Weyrauch, II, pp. 320 and 428. The classic case, however, is Lord Kelvin's. Having employed the caloric theory in his research and writing until 1850, he opens his famous paper "On the Dynamical Theory of Heat" (*Mathematical and Physical Papers* [Cambridge, 1882], 1:174–75) with a series of remarks on Davy's having "established" the dynamical theory fifty-three years before. Then he continues, "The recent discoveries made by Mayer and Joule . . . afford, *if required*, a perfect confirmation of Sir Humphry Davy's views" (italics mine). But if Davy established the dynamical theory in 1799 and if the rest of conservation follows from it, as Kelvin implies, what had Kelvin himself been doing before 1852?

99. The abstract theories of dynamical engines have no beginning in time. I pick 1760 because of its relation to the important and widely cited works of Smeaton and Borda (notes 50 and 51).

peak in the first two decades of the nineteenth century.[100] Furthermore, all three of these ingredients, except possibly the last, played important roles in the research of at least half the pioneers. That does not mean that these factors explain either the individual or collective discoveries of energy conservation. Many old discoveries and concepts were essential to the work of all the pioneers; many new ones played significant roles in the work of individuals. We have not and shall not reconstruct the causes of all that occurred. But the three factors discussed above may still provide the fundamental constellation, given the question from which we began: Why, in the years 1830–50, did so many of the experiments and concepts required for a full statement of energy conservation lie so close to the surface of scientific consciousness?

100. Merz, *European Thought*, 1:178, n. 1.

5 The History of Science

Reprinted by permission from *International Encyclopedia of the Social Sciences,* vol. 14 (New York: Crowell Collier and Macmillan, 1968), pp. 74–83. © 1968 by Crowell Collier and Macmillan.

As an independent professional discipline, the history of science is a new field still emerging from a long and varied prehistory. Only since 1950, and initially only in the United States, has the majority of even its youngest practitioners been trained for, or committed to, a full-time scholarly career in the field. From their predecessors, most of whom were historians only by avocation and thus derived their goals and values principally from some other field, this younger generation inherits a constellation of sometimes irreconcilable objectives. The resulting tensions, though they have relaxed with the increasing maturation of the profession, are still perceptible, particularly in the varied primary audiences to which the literature of the history of science continues to be addressed. Under the circumstances any brief report on development and current state is inevitably more personal and prognostic than for a longer-established profession.

Development of the Field

Until very recently most of those who wrote the history of science were practicing scientists, sometimes eminent ones. Usually history was for them a by-product of pedagogy. They saw in it, besides intrinsic appeal, a means to elucidate the concepts of their specialty, to establish its tradition, and to attract students. The historical section with which so many technical treatises and monographs still open is contemporary illustration of what was for many centuries the primary form and exclusive source for the history of science. That traditional genre appeared in classical antiquity both in historical sections of technical treatises and in a few independent histories of the most developed ancient sciences, astronomy and mathematics. Similar works—together with a growing body of heroic biography—had a continuous history from the Renaissance through the eighteenth century, when their production was much stimulated by the Enlightenment's vision of science as at once the source and the exemplar of progress. From the last fifty years of that period come the earliest historical studies that are sometimes still used as such, among them the historical narratives embedded in the technical works of Lagrange (mathematics) as well as the imposing separate treatises by Montucla (mathematics and physical science), Priestley (electricity and optics), and Delambre (astronomy). In the nineteenth and early twentieth centuries, though alternative approaches had begun to develop, scientists continued to produce both occasional biographies and magistral histories of their own specialties, for example, Kopp (chemistry), Poggendorff (physics), Sachs (botany), Zittel and Geikie (geology), and Klein (mathematics).

A second main historiographic tradition, occasionally indistinguishable from the first, was more explicitly philosophical in its objectives. Early in the seventeenth century Francis Bacon proclaimed the utility of histories of learning to those who would discover the nature and proper use of human reason. Condorcet and Comte are only the most famous of the philosophically inclined writers who, following Bacon's lead, attempted to base normative descriptions of true rationality on historical surveys of Western scientific thought. Before the nineteenth century this tradition remained predominantly programmatic, producing little significant historical research. But then, particularly in the writings of Whewell, Mach, and Duhem, philosophical concerns became a primary motive for creative activity in the history of science, and they have remained important since.

Both of these historiographic traditions, particularly when controlled by the textual-critical techniques of nineteenth-century German political history, produced occasional monuments of scholarship, which the contemporary historian ignores at his peril. But they simultaneously reinforced a concept of the field that has today been largely rejected by the nascent profession. The objective of these older histories of science was to clarify and deepen an understanding of *contemporary* scientific methods or concepts by displaying their evolution. Committed to such goals, the historian characteristically chose a single established science or branch of science—one whose status as sound knowledge could scarcely be doubted—and described when, where, and how the elements that in his day constituted its subject matter and presumptive method had come into being. Observations, laws, or theories which contemporary science had set aside as error or irrelevancy were seldom considered unless they pointed a methodological moral or explained a prolonged period of apparent sterility. Similar selective principles governed discussion of factors external to science. Religion, seen as a hindrance, and technology, seen as an occasional prerequisite to advance in instrumentation, were almost the only such factors which received attention. The outcome of this approach has recently been brilliantly parodied by the philosopher Joseph Agassi.

Until the early nineteenth century, of course, characteristics very much like these typified most historical writing. The romantics' passion for distant times and places had to combine with the scholarly standards of biblical criticism before even general historians could be brought to recognize the interest and integrity of value systems other than their own. (The nineteenth century is, for example, the period when the Middle Ages were first observed to have a history.) That transformation of sensibility which most contemporary historians would suppose essential to their field was not, however, at once reflected in the history of science. Though they agreed about nothing else, both the romantic and the scientist-historian continued to view the development of science as a quasi-mechanical march of the intellect, the successive surrender of nature's secrets to sound methods skillfully deployed. Only in this century have historians of science gradually learned to see their subject matter as something different from a chronology of accumulating positive achievement in a technical specialty defined by hindsight. A number of factors contributed to this change.

Probably the most important was the influence, beginning in late nineteenth century, of the history of philosophy. In that fi only the most partisan could feel confident of his ability to d tinguish positive knowledge from error and superstition. Dealing with ideas that had since lost their appeal, the historian could scarcely escape the force of an injunction which Bertrand Russell later phrased succinctly: "In studying a philosopher, the right attitude is neither reverence nor contempt, but first a kind of hypothetical sympathy, until it is possible to know what it feels like to believe in his theories." That attitude toward past thinkers came to the history of science from philosophy. Partly it was learned from men like Lange and Cassirer who dealt historically with people or ideas that were also important for scientific development. (Burtt's *Metaphysical Foundations of Modern Physical Science* and Lovejoy's *Great Chain of Being* were, in this respect, especially influential.) And partly it was learned from a small group of neo-Kantian epistemologists, particularly Brunschvicg and Meyerson, whose search for quasi-absolute categories of thought in older scientific ideas produced brilliant genetic analyses of concepts which the main tradition in the history of science had misunderstood or dismissed.

These lessons were reinforced by another decisive event in the emergence of the contemporary profession. Almost a century after the Middle Ages had become important to the general historian, Pierre Duhem's search for the sources of modern science disclosed a tradition of medieval physical thought which, in contrast to Aristotle's physics, could not be denied an essential role in the transformation of physical theory that occurred in the seventeenth century. Too many of the elements of Galileo's physics and method were to be found there. But it was not possible, either, to assimilate it quite to Galileo's physics or to Newton's, leaving the structure of the so-called Scientific Revolution unchanged but extending it greatly in time. The essential novelties of seventeenth-century science would be understood only if medieval science were explored first on its own terms and then as the base from which the "new science" sprang. More than any other, that challenge has shaped the modern historiography of science. The writings which it has evoked since 1920, particularly those of E. J. Dijksterhuis, Anneliese Maier, and especially Alexandre Koyré, are the models which many contemporaries aim to emulate. In addition, the discovery of medieval science and its Renaissance role has disclosed an area in which the history of science can and must be integrated with more traditional types of history. That task has barely begun, but the pioneering synthesis by Butterfield and the special studies by Panofsky and Frances Yates mark a path which will surely be broadened and followed.

A third factor in the formation of the modern historiography of science has been a repeated insistence that the student of scientific development concern himself with positive knowledge as a whole and that general histories of science replace histories of special sciences. Traceable as a program to Bacon, and more particularly to Comte, that demand scarcely influenced scholarly performance before the beginning of this century, when it was forcefully reiterated by the universally venerated Paul Tannery and then put to practice in the monumental researches of George Sarton. Subsequent experience has suggested that the sciences are not, in fact, all of a piece and that even the superhuman erudition required for a general history of science could scarcely tailor their joint evolution to a coherent narrative. But the attempt has been crucial, for it has highlighted the impossibility of attributing to the past the divisions of knowledge embodied in contemporary science curricula. Today, as historians increasingly turn back to the detailed investigation of individual branches of science, they study fields which actually existed in the periods that concern them, and they do so with an awareness of the state of other sciences at the time.

Still more recently, one other set of influences has begun to shape contemporary work in the history of science. Its result is an increased concern, deriving partly from general history and partly from German sociology and Marxist historiography, with the role of nonintellectual, particularly institutional and socioeconomic, factors in scientific development. Unlike the ones discussed above, however, these influences and the works responsive to them have to date scarcely been assimilated by the emerging profession. For all its novelties, the new historiography is still directed predominantly to the evolution of scientific ideas and of the tools (mathematical, observational, and experimental) through which these interact with each other and with nature. Its best practitioners have, like Koyré, usually minimized the importance of nonintellectual aspects of culture to the historical developments they consider. A few have acted as though the obtrusion of economic or institutional considerations into the history of science would be a denial of the integrity of science itself. As a result, there seems at times to be two distinct sorts of history of science, occasionally appearing between the same covers but rarely making firm or fruitful contact. The still dominant form, often called the "internal approach," is concerned with the substance of science as knowledge. Its newer rival, often called the "external approach," is concerned with the activities of scientists as a social group within a larger culture. Putting the two together is perhaps the greatest challenge now faced by the profession, and there are increasing signs of a response. Nevertheless, any survey of the field's present state must unfortunately still treat the two as virtually separate enterprises.

Internal History

What are the maxims of the new internal historiography? Insofar as possible (it is never entirely so, nor could history be written if it were), the historian should set aside the science that he knows. His science should be learned from the textbooks and journals of the period he studies, and he should master these and the indigenous traditions they display before grappling with innovators whose discoveries or inventions changed the direction of scientific advance. Dealing with innovators, the historian should try to think as they did. Recognizing that scientists are often famous for results they did not intend, he should ask what problems his subject worked at and how these became problems for him. Recognizing that a historic discovery is rarely quite the one attributed to its author in later textbooks (pedagogic goals inevitably transform a narrative), the historian should ask what his subject thought he had discovered and what he took the basis of that discovery to be. And in this process of reconstruction the historian should pay particular attention to his subject's apparent errors, not for their own sake but because they reveal far more of the mind at work than do the passages in which a scientist seems to record a result or an argument that modern science still retains.

For at least thirty years the attitudes which these maxims are designed to display have increasingly guided the best interpretive scholarship in the history of science, and it is with scholarship of that sort that this article is predominantly concerned. (There are other types, of course, though the distinction is not sharp, and much of the most worthwhile effort of historians of science is devoted to them. But this is not the place to consider work like that of, say, Needham, Neugebauer, and Thorndike, whose indispensable contribution has been to establish and make accessible texts and traditions previously known only through myth.) Nevertheless, the subject matter is immense; there have been few professional historians of science (in 1950 scarcely more than half a dozen in the United States); and their choice of topic has been far from random. There remain vast areas for which not even the basic developmental lines are clear.

Probably because of their special prestige, physics, chemistry, and astronomy dominate the historical literature of science. But even in these fields effort has been unevenly distributed, particularly in this century. Because they sought contemporary knowledge in the past, the nineteenth-century scientist-historians compiled surveys which often ranged from antiquity to their own day or close to it. In the twentieth century a few scientists, like Dugas, Jammer, Partington, Truesdell, and Whittaker, have written from a similar viewpoint, and some of their surveys carry the history of special fields close to the present. But few practitioners of the most developed sciences still write histories, and the members of the emerging profession have up to this time been far more systematically and narrowly selective, with a number of unfortunate consequences. The deep and sympathetic immersion in the sources which their work demands virtually prohibits wide-ranging surveys, at least until more of the field has been examined in depth. Starting with a clean slate, as they at least feel they are, this group naturally tries first to establish the early phases in the development of a science, and few get beyond that point. Besides, until the last few years almost no member of the new group has had sufficient command of the science (particularly mathematics, usually the decisive hurdle) to become a vicarious participant in the more recent research of the technically most developed disciplines.

As a result, though the situation is now changing rapidly with the entry both of more and of better-prepared people into the field, the recent literature of the history of science tends to end at the point where the technical source materials cease to be accessible to a man with elementary college scientific training. There are fine studies of mathematics to Leibniz (Boyer, Michel); of astronomy and mechanics to Newton (Clagett, Costabel, Dijksterhuis, Koyré, and Maier), of electricity to Coulomb (Cohen), and of chemistry to Dalton (Boas, Crosland, Daumas, Guerlac, Metzger). But almost no work within the new tradition has as yet been published on the mathematical physical science of the eighteenth century or on any physical science in the nineteenth.

For the biological and earth sciences, the literature is even less well developed, partly because only those subspecialties which, like physiology, relate closely to medicine had achieved professional status before the late nineteenth century. There are few of the older surveys by scientists, and the members of the new profession are only now beginning in any number to explore these fields. In biology at least there is prospect of rapid change, but up to this point the only areas much studied are nineteenth-century Darwinism and the anatomy and physiology of the sixteenth and seventeenth centuries. On the second of these topics, however, the best of the book-length studies (e.g., O'Malley and Singer) deal usually with special problems and persons and thus scarcely display an evolving scientific tradition. The literature on evolution, in the absence of adequate histories of the technical specialties which provided Darwin with both data and problems, is written at a level of philosophical generality which makes it hard to see how his *Origin of Species* could have been a major achievement, much less an achievement in the sciences. Dupree's model study of the botanist Asa Gray is among the few noteworthy exceptions.

As yet the new historiography has not touched the social sciences. In these fields the historical literature, where it exists, has been produced entirely by practitioners of the science concerned, Boring's *History of Experimental Psychology* being perhaps the outstanding example. Like the older histories of the physical sciences, this literature is often indispensable, but as history it shares their limitations. (The situation is typical for relatively new sciences: practitioners in these fields are ordinarily expected to know about the development of their specialties, which thus regularly acquire a quasi-official history; thereafter something very like Gresham's law applies.) This area therefore offers particular opportunities both to the historian of science and, even more, to the general intellectual or social historian, whose background is often especially appropriate to the demands of these fields. The preliminary publications of Stocking on the history of American anthropology provide a particularly fruitful example of the perspective which the general historian can apply to a scientific field whose concepts and vocabulary have only very recently become esoteric.

External History

Attempts to set science in a cultural context which might enhance understanding both of its development and of its effects have taken three characteristic forms, of which the oldest is the study of scientific institutions. Bishop Sprat prepared his pioneering history of the Royal Society of London almost before that organization had received its first charter, and there have since been innumerable in-house histories of individual scientific societies. These books are, however, useful principally as source materials for the historian, and only in this century have students of scientific development started to make use of them. Simultaneously they have begun seriously to examine the other types of institutions, particularly educational, which may promote or inhibit scientific advance. As elsewhere in the history of science, most of the literature on institutions deals with the seventeenth century. The best of it is scattered through periodicals (the once standard book-length accounts are regrettably out-of-date) from which it can be retrieved, together with much else concerning the history of science, through the annual "Critical Bibliography" of the journal *Isis* and through the quarterly *Bulletin signalétique* of the Centre National de la Recherche Scientifique, Paris. Guerlac's classic study on the professionalization of French chemistry, Schofield's history of the Lunar Society, and a recent collaborative volume (Taton) on scientific education in France are among the very few works on eighteenth-century scientific institutions. For the nineteenth, only Cardwell's study of England, Dupree's of the United States, and Vucinich's of Russia begin to replace the fragmentary but immensely suggestive remarks scattered, often in footnotes, through the first volume of Merz's *History of European Thought in the Nineteenth Century.*

Intellectual historians have frequently considered the impact of science on various aspects of Western thought, particularly during the seventeenth and eighteenth centuries. For the period since 1700, however, these studies are peculiarly unsatisfying insofar as they aim to demonstrate the influence, and not merely the prestige, of science. The name of a Bacon, a Newton, or a Darwin is a potent symbol: there are many reasons to invoke it besides recording a substantive debt. And the recognition of isolated conceptual parallels, for example, between the forces that keep a planet in its orbit and the system of checks and balances in the U.S. constitution, more often demonstrates interpretive ingenuity than the influence of science on other areas of life. No doubt scientific concepts, particularly those of broad scope, do help to change extra-scientific ideas. But the analysis of their role in producing this kind of change demands immersion in the literature of science. The older historiography of science does not, by its nature, supply what is needed, and the new historiography is too recent and its products too fragmentary to have had much effect. Though the gap seems small, there is no chasm that more needs bridging than that between the historian of ideas and the historian of science. Fortunately there are a few works to point the way. Among the more recent are Nicolson's pioneering studies of science in seventeenth- and eighteenth-century literature, Westfall's discussion of natural religion, Gillispie's chapter on science in the Enlightenment, and Roger's monumental survey of the role of the life sciences in eighteenth-century French thought.

The concern with institutions and that with ideas merge naturally in a third approach to scientific development. This is the study of science in a geographical area too small to permit concentration on the evolution of any particular technical specialty but sufficiently homogeneous to enhance an understanding of the social role and setting of science. Of all the types of external history, this is the newest and most revealing, for it calls forth the widest range of historical and sociological experience and skill. The small but rapidly growing literature on science in America (Dupree, Hindle, Shryock) is a prominent example of this approach, and there is promise that current studies of science in the French Revolution may yield similar illumination. Merz, Lilley, and Ben-David point to aspects of the nineteenth century on which much similar effort must be expended. The topic which has, however, evoked the greatest activity and attention is the development of science in seventeenth-century England. Because it has become the center of vociferous debate both about the origin of modern science and about the nature of the history of science, this literature is an appropriate focus for separate discussion. Here it stands for a type of research: the problems it presents will provide perspective on the relations between the internal and external approaches to the history of science.

The Merton Thesis

The most visible issue in the debate about seventeenth-century science has been the so-called Merton thesis, really two overlapping theses with distinguishable sources. Both aim ultimately to account for the special productiveness of seventeenth-century science by correlating its novel goals and values—summarized in the program of Bacon and his followers—with other aspects of contemporary society. The first, which owes something to Marxist historiography, emphasizes the extent to which the Baconians hoped to learn from the practical arts and in turn to make science useful. Repeatedly they studied the techniques of contemporary craftsmen—glassmakers, metallurgists, mariners, and the like—and many also devoted at least a portion of their attention to pressing practical problems of the day, for example, those of navigation, land drainage, and deforestation. The new problems, data, and methods fostered by these novel concerns are, Merton supposes, a principal reason for the substantive transformation experienced by a number of sciences during the seventeenth century. The second thesis points to the same novelties of the period but looks to Puritanism as their primary stimulant. (There need be no conflict. Max Weber, whose pioneering suggestion Merton was investigating, had argued that Puritanism helped to legitimize a concern with technology and the useful arts.) The values of settled Puritan communities—for example, an emphasis upon justification through works and on direct communion with God through nature—are said to have fostered both the concern with science and the empirical, instrumental, and utilitarian tone which characterized it during the seventeenth century.

Both of these theses have since been extended and also attacked with vehemence, but no consensus has emerged. (An important confrontation, centering on papers by Hall and de Santillana, appears in the symposium of the Institute for the History of Science edited by Clagett; Zilsel's pioneering paper on William Gilbert can be found in the collection of relevant articles from the *Journal of the History of Ideas* edited by Wiener and Noland. Most of the rest of the literature, which is voluminous, can be traced through the footnotes in a recently published controversy over the work of Christopher Hill.) In this literature the most persistent criticisms are those directed to Merton's definition and application of the label "Puritan," and it now seems clear that no term so narrowly doctrinal in its implications will serve. Difficulties of this sort can surely be eliminated, however, for the Baconian ideology was neither restricted to scientists nor uniformly spread through all classes and areas of Europe. Merton's label may be inadequate, but there is no doubt that the phenomenon he describes did exist. The more significant arguments against his position are the residual ones which derive from the recent transformation in the history of science. Merton's image of the Scientific Revolution, though long-standing, was rapidly being discredited as he wrote, particularly in the role it attributed to the Baconian movement.

Participants in the older historiographic tradition did sometimes declare that science as they conceived it owed nothing to economic values or religious doctrine. Nevertheless, Merton's emphases on the importance of manual work, experimentation, and the direct confrontation with nature were familiar and congenial to them. The new generation of historians, in contrast, claims to have shown that the radical sixteenth- and seventeenth-century revisions of astronomy, mathematics, mechanics, and even optics owed very little to new instruments, experiments, or observations. Galileo's primary method, they argue, was the traditional thought experiment of scholastic science brought to a new perfection. Bacon's naive and ambitious program was an impotent delusion from the start. The attempts to be useful failed consistently; the mountains of data provided by new instruments were of little assistance in the transformation of existing scientific theory. If cultural novelties are required to explain why men like Galileo, Descartes, and Newton were suddenly able to see well-known phenomena in a new way, those novelties are predominantly intellectual and include Renaissance Neoplatonism, the revival of ancient atomism, and the rediscovery of Archimedes. Such intellectual currents were, however, at least as prevalent and productive in Roman Catholic Italy and France as in Puritan circles in Britain or Holland. And nowhere in Europe, where these currents were stronger among courtiers than among craftsmen, do they display a significant debt to technology. If Merton were right, the new image of the Scientific Revolution would apparently be wrong.

In their more detailed and careful versions, which include essential qualification, these arguments are entirely convincing, up to a point. The men who transformed scientific theory during the seventeenth century sometimes talked like Baconians, but it has yet to be shown that the ideology which a number of them embraced had a major effect, substantive or methodological, on their central contributions to science. Those contributions are best understood as the result of the internal evolution of a cluster of fields which, during the sixteenth and seventeenth centuries, were pursued with renewed vigor and in a new intellectual milieu. That point, however, can be relevant only to the revision of the Merton thesis, not to its rejection. One aspect of the ferment which historians have regularly labeled "the Scientific Revolution" was a radical programmatic movement centering in England and the Low Countries, though it was also visible for a time in Italy and France. That movement, which even the present form of Merton's argument does make more comprehensible, drastically altered the appeal, the locus, and the nature of much scientific research during the seventeenth century, and the changes have been permanent. Very likely, as contemporary historians argue, none of these novel features played a large role in transforming scientific concepts during the seventeenth century, but historians must learn to deal with them nonetheless. Perhaps the following suggestions, whose more general import will be considered in the next section, may prove helpful.

Omitting the biological sciences, for which close ties to medical crafts and institutions dictate a more complex developmental pattern, the main branches of science transformed during the sixteenth and seventeenth centuries were astronomy, mathematics, mechanics, and optics. It is their development which makes the Scientific Revolution seem a revolution in concepts. Significantly, however, this cluster of fields consists exclusively of classical sciences. Highly developed in antiquity, they found a place in the medieval university curriculum where several of them were significantly further developed. Their seventeenth-century metamorphosis, in which university-based men continued to play a significant role, can reasonably be portrayed as primarily an extension of an ancient and medieval tradition developing in a new conceptual environment. Only occasionally need one have recourse to the Baconian programmatic movement when explaining the transformation of these fields.

By the seventeenth century, however, these were not the only areas of intense scientific activity, and the others—among them the study of electricity and magnetism, of chemistry, and of thermal phenomena—display a different pattern. As sciences, as fields to be scrutinized systematically for an increased understanding of nature, they were all novelties during the Scientific Revolution. Their main roots were not in the learned university tradition but often in the established crafts, and they were all critically dependent both on the new program of experimentation and on the new instrumentation which craftsmen often helped to introduce. Except occasionally in medical schools, they rarely found a place in universities before the nineteenth century, and they were meanwhile pursued by amateurs loosely clustered around the new scientific societies that were the institutional manifestation of the Scientific Revolution. Obviously these are the fields, together with the new mode of practice they represent, which a revised Merton thesis may help us understand. Unlike that in the classical sciences, research in these fields added little to man's understanding of nature during the seventeenth century, a fact which has made them easy to ignore when evaluating Merton's viewpoint. But the achievements of the late eighteenth and of the nineteenth centuries will not be comprehensible until they are taken fully into account. The Baconian program, if initially barren of conceptual fruits, nevertheless inaugurated a number of the major modern sciences.

Internal and External History

Because they underscore distinctions between the earlier and later stages of a science's evolution, these remarks about the Merton thesis illustrate aspects of scientific development recently discussed in a more general way by Kuhn. Early in the development of a new field, he suggests, social needs and values are a major determinant of the problems on which its practitioners concentrate. Also during this period, the concepts they deploy in solving problems are extensively conditioned by contemporary common sense, by a prevailing philosophical tradition, or by the most prestigious contemporary sciences. The new fields which emerged in the seventeenth century and a number of the modern social sciences provide ples. Kuhn argues, however, that the later evolution of a tech- specialty is significantly different in ways at least foreshad- d by the development of the classical sciences during the Sci- fic Revolution. The practitioners of a mature science are men trained in a sophisticated body of traditional theory and of instrumental, mathematical, and verbal technique. As a result, they constitute a special subculture, one whose members are the exclusive audience for, and judges of, each other's work. The problems on which such specialists work are no longer presented by the external society but by an internal challenge to increase the scope and precision of the fit between existing theory and nature. And the concepts used to resolve these problems are normally close relatives of those supplied by prior training for the specialty. In short, compared with other professional and creative pursuits, the practitioners of a mature science are effectively insulated from the cultural milieu in which they live their extraprofessional lives.

That quite special, though still incomplete, insulation is the presumptive reason why the internal approach to the history of science, conceived as autonomous and self-contained, has seemed so nearly successful. To an extent unparalleled in other fields, the development of an individual technical specialty can be understood without going beyond the literature of that specialty and a few of its near neighbors. Only occasionally need the historian take note of a particular concept, problem, or technique which entered the field from outside. Nevertheless, the apparent autonomy of the internal approach is misleading in essentials, and the passion sometimes expended in its defense has obscured important problems. The insulation of a mature scientific community suggested by Kuhn's analysis is an insulation primarily with respect to concepts and secondarily with respect to problem structure. There are, however, other aspects of scientific advance, such as its timing. These do depend critically on the factors emphasized by the external approach to scientific development. Particularly when the sciences are viewed as an interacting group rather than as a collection of specialties, the cumulative effects of external factors can be decisive.

Both the attraction of science as a career and the differential appeal of different fields are, for example, significantly conditioned by factors external to science. Furthermore, since progress in one field is sometimes dependent on the prior development of another, differential growth rates may affect an entire evolutionary pattern. Similar considerations, as noted above, play a major role in the inauguration and initial form of new sciences. In addition, a new technology or some other change in the conditions of society may selectively alter the felt importance of a specialty's problems or even create new ones for it. By doing so they may sometimes accelerate the discovery of areas in which an established theory ought to work but does not, thereby hastening its rejection and replacement by a new one. Occasionally, they may even shape the substance of that new theory by ensuring that the crisis to which it responds occurs in one problem area rather than another. Or again, through the crucial intermediary of institutional reform, external conditions may create new channels of communication between previously disparate specialties, thus fostering cross-fertilization which would otherwise have been absent or long delayed.

There are numerous other ways, including direct subsidy, in which the larger culture impinges on scientific development, but the preceding sketch should sufficiently display a direction in which the history of science must now develop. Though the internal and external approaches to the history of science have a sort of natural autonomy, they are, in fact, complementary concerns. Until they are practiced as such, each drawing from the other, important aspects of scientific development are unlikely to be understood. That mode of practice has hardly yet begun, as the response to the Merton thesis indicates, but perhaps the analytic categories it demands are becoming clear.

The Relevance of the History of Science

Turning in conclusion to the question about which judgments must be the most personal of all, one may ask about the potential harvest to be reaped from the work of this new profession. First and foremost will be more and better histories of science. Like any other scholarly discipline, the field's primary responsibility must be to itself. Increasing signs of its selective impact on other enterprises may, however, justify brief analysis.

Among the areas to which the history of science relates, the one least likely to be significantly affected is scientific research itself. Advocates of the history of science have occasionally described their field as a rich repository of forgotten ideas and methods, a few of which might well dissolve contemporary scientific dilemmas. When a new concept or theory is successfully deployed in a science, some previously ignored precedent is usually discovered in the earlier literature of the field. It is natural to wonder whether attention to history might not have accelerated the innovation. Almost certainly, however, the answer is no. The quantity of material to be searched, the absence of appropriate indexing categories, and the subtle but usually vast differences between the anticipation and the effective innovation, all combine to suggest that reinvention rather than rediscovery will remain the most efficient source of scientific novelty.

The more likely effects of the history of science on the fields it chronicles are indirect, providing increased understanding of the scientific enterprise itself. Though a clearer grasp of the nature of scientific development is unlikely to resolve particular puzzles of research, it may well stimulate reconsideration of such matters as science education, administration, and policy. Probably, however, the implicit insights which historical study can produce will first need to be made explicit by the intervention of other disciplines, of which three now seem particularly likely to be effective.

Though the intrusion still evokes more heat than light, the philosophy of science is today the field in which the impact of the history of science is most apparent. Feyerabend, Hanson, Hesse, and Kuhn have all recently insisted on the inappropriateness of the traditional philosopher's ideal image of science, and in search of an alternative they have all drawn heavily from history. Following directions pointed by the classic statements of Norman Campbell and Karl Popper (and sometimes also significantly influenced by Ludwig Wittgenstein), they have at least raised problems that the philosophy of science is no longer likely to ignore. The resolution of those problems is for the future, perhaps for the indefinitely distant future. There is as yet no developed and matured "new philosophy" of science. But already the questioning of older stereotypes, mostly positivistic, is proving a stimulus and release to some practitioners of those newer sciences which have most depended upon explicit canons of scientific method in their search for professional identity.

A second field in which the history of science is likely to have increasing effect is the sociology of science. Ultimately neither the concerns nor the techniques of that field need be historical. But in the present underdeveloped state of their specialty, sociologists of science can well learn from history something about the shape of the enterprise they investigate. The recent writings of Ben-David, Hagstrom, Merton, and others give evidence that they are doing so.

Very likely it will be through sociology that the history of science has its primary impact on science policy and administration.

Closely related to the sociology of science (perhaps equivalent to it if the two are properly construed) is a field that, though it scarcely yet exists, is widely described as "the science of science." Its goal, in the words of its leading exponent, Derek Price, is nothing less than "the theoretic analysis of the structure and behavior of science itself," and its techniques are an eclectic combination of the historian's, the sociologist's, and the econometrician's. No one can yet guess to what extent that goal is attainable, but any progress toward it will inevitably and immediately enhance the significance both to social scientists and to society of continuing scholarship in the history of science.

Bibliography

Other relevant material may be found in the biographies of Koyré and Sarton.

Agassi, Joseph. 1963. *Towards an Historiography of Science.* History and Theory, vol. 2. The Hague: Mouton.

Ben-David, Joseph. 1960. "Scientific Productivity and Academic Organization in Nineteenth-century Medicine." *American Sociological Review* 25:828–43.

Boas, Marie. 1958. *Robert Boyle and Seventeenth-Century Chemistry.* Cambridge: Cambridge University Press.

Boyer, Carl B. 1949. *The Concepts of the Calculus: A Critical and Historical Discussion of the Derivative and the Integral.* New York: Hafner.

A paperback edition was published in 1959 by Dover as *The History of the Calculus and Its Conceptual Development.*

Butterfield, Herbert. 1957. *The Origins of Modern Science, 1300–1800.* 2d. ed., rev. New York: Macmillan.

A paperback edition was published in 1962 by Collier.

Cardwell, Donald S. L. 1957. *The Organisation of Science in England: A Retrospect.* Melbourne and London: Heinemann.

Clagett, Marshall. 1959. *The Science of Mechanics in the Middle Ages.* Madison: University of Wisconsin Press.

Cohen, I. Bernard. 1956. *Franklin and Newton: An Inquiry into Speculative Newtonian Experimental Science and Franklin's*

Work in Electricity as an Example Thereof. American Philosophical Society. Memoirs, vol. 43. Philadelphia: The Society.

Costabel, Pierre. 1960. *Leibniz et la dynamique: Les textes de 1692*. Paris: Hermann.

Crosland, Maurice. 1963. "The Development of Chemistry in the Eighteenth Century." *Studies on Voltaire and the Eighteenth Century* 24:369–441.

Daumas, Maurice. 1955. *Lavoisier: Théoricien et expérimentateur*. Paris: Presses Universitaires de France.

Dijksterhuis, Edward J. 1961. *The Mechanization of the World Picture*. Oxford: Clarendon.
First published in Dutch in 1950.

Dugas, René. 1955. *A History of Mechanics*. Neuchâtel: Editions du Griffon; New York: Central Book.
First published in French in 1950.

Duhem, Pierre. 1906–13. *Etudes sur Léonard de Vinci*. 3 vols. Paris: Hermann.

Dupree, A. Hunter. 1957. *Science in the Federal Government: A History of Policies and Activities to 1940*. Cambridge, Mass.: Belknap.

Dupree, A. Hunter. 1959. *Asa Gray: 1810–1888*. Cambridge, Mass.: Harvard University Press.

Feyerabend, P. K. 1962. "Explanation, Reduction and Empiricism." In Herbert Feigl and Grover Maxwell, eds., *Scientific Explanation, Space, and Time*, pp. 28–97. Minnesota Studies in the Philosophy of Science, vol. 3. Minneapolis: University of Minnesota Press.

Gillispie, Charles C. 1960. *The Edge of Objectivity: An Essay in the History of Scientific Ideas*. Princeton, N.J.: Princeton University Press.

Guerlac, Henry. 1959. "Some French Antecedents of the Chemical Revolution." *Chymia* 5:73–112.

———. 1961. *Lavoisier; the Crucial Year: The Background and Origin of His First Experiments on Combustion in 1772*. Ithaca, N.Y.: Cornell University Press.

Hagstrom, Warren O. 1965. *The Scientific Community*. New York: Basic Books.

Hanson, Norwood R. 1961. *Patterns of Discovery: An Inquiry into the Conceptual Foundations of Science*. Cambridge: Cambridge University Press.

Hesse, Mary B. 1963. *Models and Analogies in Science.* London: Sheed & Ward.

Hill, Christopher. 1965. "Debate: Puritanism, Capitalism and the Scientific Revolution." *Past and Present*, no. 29:68–97.
Articles relevant to the debate may also be found in numbers 28, 31, 32, and 33.

Hindle, Brooke. 1956. *The Pursuit of Science in Revolutionary America, 1735–1789.* Chapel Hill: University of North Carolina Press.

Institute for the History of Science, University of Wisconsin, 1957. 1959. *Critical Problems in the History of Science: Proceedings.* Edited by Marshall Clagett. Madison: University of Wisconsin Press.

Jammer, Max. 1961. *Concepts of Mass in Classical and Modern Physics.* Cambridge, Mass.: Harvard University Press.

Journal of the History of Ideas. 1957. *Roots of Scientific Thought: A Cultural Perspective.* Edited by Philip P. Wiener and Aaron Noland. New York: Basic Books.
Selections from the first 18 volumes of the *Journal.*

Koyré, Alexandre. 1939. *Etudes galiléennes.* 3 vols. Actualités scientifiques et industrielles, nos. 852, 853, and 854. Paris: Hermann.
Volume 1: *À l'aube de la science classique.* Volume 2: *La loi de la chute des corps: Descartes et Galilée.* Volume 3: *Galilée et la loi d'inertie.*

———. 1961. *La révolution astronomique: Copernic, Kepler, Borelli.* Paris: Hermann.

Kuhn, Thomas S. 1962. *The Structure of Scientific Revolutions.* Chicago: University of Chicago Press.
A paperback edition was published in 1964.

Lilley, S. 1949. "Social Aspects of the History of Science." *Archives internationales d'histoire des sciences* 2:376–443.

Maier, Anneliese. 1949–58. *Studien zur Naturphilosophie der Spätscholastik.* 5 vols. Rome: Edizioni de "Storia e Letteratura."

Merton, Robert K. 1967. *Science, Technology and Society in Seventeenth-Century England.* New York: Fertig.

———. 1957. "Priorities in Scientific Discovery: A Chapter in the Sociology of Science." *American Sociological Review* 22: 635–59.

Metzger, Hélène. 1930. *Newton, Stahl, Boerhaave et la doctrine chimique.* Paris: Alcan.

Meyerson, Emile. 1964. *Identity and Reality.* London: Allen & Unwin.

First published in French in 1908.

Michel, Paul-Henri. 1950. *De Pythagore à Euclide.* Paris: Édition "Les Belles Lettres."

Needham, Joseph. 1954–1965. *Science and Civilisation in China.* 4 vols. Cambridge: Cambridge University Press.

Neugebauer, Otto. 1957. *The Exact Sciences in Antiquity.* 2d ed. Providence, R.I.: Brown University Press.

A paperback edition was published in 1962 by Harper.

Nicolson, Marjorie H. 1960. *The Breaking of the Circle: Studies in the Effect of the "New Science" upon Seventeenth-Century Poetry.* Rev. ed. New York: Columbia University Press.

A paperback edition was published in 1962.

O'Malley, Charles D. 1964. *Andreas Vesalius of Brussels, 1514–1564.* Berkeley and Los Angeles: University of California Press.

Panofsky, Erwin. 1954. *Galileo as a Critic of the Arts.* The Hague: Nijhoff.

Partington, James R. 1962–. *A History of Chemistry.* New York: St. Martins.

Volumes 2–4 were published from 1962 to 1964; Volume 1 is in preparation.

Price, Derek J. de Solla. 1966. "The Science of Scientists." *Medical Opinion and Review* 1:81–97.

Roger, Jacques. 1963. *Les sciences de la vie dans la pensée française du XVIIIe siècle: La génération des animaux de Descartes à* l'Encyclopédie. Paris: Colin.

Sarton, George. 1927–48. *Introduction to the History of Science.* 3 vols. Baltimore: Williams & Wilkins.

Schofield, Robert E. 1963. *The Lunar Society of Birmingham: A Social History of Provincial Science and Industry in Eighteenth-Century England.* Oxford: Clarendon.

Shryock, Richard H. 1947. *The Development of Modern Medicine.* 2d ed. New York: Knopf.

Singer, Charles J. 1922. *The Discovery of the Circulation of the Blood.* London: Bell.

Stocking, George W. Jr. 1966. "Franz Boas and the Culture Concept in Historical Perspective." *American Anthropologist* New Series 68:867–82.

Taton, René, ed. 1964. *Enseignement et diffusion des sciences en France au XVIIIe siècle*. Paris: Hermann.

Thorndike, Lynn. 1959–64. *A History of Magic and Experimental Science*. 8 vols. New York: Columbia University Press.

Truesdell, Clifford A. 1960. *The Rational Mechanics of Flexible or Elastic Bodies 1638–1788: Introduction to Leonhardi Euleri* Opera omnia *Vol. X et XI seriei secundae*. Leonhardi Euleri Opera omnia, Ser. 2, Vol. 11, part 2. Turin: Fussli.

Vucinich, Alexander S. 1963. *Science in Russian Culture*. Volume 1: *A History to 1860*. Stanford University Press.

Westfall, Richard S. 1958. *Science and Religion in Seventeenth-Century England*. New Haven: Yale University Press.

Whittaker, Edmund, 1951–53. *A History of the Theories of Aether and Electricity*. 2 vols. London: Nelson.

Volume 1: *The Classical Theories*. Volume 2: *The Modern Theories, 1900–1926*.

Volume 1 is a revised edition of *A History of the Theories of Aether and Electricity from the Age of Descartes to the Close of the Ninteenth Century*, published in 1910. A paperback edition was published in 1960 by Harper.

Yates, Frances A. 1964. *Giordano Bruno and the Hermetic Tradition*. Chicago: University of Chicago Press.

6

The Relations between History and the History of Science

Reprinted by permission from *Daedalus* 100 (1971): 271–304.

The invitation to write this essay asks that I address myself to the relations between my own field and other sorts of history. "For several decades," it points out, "the history of science has seemed a discipline apart with only very tenuous links with other kinds of historical study." That generalization, which errs only in supposing that the separation is but a few decades old, isolates a problem with which I have struggled, both intellectually and emotionally, since I first began to teach the history of science, twenty years ago. My colleagues and my students are no less aware of it than I, and its existence does much to determine both the scale and direction of our discipline's development. Strangely enough, however, though we repeatedly gnaw at it among ourselves, no one has previously made the problem a matter for public scrutiny and discussion. The opportunity to do so here is correspondingly welcome. Historians of science, if they must act alone, are unlikely to succeed in resolving the central dilemma of their field.

In revising this essay I have profited from occasional comments made at the conference for which it was prepared, particularly those of M. I. Finley. Even more helpful have been criticisms of my draft by several colleagues: T. M. Brown, Roger Hahn, J. L. Heilbron, and Carl Schorske. None of them agrees altogether with the views here expressed, but the paper is better for their intervention.

That perception of my assignment determines my approach. My topic is one I have lived with rather than studied. The data I bring to its analysis are correspondingly personal and impressionistic rather than systematic, with the result, among others, that I shall consider only the situation in the United States. Partisanship I shall try to avoid, but without hope of entire success, for I take up the subject as an advocate, a man much concerned with some central impediments to the development and exploitation of his special field.

Despite the universal lip service paid by historians to the special role of science in the development of Western culture during the past four centuries, the history of science is for most of them still foreign territory. In many cases, perhaps in most, such resistance to foreign travel does no obvious harm, for scientific development has little apparent relevance to many of the central problems of modern Western history. But men who consider socioeconomic development or who discuss changes in values, attitudes, and ideas have regularly adverted to the sciences and must presumably continue to do so. Even they, however, regularly observe science from afar, balking at the border which would give access to the terrain and the natives they discuss. That resistance is damaging, both to their own work and to the development of the history of science.

To identify the problem more clearly, I shall begin this essay by mapping the border which has heretofore separated the traditional fields of historical studies from history of science. Conceding that part of the separation is due simply to the intrinsic technicality of science, I shall next try to isolate and to examine the consequences of the still substantial division which will need to be explained in other ways. Seeking such explanations, I shall first discuss some aspects of a traditional historiography of science that have characteristically repelled and sometimes also misled historians. Since that tradition has, however, been largely out of date for a quarter of a century, it cannot entirely explain the historians' contemporary stance. Fuller understanding must depend as well upon an examination of selected aspects of the traditional structure and ideology of the historical profession, topics to be examined briefly in the penultimate section below. To me, at least, the more sociological sources of division there discussed seem central, and it is hard to see how they are to be entirely overcome. Nevertheless, I shall consider in closing a few recent developments, primarily within my own discipline, which suggest that an at least partial rapprochement may characterize the decade immediately ahead.

What does one have in mind when speaking of the history of science as "a discipline apart"? Partly that almost no students of history pay any attention to it. Since 1956 my own courses in the history of science have regularly been listed among history courses under the masthead of the department of which I was a member. Yet in those courses only about one student in twenty has been an undergraduate history major or a graduate student of history, excepting history of science. The majority of those enrolled have regularly been scientists or engineers. Among the remainder, philosophers and social scientists outnumber historians, and students of literature are not far behind. Again, in both of the history departments to which I have belonged, a history of science area has been an available minor field option for historians taking graduate general examinations. I can think, however, of only five students who have elected it in fourteen years, a particular misfortune because these examinations provide an especially effective route to rapprochement. For some time I feared that the fault was my own, since my training was in physics rather than history and my teaching probably embodies residues. But all the colleagues to whom I have bemoaned the situation, many of them trained as historians, report identical experiences. Furthermore, the subject they teach appears not to matter. Courses on the Scientific Revolution or on science in the French Revolution seem no more attractive to prospective historians than courses on the development of modern physics. Apparently the word "science" in a title is sufficient to turn students of history away.

Those phenomena have a corollary which is equally revealing. Though the history of science remains a small field, it has expanded more than tenfold in the last fifteen years, especially during the last eight. Most new members of the discipline are placed in history departments, which is, I shall later urge, where they belong. But the pressure to employ them almost always comes from outside rather than from within the department to which they are [illegible] mately attached. Usually the initiative is taken by scientist philosophers who must persuade the university administratio add a new slot in history. Only after that condition is met m historian of science be appointed. Thereafter, he is usually tr with complete cordiality within his new department; no group has received me more warmly nor supplied more of my close friends than my history colleagues. Nevertheless, in subtle ways the historian of science is sometimes asked to maintain intellectual distance. I have, for example, occasionally had to defend the work of a colleague or student from a historian's charge that it was not really history of science at all but just history. In ways that are obscure, and perhaps correspondingly important, a historian of science is expected, occasionally even by older historians of science, to be not quite a historian.

The preceding remarks are directed to the social indices of separatism. Look now at some of its pedagogic and intellectual consequences. These seem to be primarily of two sorts, neither of which can be considered in much detail until I discuss, below, the extent to which they are merely the inevitable results of the intrinsic technicality of scientific source materials. Even a sketchy description at this point will, however, point the direction of my argument.

One overall consequence of separatism has, I think, been the abdication by historians of responsibility for evaluating and portraying the role of science in the development of Western culture since the end of the Middle Ages. To those tasks the historian of science can and must make essential contributions, at least by providing the books, monographs, and articles which will be the main sources for other sorts of historians. But insofar as his first commitment is to his specialty, the student of scientific development is no more responsible for the task of integration than the historian of ideas or of socioeconomic development, and he has generally been less well equipped than they to perform it. What is needed is a critical interpenetration of the concerns and achievements of historians of science with those of men tilling certain other historical fields, and such interpenetration, if it has occurred at all, is not evident in the work of most current historians. The usual global acknowledgments that science has somehow been vastly important to the development of modern Western society provide no substitute. Taken in conjunction with the few traditional examples used to illustrate them, they often exaggerate and regularly distort the nature, extent, and timing of the sciences' role.

Surveys of the development of Western civilization illustrate the main consequences of the failure to interpenetrate. Perhaps the most striking of these is the almost total neglect of scientific development since 1750, the period during which science assumed its main role as a historical prime mover. A chapter on the Industrial Revolution—the relation of which to science is at once interesting, obscure, and undiscussed—is sometimes succeeded by a section on Darwinism, mostly social. Often that is all! The overwhelming majority of the space devoted to science in all but a very few general history books is reserved for the years before 1750, an imbalance with disastrous consequences to which I shall return below.[1]

Neglect of science, though less extreme, used also to characterize discussions of European history before 1750. With respect to space allocation, however, that oversight has been generously rectified since the appearance in 1949 of Herbert Butterfield's admirable *Origins of Modern Science.* By now almost all surveys have come to include a chapter or major section on the Scientific Revolution of the sixteenth and seventeenth centuries. But those chapters often fail to recognize, much less to confront, the principal historiographic novelty which Butterfield discovered in the current specialists' literature and made available to a wider audience—the relatively minor role played by new experimental methods in the substantive changes of scientific theory during the Scientific Revolution. They are still dominated by old myths about the role of method, to the consequences of which I shall return below.[2]

Perhaps it is some sense of that inadequacy which often makes historians reluctant to give lectures accompanying the reading on the birth of modern science. Occasionally, if unable to co-opt a historian of science to fill the gap, they simply assign chapters in

1. Roger Hahn persuades me that a few very recent textbooks show signs of change. Perhaps I am *merely* impatient. But the progress of the last half-dozen years, if it is real, still seems to me belated, scattered, and incomplete. Why, for example, has J. H. Randall's *Making of the Modern Mind,* a book first published in 1926 and long out of date, yet to be surpassed as a balanced survey of science's role in the development of Western thought?

2. One aspect of Butterfield's discussion has, in fact, helped to preserve the myths. The historiographic novelties accessible through his book are concentrated in chaps. 1, 2, and 4, which deal with the development of astronomy and mechanics. These are, however, juxtaposed with essentially traditional accounts of the methodological views of Bacon and Descartes, illustrated in application by a chapter on William Harvey. The two resulting versions of the requisites for a transformed science are hard to reconcile, a fact which Butterfield's subsequent discussion of the chemical revolution makes particularly apparent.

Butterfield as a supplement and reserve discussion for section meetings. Butterfield or the bomb has persuaded historians that they must take some account of the role of science, and they attempt to discharge that obligation with a block of material on the Scientific Revolution. But the chapters they then produce seldom reflect an awareness of the problems with which their subject has confronted recent generations of academic specialists. Students must usually look elsewhere for examples of the critical standards ordinarily defended by the profession.

Neglect of the current specialists' literature is, however, only one part of the problem and perhaps not the most serious. More central is the peculiar selectivity with which historians approach the sciences, whether through primary or secondary sources. Dealing with, say, music or the arts, the historian may read program notes and the catalogs of exhibits, but he also listens to symphonies and looks at paintings, and his discussion, whatever its sources, is directed to them. Dealing with the sciences, however, he reads *and discusses* programmatic works almost exclusively: Bacon's *Novum Organum*, but usually book 1 (the idols) rather than book 2 (heat as motion); Descartes' *Discourse on Method*, but not the three substantive essays to which it provides the introduction; Galileo's *Assayer*, but only the introductory pages of his *Two New Sciences*; and so on. The same selectivity shows in the historian's attention to secondary works: Alexandre Koyré's *From the Closed World to the Infinite Universe*, but not his *Etudes galiléennes* or *The Problem of Fall*; E. A. Burtt's *Metaphysical Foundations of Modern Physical Science*, but not E. J. Dijksterhuis's magistral *Mechanization of the World Picture*.[3] Even within individual works there is a

3. The following observation may strengthen the point at which I aim. In the arts the men who create and those who criticize belong to separate, often hostile, groups. Historians may sometimes rely excessively on the latter, but they know the difference between critics and artists, and they are careful to acquaint themselves with works of art as well. In the sciences, on the other hand, the nearest equivalents to the works of critics are written by scientists themselves, usually in prefatory chapters or separate essays. Historians usually rely *exclusively* on these works of "criticism," failing to note, because their authors were also creative scientists, that that selection leaves the science out. On the significance of the different role of the critic in science and in art see my "Comment [on the Relation of Science and Art]," in *Comparative Studies in Society and History* 11 (1969): 403–12 (pp. 340–51 below).

marked tendency, which I shall illustrate below, to skip the chapters that deal with technical contributions.

I do not suggest that what scientists say about what they do is irrelevant to their performance and their concrete achievements. Nor am I suggesting that historians ought not to read and discuss programmatic works. But, as the parallel to program notes should indicate, the relation of prefaces and programmatic writings to substantive science is seldom literal and always problematic. The former must, of course, be read, for they are frequently the media through which scientific ideas reach a larger public. But they are often decisively misleading with respect to a whole series of issues that the historian ought, and often pretends, to deal with: Where do influential scientific ideas come from? What gives them their special authority and appeal? To what extent do they remain the same ideas as they become effective in the larger culture? And, finally, if their influence is not literal, in what sense is it really due to the science to which it is imputed?[4] The intellectual impact of the sciences on extrascientific thought will not, in short, be understood without attention also to the technical core of science. That historians regularly attempt such a sleight of hand suggests that one essential part of what has to this point been described as a gap between history and the history of science might more accurately be seen as a barrier between historians as a group and the sciences. To that point, also, I shall return below.

Before looking more closely at the manner in which historians approach the sciences, I must, however, ask how much may reasonably be expected of them. That question, in turn, demands a sharp separation between the problems of intellectual history, on the one hand, and those of socioeconomic history, on the other. Let me consider them in order.

Intellectual history is the area in which the historian's selectivity with respect to sources has its primary effect, and one may well wonder whether he has an alternative. Excepting historians of sci-

4. For an example of the sort of illumination that can be provided by someone who knows the science and its history see the discussion of the role of science in the Enlightenment by C. C. Gillispie, *The Edge of Objectivity* (Princeton: Princeton University Press, 1960), chap. 5.

ence, among whom the requisite skills are also relatively rare, almost no historians have the training required to read, say, the works of Euler and Lagrange, Maxwell and Boltzmann, or Einstein and Bohr. But that is a very special list in several respects. All the men on it are mathematical physicists; the oldest of them was not born until the first decade of the eighteenth century; and none of them, so far as I can see, has had more than the most tenuous and indirect impact upon the development of extrascientific thought.

The last point, which is the crucial one, may be debatable and ultimately wrong with respect to Einstein and Bohr. Discussions of the contemporary intellectual scene often refer to relativity and the quantum theory when discussing such issues as the limitations of science and of reason. Yet the arguments for direct influence—as against the appeal to authority in support of views held for other reasons—have so far been extremely forced. My own suspicion, which provides at least a reasonable working hypothesis, is that after a science has become thoroughly technical, particularly mathematically technical, its role as a force in intellectual history becomes relatively insignificant. Probably there are occasional exceptions, but if Einstein and Bohr provide them, then the exceptions prove the rule. Whatever their role may have been, it is very different from that of, say, Galileo or Descartes, of Lyell, Playfair, or Darwin, or, for that matter, of Freud, all of whom were read by laymen. If the intellectual historian must consider scientists, they are generally the early figures in the development of their fields.

Not surprisingly, just because the figures he must treat are the early ones, the intellectual historian could handle them in depth if he wished to do so. The job would not be easy: I am not arguing that no significant effort is required—only that there is no other way. Nor would every historian be responsible for undertaking it regardless of his interests. But the man whose concerns include ideas affected by scientific development could study the technical scientific source materials to which he currently only makes reference. Very little of the technical literature written before 1700 is in principle inaccessible to anyone with sound high school scientific training, at least not if he is willing to undertake a modicum of additional work as he goes along. For the eighteenth century the same background in science is adequate to the literature of chemistry, experimental physics (particularly electricity, optics, and heat), geology, and biology—all of science, in short, excepting mathematical mechanics and astronomy. For the nineteenth century most of physics and much of chemistry becomes excessively technical, but men with high school science have access to almost the whole literature of geology, biology, and psychology. I do not suggest that the historian should become a historian of science whenever scientific development becomes relevant to the topic he studies. Here, as elsewhere, specialization is inevitable. But he could in principle do so, and he can therefore certainly command the specialists' secondary literature on his topic. By failing to do even that, he ignores constitutive elements and problems of scientific advance, and the result as I shall shortly indicate shows in his work.

The preceding list of topics potentially accessible to the intellectual historian is revealing in two respects. First, as already indicated, it includes all the technical subject matters with which, *qua* intellectual historian, he is likely to wish to deal. Second, it is coextensive with the list of fields which have been most and best discussed by historians of science. Contrary to a widespread impression, historians of science have seldom dealt in depth with the development of the technically most advanced subjects. Studies of the history of mechanics are sparse from the eve of the publication of Newton's *Principia*, histories of electricity break off with Franklin or at most with Charles Coulomb, of chemistry with Antoine Lavoisier or John Dalton, and so on. The main exceptions, though not the only ones, are Whiggish compendia by scientists, sometimes invaluable as reference works, but otherwise virtually useless to the man whose interests include the development of ideas. However regrettable, that imbalance in favor of relatively nontechnical subjects should surprise no one. Most of the men who have produced the models which contemporary historians of science aim to emulate have not been scientists nor have they had much scientific training. Interestingly enough, however, their background has not been in history either, though historians might have done the job and even done it better since their concerns would not have been so narrowly focused on the conceptual. Instead, they have come from philosophy, though mostly, like Koyré, from Continental schools where the divide between history and philosophy is by no means so deep as in the English-speaking world. All of which suggects once more that a central part of the problem to which this paper is addressed arises from the attitudes of historians toward science.

I shall explore these attitudes further near the end of this essay, but must first ask whether they make any difference to the performance of the tasks which intellectual historians undertake. Obviously they do not in the large proportion of cases which involve scientific ideas only marginally or not at all. In numerous other cases, however, characteristic infirmities result from what I have previously described as history derived predominantly from prefaces and programmatic works. When scientific ideas are discussed without reference to the concrete technical problems against which they were forged, what results is a decidedly misleading notion of the way in which scientific theories develop and impinge on their extrascientific environment.

One form which the systematic misdirection takes is particularly clear in discussions of the Scientific Revolution, including many by older historians of science: an excessive emphasis on the role of new methods, particularly on the power of experiment to create, by itself, new scientific theories. Reading the continuing argument over the so-called Merton thesis, for example, I am constantly depressed by the almost universal misstatement of what that debate is about. What is really at issue, I take it, is an explanation of the rise and dominion of the Baconian movement in England. Both proponents and critics of the Merton thesis simply take it for granted that an explanation of the rise of the new experimental philosophy is tantamount to an explanation of scientific development. On that view, if Puritanism or some other new trend in religion increased the dignity of manual manipulation and fostered the search for God in His works, then, ipso facto, it fostered science. Conversely, if first-rate science was done in Catholic countries, then no Protestant religious movements could be responsible for the rise of seventeenth-century science.

That all-or-nothing polarization is, however, unnecessary and it may well be false. A strong case can be made for the thesis that Baconian experimentalism had comparatively little to do with the main changes of scientific theory which marked the Scientific Revolution. Astronomy and mechanics were transformed with little recourse to experiment and none to new sorts of experimentation. In optics and physiology experiment played a larger role, but the models were not Baconian but rather classic and medieval: Galen in physiology, Ptolemy and Alhazen in optics. These fields plus mathematics exhaust the list of those in which theory was radically transformed during the Scientific Revolution. With respect to their practice neither experimentalism nor its putative religious correlate should be expected to make much difference.

That view, if correct, does not, however, render either Baconianism or new religious movements unimportant to scientific development. What it does suggest is that the role of the new Baconian methods and values was not to produce new theories in previously established sciences but rather to make new fields, often those with roots in the prior crafts, available for scientific scrutiny (for example, magnetism, chemistry, electricity, and the study of heat). Those fields, however, received little significant theoretical reordering before the mid-eighteenth century, the time through which one must wait to discover that the Baconian movement in the sciences was by no means a fraud. That Britain rather than Catholic France, especially after the revocation of the Edict of Nantes, played the dominant role in bringing order to these newer, more Baconian fields may indicate that a revised Merton thesis will yet prove immensely informative. Perhaps it will even help us understand why one old saw about science continues to withstand close scrutiny: at least from 1700 to 1850, British science was predominantly experimental and mechanical, French mathematical and rationalistic. In addition, it may tell us something about the quite special roles played by Scotland and Switzerland in the scientific developments of the eighteenth century.

That historians have had such difficulty in even imagining possibilities like these is, I think, at least partly due to a widespread conviction that scientists discover truth by the quasi-mechanical (and perhaps not very interesting) application of scientific method. Having accounted for the seventeenth-century discovery of method, the historian may, and indeed does, leave the sciences to shift for themselves. That attitude, however, cannot be quite conscious, for another main by-product of preface history is incompatible with it. On the rare occasions when they turn from scientific methods to the substance of new scientific theories, historians seem invariably to give excessive emphasis to the role of the surrounding climate of extrascientific ideas. I would not argue for a moment that that climate is unimportant to scientific development. But, except in the rudimentary stages of the development of a field, the ambient intellectual milieu reacts on the theoretical structure of a science only to the extent that it can be made relevant to the concrete technical problems with which the practitioners of that field engage. Historians of science may, in the past, have been excessively concerned with this technical core, but historians have usually ignored its existence entirely. They know it is there, but they act as though it were the mere product of science—of proper method acting in a suitable environment—rather than being the most essential of all the various determinants of a science's development. What results from this approach is reminiscent of the story of the emperor's new clothes.

Let me give two concrete examples. Both intellectual historians and historians of art often describe the novel intellectual currents of the Renaissance, especially Neoplatonism, which made it possible for Kepler to introduce the ellipse to astronomy, thus breaking the traditional hold of orbits compounded from perfect circular motions. On this view, Tycho's neutral observations plus the Renaissance intellectual milieu yield Kepler's laws. What is regularly ignored, however, is the elementary fact that elliptical orbits would have been useless if applied to any geocentric astronomical scheme. Before the use of ellipses could transform astronomy, the sun had to replace the earth at the center of the universe. That step was not, however, taken until just over a half-century before Kepler's work, and to it the novel *intellectual* climate of the Renaissance made only equivocal contributions. It is an open question, as well as an interesting and important one, whether Kepler might not equally easily have been led to ellipses without benefit of Neoplatonism.[5] To tell the story without reference to any of the technical factors on which the answer to that question depends is to misrepresent the manner in which scientific laws and theories enter the realm of ideas at large.

A more important example to the same effect is provided by countless standard discussions of the origin of Darwin's theory of evolution.[6] What was required, we are told, to transform the static

5. T. S. Kuhn, *The Copernican Revolution* (Cambridge, Mass.: Harvard University Press, 1957), pp. 135–43. N. R. Hanson, *Patterns of Discovery* (Cambridge: University Press, 1958), chap. 4. Note that there are other aspects of Kepler's thought to which the relevance of Neoplatonism is beyond doubt.

6. See, for example, R. M. Young, "Malthus and the Evolutionists: The Common Context of Biological and Social Theory," *Past and Present*, no 43 (1969), pp. 109–45, an essay which includes much useful guidance to the

Chain of Being into an ever-moving escalator was the currency of such ideas as infinite perfectability and progress, the laissez faire competitive economy of an Adam Smith, and, above all, the population analyses of Malthus. I cannot doubt that factors of this sort were vitally important; anyone who does would do well to ask how, in their absence, the historian is to understand the proliferation, particularly in England, of pre-Darwinian evolutionary theories like those of Erasmus Darwin, Spencer, and Robert Chambers. Yet these speculative theories were uniformly anathema to the scientists whom Charles Darwin managed to persuade in the course of making evolutionary theory a standard ingredient of the Western intellectual heritage. What Darwin did, unlike these predecessors, was to show how evolutionary concepts should be applied to a mass of observational materials which had accumulated only during the first half of the nineteenth century and were, quite independently of evolutionary ideas, already making trouble for several recognized scientific specialties. This part of the Darwin story, without which the whole cannot be understood, demands analysis of the changing state, during the decades before the *Origin of Species*, of fields like stratigraphy and paleontology, the geographical study of plant and animal distribution, and the increasing success of classificatory systems which substituted morphological resemblances for Linneaus's parallelisms of function. The men who,

recent literature on Darwinism. Note, however, one irony which illustrates the problems of perception now under discussion. Young opens by deploring the assumptions, widespread among "both historians of science and other sorts of historians . . . that scientific ideas and findings can be dealt with as relatively unequivocal units with fairly sharply defined boundaries . . . [and] that 'non-scientific' factors [have] played relatively little part in shaping the development of scientific ideas." His paper is intended as "a case study which attempts to break down barriers in one small area between the history of science and other branches of history." Obviously, this is just the sort of contribution which I too would particularly welcome. Yet Young cites almost no literature which has attempted to explain the emergence of Darwinism as a response to the development of *scientific* ideas or techniques, and indeed there is very little to cite. Nor does his own paper make any attempt to deal with the technical issues which may have helped to shape Darwin's thought. Very likely it will be for some time the standard account of Malthus's influence on evolutionary thought, for it is admirably thorough, erudite, and perceptive. But far from being a barrier breaker, it belongs to a standard historiographic tradition which has done much to preserve the very separation Young deplores.

in developing natural systems of classification, first spoke of tendrils as "aborted" leaves or who accounted for the different number of ovaries in closely related plant species by referring to the "adherence" in one species of organs separate in the other were not evolutionists by any means. But without their work, Darwin's *Origin* could not have achieved either its final form or its impact on the scientific and the lay public.

One last point will conclude this portion of my argument. I said earlier that, in accounting for the genesis of novel scientific theories, the emphasis on method and the emphasis on extrascientific intellectual milieu were not quite compatible. I would now add that, at the most fundamental level, the two prove to be identical in their effects. Both induce an apparently incurable Whiggishness which permits the historian to dismiss as superstition all the scientific forebears of the ideas with which he deals. The hold of the circle on the astronomical imagination is to be understood as a product of the Platonic infatuation with geometric perfection, perpetuated by medieval dogmatism; the endurance in biology of the idea of fixed species is to be understood as the result of an excessively literal reading of Genesis. What is missing from the first account, however, is any reference to the elegant and predictively successful astronomical systems built from circles, an achievement on which Copernicus did not himself improve. What is missing from the second is any recognition that the observed existence of discrete species, without which there could be no taxonomic enterprise, becomes extremely difficult to understand unless the current members of each descend from some original pair. Since Darwin the definition of basic taxonomic categories like species and genus have necessarily become and remained relatively arbitrary and extraordinarily problematic. Conversely, one technical root of Darwin's work is the increasing difficulty, during the early nineteenth century, of applying these standard classificatory tools to a body of data vastly expanded by, among other things, exploration of the New World and the Pacific. In short, ideas which the historian dismisses as superstitions usually prove to have been crucial elements in highly successful older scientific systems. When they do, the emergence of novel replacements will not be understood as the consequence merely of good method applied in a favorable intellectual milieu.

I have spoken so far of the effect of preface history on the man concerned to place science in intellectual history. Turning now to standard views about the socioeconomic role of science, one encounters a very different situation. What the historian lacks in this area is not so much a knowledge of technical sources, which would in any case be largely irrelevant, as a command of conceptual distinctions essential to the analysis of science as a social force. Some of those distinctions would generate themselves if the socioeconomic historian possessed a better understanding of the nature of science as an enterprise and of its changes over time. Concerned with the role of the sciences, he requires at least a global sense of how men gain membership in scientific communities, of what they then do, where their problems come from, and what they receive as solutions. To this extent, his needs overlap the intellectual historian's, though they are technically far less demanding. But the socioeconomic historian also has needs which the intellectual historian does not: some knowledge of the nature of technology as an enterprise, an ability to distinguish it from science, both socially and intellectually, and above all a sensitivity to the various modes of interaction between the two.

Science, when it affects socioeconomic development at all, does so through technology. Historians tend frequently to conflate the two enterprises, abetted by prefaces which, since the seventeenth century, have regularly proclaimed the utility of science and have often then illustrated it with explanations of existing machines and modes of production.[7] On these issues, too, Bacon has been taken

7. The historian's difficulties with science-*cum*-technology are nowhere better illustrated than in discussions of the Industrial Revolution. The long-standard attitude is that of T. S. Ashton, *The Industrial Revolution, 1760–1830* (London and New York: Oxford University Press, 1948), p. 15: "The stream of English scientific thought, issuing from the teaching of Francis Bacon, and enlarged by the genius of Boyle and Newton, was one of the main tributaries of the industrial revolution." Roland Mousnier's *Progrès scientifique et technique au XVIII*^e^ *siècle* (Paris: Plon, 1958) takes the opposite position in an even more extreme form, arguing for total independence of the two enterprises. As a corrective to the view that the Industrial Revolution was applied Newtonian science, Mousnier's version is an improvement, but it entirely misses the significant methodological and i
logical interactions of eighteenth-century science and technology. For t
see below or the excellent sketch in the chapter "Science" in E. J. Hobsba

not only seriously, as he should be, but literally, as he should not. The methodological innovations of the seventeenth century are thus seen as the source of a useful as well as sound science. Explicitly or implicitly, science is portrayed as having played a steadily increasing socioeconomic role ever since. In fact, however, despite the hortatory claims of Bacon and his successors for three centuries, technology flourished without significant substantive inputs from the sciences until about one hundred years ago. The emergence of science as a prime mover in socioeconomic development was not a gradual but a sudden phenomenon, first significantly foreshadowed in the organic-chemical dye industry in the 1870s, continued in the electric power industry from the 1890s, and rapidly accelerated since the 1920s. To treat these developments as the emergent consequences of the Scientific Revolution is to miss one of the radical historical transformations constitutive of the contemporary scene. Many current debates over science policy would be more fruitful if the nature of this change were better understood.

To that transformation I shall return but must first sketch, however simplistically and dogmatically, some background for it. Science and technology had been separate enterprises before Bacon announced their marriage in the beginning of the seventeenth century, and they continued separate for almost three centuries more. Until late in the nineteenth century, significant technological innovations almost never came from the men, the institutions, or the social groups that contributed to the sciences. Though scientists sometimes tried and though their spokesmen often claimed success, the effective improvers of technology were predominantly craftsmen, foremen, and ingenious contrivers, a group often in sharp conflict with their contemporaries in the sciences.[8] Scorn for in-

The Age of Revolution, 1789–1848 (Cleveland: World Publishing Company, 1962).

8. R. P. Multhauf, "The Scientist and the 'Improver' of Technology," *Technology and Culture* 1 (1959): 38–47; C. C. Gillispie, "The *Encyclopédie* and the Jacobin Philosophy of Science," in M. Clagett, ed., *Critical Problems in the History of Science* (Madison: University of Wisconsin Press, 1959), pp. 255–89. For hints at an explanation of the dichotomy, see my "Comments" in R. R. Nelson, ed., *The Rate and Direction of Inventive Activity*, a Report of the National Bureau of Economic Research (Princeton: Princeton University Press, 1962), pp. 379–84, 450–57, and the

ventors shows repeatedly in the literature of science, and hostility to the pretentious, abstract, and wool-gathering scientist is a persistent theme in the literature of technology. There is even evidence that this polarization of science and technology has deep sociological roots, for almost no historical society has managed successfully to nurture both at the same time.

Greece, when it came to value its science, viewed technology as a finished heritage from its ancient gods; Rome, on the other hand, famous for its technology, produced no notable science. The series of late-medieval and Renaissance technological innovations which made possible the emergence of modern European culture had largely ceased before the Scientific Revolution began. Britain, though it produced a significant series of isolated innovators, was generally backward in at least the abstract and developed sciences during the century which embraces the Industrial Revolution, while technologically second-rate France was the world's preeminent scientific power. With the possible exceptions (it is too early to be sure) of the United States and the Soviet Union since about 1930, Germany during the century before World War II is the only nation that has managed simultaneously to support first-rate traditions in both science and technology. Institutional separation—the universities for *Wissenschaft* and the Technische Hochschulen for industry and the crafts—is a likely cause of that unique success. As a first approximation, the historian of socioeconomic development would do well to treat science and technology as radically distinct enterprises, not unlike the sciences and the arts. That technologies have, between the Renaissance and the late nineteenth century, usually been classified as arts is not an accident.

Starting from this perspective one can ask, as the socioeconomic historian must, about interactions between the two enterprises, now seen as distinct. Such interactions have characteristically been of three sorts, one dating from antiquity, the second from the mid-eighteenth century, and the third from the late nineteenth. The longest lasting, now probably finished except in the social sciences, is the impact of preexisting technologies, whatever their source, on

epilogue of my paper, "The Essential Tension: Tradition and Innovation in Scientific Research," in C. W. Taylor and Frank Barron, eds., *Scientific Creativity: Its Recognition and Development* (New York: Wiley, 1963), pp. 341–54, pp. 237–39, below.

the sciences. Ancient statics, the new sciences of the seventeenth century like magnetism and chemistry, and the development of thermodynamics in the nineteenth century all provide examples. In each of these cases and countless others, critically important advances in the understanding of nature resulted from the decision of scientists to study what craftsmen had already learned how to do. There are other main sources of novelty in the sciences, but this one has too often been underrated, except perhaps by Marxists.

In all these cases, however, the resulting benefits have accrued to science not to technology, a point which Marxist historians repeatedly miss. When Kepler studied the optimum dimensions of wine casks, the proportions which would yield maximum content for the least consumption of wood, he helped to invent the calculus of variations, but existing wine casks were, he found, already built to the dimensions he derived. When Sadi Carnot undertook to produce the theory of the steam engine, a prime mover to which, as he emphasized, science had contributed little or nothing, the result was an important step toward thermodynamics; his prescription for engine improvement, however, had been embodied in engineering practice before his study began.[9] With few exceptions, none of much significance, the scientists who turned to technology for their problems succeeded merely in validating and explaining, not in improving, techniques developed earlier and without the aid of science.

A second mode of interaction, visible from the mid-eighteenth century, was the increasing deployment in the practical arts of methods borrowed from science and sometimes of scientists themselves.[10] The effectiveness of the movement remains uncertain. It has, for example, no apparent role in the development of the new textile machinery and iron fabricating techniques so important to the Industrial Revolution. But the "experimental farms" of eighteenth-century Britain, the record books of the stock breeders, and

9. W. C. Unwin, "The Development of the Experimental Study of Heat Engines," *The Electrician* 35 (1895): 46–50, 77–80, is a striking account of the difficulties encountered when attempting to use Carnot's theory and its successors for practical engineering design.

10. C. C. Gillispie, "The Natural History of Industry," *Isis* 48 (1957): 398–407; R. E. Schofield, "The Industrial Orientation of the Lunar Society of Birmingham," *Isis* 48 (1957): 408–15. Note the extent to which both authors, while disagreeing vehemently, are nevertheless defending the same thesis in different words.

the experiments on steam that Watt performed in developing the separate condenser are all plausibly seen as a conscious attempt to employ scientific methods in the crafts, and such methods were on occasions productive. The men who used them were seldom, however, contributors to contemporary science which, in any case, few of them knew. When they succeeded, it was not by applying existing science but by a frontal attack, however methodologically sophisticated, on a recognized social need.

Only in chemistry is the situation significantly more equivocal.[11] Particularly in France, distinguished chemists, including both Lavoisier and C. L. Berthollet, were employed to supervise and improve such industries as dyeing, ceramics, and gunpowder. Their regimens, furthermore, were an apparent success. But the changes they introduced were neither dramatic nor, in any obvious way, dependent on contemporary chemical theory and discovery. Lavoisier's new chemistry is a case in point. It undoubtedly provided a more profound understanding of the previously developed technology of ore reduction, acids manufacture, and so on. In addition it permitted the gradual elaboration of better techniques of quality control. But it was responsible for no fundamental changes in these established industries, nor did it have an observable role in the nineteenth-century development of such new technologies as sulfuric acid, soda, or wrought iron and steel. If one looks for important new processes which result from the development of scientific knowledge, one must wait for the maturation of organic chemistry, current electricity, and thermodynamics during the generations from 1840 to 1870.

Products and processes derived from prior scientific research and dependent for their development on additional research by men with scientific training display a third mode of interaction between science and technology.[12] Since its emergence in the organic dye industry a century ago, it has transformed communication, the

11. H. Guerlac, "Some French Antecedents of the Chemical Revolution," *Chymia* 5 (1968): 73–112; Archibald Clow and N. L. Clow, *The Chemical Revolution* (London: Batchworth Press, 1952); and L. F. Haber, *The Chemical Industry during the Nineteenth Century* (Oxford: Clarendon Press, 1958).

12. John Beer, *The Emergence of the German Dye Industry*, Illinois Studies in the Social Sciences, vol. 44 (Urbana: University of Illinois Press, 1959); H. C. Passer, *The Electrical Manufacturers, 1875–1900* (Cambridge, Mass.: Harvard University Press, 1953).

generation and distribution of power (twice), the materials both of industry and of everyday life, and also both medicine and warfare. Today its omnipresence and importance disguise the still real cleavage between science and technology. In the process, they make it difficult to realize how very recent and decisive the emergence of this kind of interaction has been. Even economic historians seldom seem aware of the qualitative divide between the forces promoting change during the Industrial Revolution and those operative in the twentieth century. Most general histories disguise even the existence of any such transformation. One need not, however, inflate the importance of history of science to suppose that since 1870 science has assumed a role which no student of modern socioeconomic development may responsibly ignore.

What are the sources of the transformation and how may the socioeconomic historian contribute to their understanding? I suggest there are two, of which he can recognize the first and participate in unraveling the second. No science, however highly developed, need have applications which will significantly alter existing technological practice. The classical sciences like mechanics, astronomy, and mathematics had few such effects even after they were recast during the Scientific Revolution. The sciences which did were those born of the Baconian movement of the seventeenth century, particularly chemistry and electricity. But even they did not reach the levels of development required to generate significant applications until the middle third of the nineteenth century. Before the maturation of these fields at mid-century, there was little of much socioeconomic importance that scientific knowledge in any field could produce. Though few socioeconomic historians are equipped to follow the technical aspects of the advances which suddenly made science productive of new materials and devices, they can surely be aware of these developments and their special role.

Internal technical development was not, however, the only requisite for the emergence of a socially significant science, and about what remained the socioeconomic historian could have a great deal
icance to say. During the nineteenth century the institu-
nd social structure of the sciences was transformed in ways
n foreshadowed in the Scientific Revolution. Beginning in
30s and continuing through the first half of the following

century, newly formed societies of specialists in individual branches of science assumed the leadership which the all-embracing national societies had previously attempted to supply. Simultaneously, private scientific journals and particularly journals of individual specialties proliferated rapidly and increasingly replaced the house organs of the national academies which had previously been the almost exclusive media of public scientific communication. A similar change is visible in scientific education and in the locus of research. Excepting in medicine and at a few military schools, scientific education scarcely existed before the foundation of the Ecole polytechnique in the last decade of the eighteenth century. That model spread rapidly, however, first to Germany, then to the United States, and finally, more equivocally, to England. With it developed other new institutional forms, especially teaching and research laboratories, like Justus von Liebig's at Giessen or the Royal College of Chemistry in London. These are the developments which first made possible and then supported what had previously scarcely existed, the professional scientific career. Like a potentially applicable science, they emerged relatively suddenly and quickly. Together with the maturation of the Baconian sciences of the seventeenth century, they are the pivot of a second scientific revolution which centered in the first half of the *nineteenth* century, a historical episode at least as crucial to an understanding of modern times as its older namesake. It is time it found its way into history books, but it is too much a part of other developments in the nineteenth century to be untangled by historians of science alone.

I have so far been describing the historian's neglect of science and its history, repeatedly implying while doing so that the blame lies exclusively with historians, scarcely at all with the specialists who have chosen science as their object of study. Today, for reasons to which I shall return, that allocation of responsibility seems to me increasingly nearly justified, if ultimately unfair. But the current situation is in part a product of the past. If the contemporary gap between history and the history of science is to be further analyzed in hope of its amelioration, the contribution to separatism made by the history of the history of science must first be recognized.

Until the early years of this century, history of science, or what little there was of it, was dominated by two main traditions.[13] One of them, with an almost continuous tradition from Condorcet and Comte to Dampier and Sarton, viewed scientific advance as the triumph of reason over primitive superstition, the unique example of humanity operating in its highest mode. Though vast scholarship, some of it still useful, was sometimes expended on them, the chronicles which this tradition produced were ultimately hortatory in intent, and they included remarkably little information about the content of science beyond who first made which positive discovery when. Except occasionally for reference or the preparation of historiographic articles, no contemporary historian of science reads them, a fact which does not yet seem to have been as widely appreciated as it should be by the historical profession at large. Though I know it will give offense to some people whose feelings I value, I see no alternative to underscoring the point. Historians of science owe the late George Sarton an immense debt for his role in establishing their profession, but the image of their specialty which he propagated continues to do much damage even though it has long since been rejected.

A second tradition, more important both for its products and because, particularly on the Continent, it still displays some life, originates with practicing scientists, sometimes eminent ones, who have from time to time prepared histories of their specialties. Their work usually began as a by-product of science pedagogy and was directed predominantly to science students. Besides intrinsic appeal, they saw in such histories a means to elucidate the contents of their specialty, to establish its tradition, and to attract students. The volumes they produced were and are quite technical, and the best of them can still be used with profit by specialists with different historiographic inclinations. But seen as history, at least from current perspectives, the tradition has two great limitations. Excepting in occasional naive asides, it produced exclusively internal histories which considered neither context for, nor external effects of, the evolution of the concepts and techniques being discussed. That limitation need not always have been a defect, for the mature sciences are regularly more insulated from the external climate, at

13. A number of the following points are developed more fully in my "History of Science," *International Encyclopedia of the Social Sciences* (New York, 1968), 14:74–83, 105–26, below.

least of ideas, than are other creative fields. But it was undoubtedly badly overdone and, in any case, made work in this mode unattractive to historians, excepting perhaps historians of ideas.

Even the purest historians of ideas were, however, repelled and on occasion seriously misled by a second and even more pronounced defect of this tradition. Scientist-historians and those who followed their lead characteristically imposed contemporary scientific categories, concepts, and standards on the past. Sometimes a specialty which they traced from antiquity had not existed as a recognized subject for study until a generation before they wrote. Nevertheless, knowing what belonged to it, they retrieved the current contents of the specialty from past texts of a variety of heterogeneous fields, not noticing that the tradition they constructed in the process had never existed. In addition, they usually treated concepts and theories of the past as imperfect approximations to those in current use, thus disguising both the structure and integrity of past scientific traditions. Inevitably, histories written in this way reinforced the impression that the history of science is a not very interesting chronicle of the triumph of sound method over careless error and superstition. If these were the only possible models available, one could criticize historians for little except being too easily misled.

But they are not the only models, nor for thirty years have they been even the ones dominant in the profession. Those derive from a more recent tradition which increasingly adapted to the sciences an approach discovered in late-nineteenth-century histories of philosophy. In that field, of course, only the most partisan could feel confident of their ability to distinguish positive knowledge from error and superstition. As a result, historians could scarcely escape the force of an injunction later phrased succinctly by Bertrand Russell: "In studying a philosopher, the right attitude is neither reverence nor contempt, but first a kind of hypothetical sympathy, until it is possible to know what it feels like to believe in his theories."[14] In the history of ideas, the resulting tradition is the one which produced both Ernst Cassirer and Arthur Lovejoy, men whose work, however profound its limitations, has had a great and fructifying influence on the subsequent treatment of ideas i
tory. What is surprising and remains to be explained is the la

14. Bertrand Russell, *A History of Western Philosophy* (New
Simon & Schuster, 1945), p. 39.

any comparable influence, even on intellectual historians, of the works of the men who, following Alexandre Koyré, have for a generation been developing the same models for the sciences. Seen through their writings, science is not the same enterprise as the one represented in either of the older traditions. For the first time it has become potentially a fully historical enterprise, like music, literature, philosophy, or law.

I say "potentially" because that model too has limitations. Though it has extended the proper subject matter of the historian of science to the entire context of ideas, it remains internal history in the sense that it pays little or no attention to the institutional or socioeconomic context within which the sciences have developed. Recent historiography has, for example, largely discredited the myth of method, but it has then had difficulty finding any significant role for the Baconian movement and has had little but scorn for either the Merton thesis or the relation between science and technology, industry, or the crafts.[15] It is time to confess that a few of the object lessons I have read to historians above could be fruitfully circulated in my own profession as well. But the areas to which these object lessons apply are the interstices between the history of science and the now standard concerns of the cultural and socioeconomic historian. They will need to be worked by both groups. Given a model of the internal development of science which provides points of entrée, historians of science are now increasingly turning to it, a movement to be discussed in my concluding section. I am aware of no comparable movement within the historical profession at large.

Clearly, historians of science must share the blame. But no catalog of their past and present sins will entirely explain the realities of their current relation to the rest of the historical profession. What currency their work has achieved has come primarily through Butterfield's book, published nearly thirty years ago, when their discipline was embryonic, and never fully assimilated since. Neglect of their subject matter, science, remains particularly acute for just the years during which it became a major historical force. Though usually placed in history departments, their courses are

15. T. S. Kuhn, "Alexandre Koyré and the History of Science," *Encounter* 34 (1970): 67–70.

seldom taken and their books seldom read by historians. About the causes of that situation I can only speculate and part of that speculation must deal with subjects that I know only through conversations with colleagues and friends. Nevertheless, the occasion of this volume may provide an excuse for speculation.

Two sorts of explanations suggest themselves, of which the first arises from what is perhaps a factor unique to history among the learned disciplines. The history of science is not in principle a narrower specialty than, say, political, diplomatic, social, or intellectual history. Nor are its methods radically distinct from the ones employed in those fields. But it is a specialty of a different sort, for it is concerned in the first instance with the activity of a special group—the scientists—rather than with a set of phenomena which must at the start be abstracted from the totality of activities within a geographically defined community. In this respect its natural kin are the history of literature, of philosophy, of music, and of the plastic arts.[16] These specialties, however, are not ordinarily offered by departments of history. Instead, they are more or less integral parts of the offering of the department responsible for the discipline of which the history is to be studied. Perhaps historians react to the history of science in the same way that they do to the history of other disciplines. Perhaps it is only the proximity created by membership in the same department which leads to a special sense of strain.

16. M. I. Finley points out that the history of law would provide an even more revealing parallel. The law, after all, is one of the obvious determinants of the sorts of political and social developments which historians have traditionally studied. But, excepting for reference to the expression of the will of society through legislation, historians seldom pay attention to its evolution as an institution. Reactions at the conference to Peter Paret's insistence that military history must be in part the history of the military establishment as an institution with a life that is in part its own suggests how deep resistance to disciplinary history sometimes lies. Participants suggested, for example, that what military history ought to be is the study of the social sources of war and of the effects of war on society. But these subjects, though they perhaps provide the main reasons for wanting to have military history done, must not be its primary focus. An understanding of wars, their development and consequences, depends in essential ways on an understanding of military establishments. In any case, the subject war-and-society is as much the responsibility of the general historian as of his colleague who specializes in military history. The parallel to history of science is very close.

Those suggestions I owe to Carl Schorske, one of the two historians with whom I and my students have interacted most closely and fruitfully since I first began to teach in a history department fourteen years ago. He has persuaded me, though not until this essay was well advanced, that many of the problems discussed under the rubric science-in-intellectual-history, above, have precise parallels in the historian's typical discussion of other intellectual, literary, and artistic pursuits. Historians are, he argues, often quite adept at retrieving from a novel, a painting, or a philosophical disquisition, themes which reflect contemporary social problems and values. What they regularly miss, however, sometimes by explaining them away, are those aspects of these artifacts that are internally determined, partly by the intrinsic nature of the discipline which produces them and partly by the special role which that discipline's past always plays in its current evolution. Artists, whether in imitation or revolt, build from past art. Like scientists, philosophers, writers, and musicians, they live and work both within a larger culture and within a quasi-independent disciplinary tradition of their own. Both environments shape their creative products, but the historian all too often considers only the first.

Excepting in my own field, my competence for evaluating these generalizations is restricted to the history of philosophy. There, however, they fit as precisely as they do in the history of science. Since they are, in addition, extremely plausible, I shall tentatively accept them. What historians generally view as historical in the development of individual creative disciplines are those aspects which reflect its immersion in a larger society. What they all too often reject, as not quite history, are those internal features which give the discipline a history in its own right.

The perception which permits that rejection seems to me profoundly unhistorical. The historian does not apply it in other realms. Why should he do so here? Consider, for example, the manner in which historians treat geographical and linguistic subdivisions. Few of them would deny the existence of problems which can be treated only on the gigantic canvas of world history. But they do not therefore deny that the study of the development of Europe or America is also historical. Nor do they resist the next step, which finds a legitimate role for national or even county histories provided that their authors remain alert to the aspects of their restricted subject which are determined by the influence of surrounding groups. When, as is inevitable, communication problems arise, for example between British and European historians, these are deplored as historiographic blinders and as likely sources of error. The feelings which are generated sometimes resemble those which historians of science or art regularly encounter, but no one would say *out loud* that French history is by definition historical in some sense in which British is not. Yet that is very often the response when the analytic units shift from geographically defined subsystems to groups whose cohesiveness—not necessarily less (or more) real than that of a national community—derives from training in a special discipline and an allegiance to its special values. Perhaps if historians could admit the existence of seams in Clio's web, they could more easily recognize that there are no rents.

The resistance to disciplinary histories is not, of course, exclusively the fault of the historians who work within history departments. With a few notable exceptions like Paul Kristeller and Erwin Panofsky, the men who study the development of a discipline from within that discipline's parent department concentrate excessively on the internal logic of the field they study, often missing both consequences and causes in the larger cultures. I remember with deep embarrassment the day on which a student found occasion to remind me that Arnold Sommerfeld's relativistic treatment of the atom was invented midway through the First World War. Institutional separations depress historical sensitivities on both sides of the barrier they create. Nor is separation the only source of difficulty. The man who teaches within the department responsible for the discipline he studies almost always addresses himself to practitioners of that discipline or, in the case of literature and the arts, to its critics. Usually the historical dimension of his work is subordinated to the function of teaching and perfecting the current discipline. The history of philosophy, as taught within philosophy departments, is often, for example, a parody of the historical. Reading a work of the past, the philosopher regularly seeks the author's positions on current problems, criticizes them with the aid of current apparatus, and interprets his text to maximize its coherence with modern doctrine. In that process the historic original is often lost. I am told, for example, of the response of a former philosophy colleague to a student who questioned his reading of a passage in Marx. "Yes," he said, "the words do seem to say what you suggest. But that cannot be what Marx meant, for it is plainly false." Why Marx should have chosen to use the words he did was not a problem worth pausing for.

Most examples of the Whiggishness enforced by placing history in the service of a parent discipline are more subtle but no less unhistorical. The damage they do is no greater, I think, than that done by the historian's rejection of disciplinary history, but it is surely as great. I have already pointed out that the history of science displayed the same unhistorical syndromes when it was taught within science departments. The forces that have increasingly transferred it to history departments in recent years have placed it where it belongs. Though a shotgun was required for the wedding and though the strains characteristic of forced marriages result, the offspring may yet be viable. I cannot doubt that similar compulsory association with the practitioners of other branches of disciplinary history would be equally fruitful. Perhaps, as my first history department chairman, the late George Guttridge, once remarked, we shall soon recognize how badly history fits the departmental organization of American universities. Some transdepartmental institutional arrangement is badly needed, perhaps a faculty or school of historical studies which would bring together all those whose concern, regardless of their departmental affiliations, is with the past in evolution.

I have been considering the suggestion that the relations between history and the history of science differ only in intensity, not in kind, from the relations between history and the study of the development of other disciplines. The parallels are, I think, clear, and they carry us a long way toward an understanding of the problem I have been asked to discuss. But they are not complete, and they do not explain everything. Treating literature, art, or philosophy, historians do, I have suggested, read sources as they do not in the sciences. The historian's ignorance of even the main developmental stages of science has no parallel for the other disciplines on which he touches. Even offered in other departments, courses in the history of literature and the arts are more likely to attract historians than courses in the history of science. Above all, there is no precedent in other disciplines for the historian's exclusive attention to a single period when discussing a science. Those historians who consider art, literature, or philosophy at all are as likely to do so when dealing with the nineteenth century as with the Renaissance. Science, on the other hand, is a topic to be discussed only between 1540 and 1700. One reason, I suspect, for the historian's characteristic emphasis on the discovery of method is that it protects him from the need to deal with the sciences after that period. With their method in hand, the sciences cease to be historical, a perception for which there is no parallel in the historian's view of other disciplines.

Contemplating these phenomena and some more personal experiences to be illustrated below, I reluctantly conclude that part of what separates historians from their colleagues in the history of science is what, in addition to personality, separates F. R. Leavis from C. P. Snow. Though I sympathize with those who believe it has been misnamed, the two-culture problem is another probable source of the difficulties we have been considering.

My basis for that conjecture is largely impressionistic, but not entirely so. Consider the following quotation from a British psychologist whose tests enable him to predict with some assurance the future specialties of high school students, even though (like I.Q. tests, which he includes) they discriminate scarcely at all between those who will do well and badly after specialization:

> The typical historian or modern linguist had, relatively speaking, rather a low I.Q., and a verbal bias of intelligence. He was prone to work erratically on the intelligence test, accurate at times and slap dash at others; and his interests tended to be cultural rather than practical. The young physical scientist often had a high I.Q., and a non verbal bias of ability; he was usually consistently accurate; his interests were usually technical, mechanical, or in life out of doors. Naturally, these rules-of-thumb were not perfect: a minority of arts specialists had scores like scientists, and vice versa. But, by and large, the predictions held surprisingly well, and at the extremes they were infallible.[17]

Together with other evidence from the same source, this passage suggests that historians and scientists, at least those of the more mathematical and abstract sort, are polar types.[18] Other studies,

17. Liam Hudson, *Contrary Imaginations: A Psychological Study* *English Schoolboy* (London: Methuen, 1966), p. 22.

18. A fuller analysis, to which Hudson's pioneering book provide fascinating leads, would recognize that there are multiple dimens polarization. For example, the same sort of scientists who are mos

though insufficiently detailed to single out historians, indicate that scientists as a group come from a lower socioeconomic stratum than their academic colleagues in other fields.[19] Personal impressions, both from my own school days and my children's, suggest that the intellectual differences appear quite early, especially in mathematics, where they are often obvious before age fourteen. I am thinking now not primarily of ability or creativity, but merely of affection. Though there are both exceptions and a large middle ground, I suggest that a passion for history is seldom compatible with even a developed liking for mathematics or laboratory science, and vice versa.

Not surprisingly, as these polarities develop and are embodied in career decisions, they often find expression in defensiveness and hostility. The historians who read this essay will not need to be told of the often overt disdain of scientists for historical studies. Unless I suppose that it is reciprocated, I cannot account for the stance, described above, of historians toward the sciences. Historians of science ought to be exceptions, but even they often prove the rule. Most of them begin in science, turning to its history only at the graduate level. Those who do frequently insist that their interest is only in the history of science, not in mere history, a field they conceive of as at once irrelevant and uninteresting. As a result, they are more easily attracted to special departments or programs than to regular history departments. Fortunately, it is usually possible to convert them once they are there.

If, however, many historians are hostile to science—as I suppose—it must be admitted that they disguise it well, far better, for example, than their colleagues in literature, language, and the arts, who are often entirely explicit. Yet that difference provides at least no counterevidence, for it could have been expected. Like philosophers and unlike most students of literature and art, historians

to disdain history are often passionately interested in music though not usually in the other main forms of artistic expression. Neither Hudson nor I is referring to a simple spectrum ranging from the artist, at one extreme, to the scientist, at the other, with the historian and artist at the same end of the spectrum.

19. C. C. Gillispie, "Remarks on Social Selection as a Factor in the Progressivism of Science," *American Scientist* 56 (1968): 439–50, underscores the phenomenon and provides relevant bibliography.

see their enterprise as somehow cognitive and thus akin to science if not of it. With scientists they share such values as impartiality, objectivity, and faithfulness to evidence. They too have tasted the forbidden fruit of the tree of knowledge, and the antiscientific rhetoric of the arts is no longer available to them. There are, however, subtler ways of expressing hostility, some of which I have suggested above. This part of my argument will therefore conclude with some evidence of a more personal sort.

The first is a memorable encounter with a much-valued friend and colleague, who has from time to time organized and led an experimental seminar at Princeton designed to acquaint first-year graduate students with ancillary methods and approaches for which the future specialist may some day find a use. When appropriate, a local or visiting specialist is asked to manage discussion and to consult about the preparatory reading. Several years ago I accepted an invitation to lead the group in the first of a pair of meetings on the history of science. The central item of the reading, selected after much talk, was an old book of mine, *The Copernican Revolution*. That choice may not have been the best, but there were reasons for it, explicit both in my conversations with my colleague and in the preface. Though not a text, the book was written so that it could be used in college courses on science for the nonscientist. It would not, therefore, present insuperable obstacles to our graduate students. More important, when it was written, the book was the only one that attempted to portray, within a single pair of covers, both the technical-astronomical and the wider intellectual-historical dimensions of the revolution. It was thus a concrete example of the point I have argued more abstractly above: the role of science in intellectual history cannot be understood without the science. How many students grasped that point I cannot be sure, but my colleague did not. Midway through a lively discussion, he interjected, "But, of course, I skipped the technical parts." Since he is a busy man, the omission may not be surprising. But what is suggested by his willingness, unsolicited, to make it public?

My second, briefer example is in the public domain. Frank Manuel's *Portrait of Isaac Newton* is surely the most brilliant and thorough study of its subject in a very long time. Excepting those offended by its psychoanalytic approach, the Newtonian experts with whom I have discussed it assure me that it will affect their work for years to come. The history of science would be far poorer if it had not been written. Nevertheless, in the present context, it raises a fundamental question. Is there any field but science in which one can imagine a historian's preparing a major biography which omits, consciously and deliberately, any attempt to deal with the creative work which made its subject's life a worthy object of study. I cannot think of a similar labor of love devoted to a major figure in the arts, philosophy, religion, or public life. Under the circumstances, I am not sure that love is the emotion involved.

These examples were introduced by the claim that they would display hostility to science. Having presented them, I confess my uncertainty that "hostility" is altogether the appropriate term. But they are examples of strange behavior. If what they illustrate must for the moment remain obscure, it may nevertheless constitute the central impediment separating history and the history of science.

Having by now said more than all I know about the barriers that divide history from the history of science, I shall conclude with some brief remarks on signs of change. One of them is the mere proliferation of historians of science and their increasing placement in departments of history. Though both numbers and proximity may be initially a source of friction, they also increase the availability of communication channels. Growth is also responsible for a second encouraging development, the increasing attention now being devoted to periods more recent than the Scientific Revolution and to previously little explored parts of science. The better secondary literature will not for much longer be restricted to the sixteenth and seventeenth centuries, nor will it continue to deal primarily with the physical sciences. The current increase in the study of the history of the life sciences may prove particularly important. These fields have, until recently, been far less technical than the main physical sciences contemporary with them. Studies that trace their development are likely to be correspondingly more accessible to the historian who would like to discover what the history of science is about.

Look next at two other developments the effects of which are now observable among many of the younger practitioners of the history of science. Led by Frances Yates and Walter Pagel, they are now finding increasingly significant roles for Hermeticism and related movements in the early stages of the Scientific Revolution.[20] The original and exciting literature which results may well have three effects which transcend its explicit subject. First, just because Hermeticism was an avowedly mystical and irrational movement, recognition of its roles should help to make science more palatable to historians repelled by what many have taken to be a quasi-mechanical enterprise, governed by pure reason and cold fact. (It would plainly be absurd to select the rational elements from Hermeticism for exclusive attention as an older generation has done with Neoplatonism.) Second, Hermeticism now appears to have affected two aspects of scientific development previously seen as mutually exclusive and defended by competing schools. On the one hand, it was an intellectual, quasi-metaphysical movement which changed man's ideas about the entities and causes underlying natural phenomena; as such it is analyzable by the usual techniques of the historian of ideas. But it was also a movement which, in the figure of the magus, prescribed new goals and methods for science. Treatises on, for example, natural magic show that the new emphasis on the power of science, on the study of crafts, and on mechanical manipulation and machines is in part the product of the same movement that changed the intellectual climate. Two disparate approaches to the history of science are thus unified in a way likely to have particular appeal to the historian. Finally, newest, and perhaps most important, Hermeticism now has begun to be studied as a class movement with a discernible social base.[21] If that development continues, the study of the Scientific Revolution will become multidimensional cultural history of the sort many historians are now also striving to create.

I turn finally to the newest movement of all, apparent primarily among graduate students and the very youngest members of the profession. Perhaps partly because of their increasing contact with historians, they are turning more and more to the study of what is often described as external history. Increasingly they emphasize

20. F. A. Yates, "The Hermetic Tradition in Renaissance Science," in C. S. Singleton, ed., *Art, Science, and History in the Renaissance* (Baltimore: Johns Hopkins University Press, 1967), pp. 255–74; Walter Pagel, *William Harvey's Biological Ideas* (New York: Karger, 1967).

21. P. M. Rattansi, "Paracelsus and the Puritan Revolution," *Ambix* 11 (1963): 24–32, and "The Helmontian-Galenist Controversy in Restoration England," *Ambix* 12 (1964): 1–23.

the effects on science not of the intellectual but of the socioeconomic milieu, effects manifest in changing patterns of education, institutionalization, communication, and values. Their efforts owe something to the older Marxist histories, but their concerns are at once broader, deeper, and less doctrinaire than those of their predecessors. Because historians will find themselves more at home with the studies that result than they have been with older histories of science, they are particularly likely to welcome the change. Indeed they may even learn from it something of more general relevance. Like literature and the arts, science is the product of a group, a community of scientists. But in the sciences, particularly in the later stages of their development, disciplinary communities are both easier to isolate and also more nearly self-contained and self-sufficient than the relevant groups in other fields. As a result the sciences provide a particularly promising area in which to explore the role of forces current in the larger society in shaping the evolution of a discipline which is simultaneously controlled by its own internal demands.[22] That study, if successful, could provide models for a variety of fields besides the sciences.

All these developments are necessarily encouraging to anyone bothered by the traditional chasm between history and the history of science. If they continue, as seems likely, it will be less deep a decade hence than it has been in the past. But it is not likely to disappear, for the new trends described above can have only indirect, partial, and long-range effects on what I take to be the fundamental source of the division. Perhaps the example of the history of science can by itself undermine the historian's resistance to disciplinary history, but I would be more confident if I knew the reasons for that resistance in the past. In any case, the history of science is, by itself, an unlikely remedy for a social malady so deep and widespread as the two-culture problem. Instead, in my most depressed moments, I sometimes fear that the history of science may yet be that problem's victim. Though I welcome the turn to the external history of science as redressing a balance which has long been seriously askew, its new popularity may not be an unmixed

22. The penultimate section of the article cited in note 13 elaborates this possibility in theoretical terms. T. M. Brown's "The College of Physicians and the Acceptance of Iatromechanism in England, 1665–1695," *Bulletin of the History of Medicine* 44 (1970): 12–30, provides a concrete example.

blessing. One reason it now flourishes is undoubtedly the increasingly virulent antiscientific climate of these times. If it becomes the exclusive approach, the history of science could be reduced to a higher-level version of the tradition which, by leaving the science out, ignored the internalities that shape the development of any discipline. That price would be too high to pay for rapprochement, but unless historians can find a place for the history of disciplines, it will be hard to avoid.

II Metahistorical Studies

7

The Historical Structure of Scientific Discovery

My object in this article is to isolate and illuminate one small part of what I take to be a continuing historiographic revolution in the study of science.[1] The structure of scientific discovery is my particular topic, and I can best approach it by pointing out that the subject itself may well seem extraordinarily odd. Both scientists and, until quite recently, historians have ordinarily viewed discovery as the sort of event which, though it may have preconditions and surely has consequences, is itself without internal structure. Rather than being seen as a complex development extended both in space and time, discovering something has usually seemed to be a unitary event, one which, like seeing something, happens to an individual at a specifiable time and place.

This view of the nature of discovery has, I suspect, deep roots in the nature of the scientific community. One of the few historical elements recurrent in the textbooks from which the prospective scientist learns his field is the attribution of particular natural phenomena to the historical personages who first discovered them. As a result of this and other aspects of their training, discovery be-

1. The larger revolution will be discussed in my forthcoming book, *The Structure of Scientific Revolutions,* to be published in the fall by the University of Chicago Press. The central ideas in this paper have been abstracted from that source, particularly from its sixth chapter, "Anomaly and the Emergence of Scientific Discoveries" [2d ed., 1970].

comes for many scientists an important goal. To make a discovery is to achieve one of the closest approximations to a property right that the scientific career affords. Professional prestige is often closely associated with these acquisitions.[2] Small wonder, then, that acrimonious disputes about priority and independence in discovery have often marred the normally placid tenor of scientific communication. Even less wonder that many historians of science have seen the individual discovery as an appropriate unit with which to measure scientific progress and have devoted much time and skill to determining what man made which discovery at what point in time. If the study of discovery has a surprise to offer, it is only that, despite the immense energy and ingenuity expended upon it, neither polemic nor painstaking scholarship has often succeeded in pinpointing the time and place at which a given discovery could properly be said to have "been made."

That failure, both of argument and of research, suggests the thesis that I now wish to develop. Many scientific discoveries, particularly the most interesting and important, are not the sort of event about which the questions "Where?" and, more particularly, "When?" can appropriately be asked. Even if all conceivable data were at hand, those questions would not regularly possess answers. That we are persistently driven to ask them nonetheless is symptomatic of a fundamental inappropriateness in our image of discovery. That inappropriateness is here my main concern, but I approach it by considering first the historical problem presented by the attempt to date and to place a major class of fundamental discoveries.

The troublesome class consists of those discoveries—including oxygen, the electric current, X rays, and the electron—which could not be predicted from accepted theory in advance and which therefore caught the assembled profession by surprise. That kind of discovery will shortly be my exclusive concern, but it will help first to note that there is another sort and one which presents very few of the same problems. Into this second class of discoveries fall the neutrino, radio waves, and the elements which filled empty places

...or a brilliant discussion of these points, see R. K. Merton, "Priorities ...ntific Discovery: A Chapter in the Sociology of Science," *American ...gical Review* 22 (1957): 635. Also very relevant, though it did not ... until this article had been prepared, is F. Reif, "The Competitive ... of the Pure Scientist," *Science* 134 (1961): 1957.

in the periodic table. The existence of all these objects had been predicted from theory before they were discovered, and the men who made the discoveries therefore knew from the start what to look for. That foreknowledge did not make their task less demanding or less interesting, but it did provide criteria which told them when their goal had been reached.[3] As a result, there have been few priority debates over discoveries of this second sort, and only a paucity of data can prevent the historian from ascribing them to a particular time and place. Those facts help to isolate the difficulties we encounter as we return to the troublesome discoveries of the first class. In the cases that most concern us here there are no benchmarks to inform either the scientist or the historian when the job of discovery has been done.

As an illustration of this fundamental problem and its consequences, consider first the discovery of oxygen. Because it has repeatedly been studied, often with exemplary care and skill, that discovery is unlikely to offer any purely factual surprises. Therefore it is particularly well suited to clarify points of principle.[4] At least three scientists—Carl Scheele, Joseph Priestley, and Antoine Lavoisier—have a legitimate claim to this discovery, and polemicists have occasionally entered the same claim for Pierre Bayen.[5]

3. Not all discoveries fall so neatly as the preceding into one or the other of my two classes. For example, Anderson's work on the positron was done in complete ignorance of Dirac's electron theory from which the new particle's existence had already been very nearly predicted. On the other hand, the immediately succeeding work by Blackett and Occhialini made full use of Dirac's theory and therefore exploited experiment more fully and constructed a more forceful case for the positron's existence than Anderson had been able to do. On this subject see N. R. Hanson, "Discovering the Positron," *British Journal for the Philosophy of Science* 12 (1961): 194; 12 (1962): 299. Hanson suggests several of the points developed here. I am much indebted to Professor Hanson for a preprint of this material.

4. I have developed a less familiar example from the same viewpoint in "The Caloric Theory of Adiabatic Compression," *Isis* 49 (1958): 132. A closely similar analysis of the emergence of a new theory is included in the early pages of my essay "Energy Conservation as an Example of Simultaneous Discovery," in *Critical Problems in the History of Science*, ed. M. Clagett (Madison: University of Wisconsin Press, 1959), pp. 321–56 (pp. 66–104 above). Reference to these papers may add depth and detail to the following discussion.

5. The still classic discussion of the discovery of oxygen is A. N. Meldrum, *The Eighteenth Century Revolution in Science: The First Phase* (Calcutta, 1930), chap. 5. A more convenient and generally quite reliable

Scheele's work, though it was almost certainly completed before the relevant researches of Priestley and Lavoisier, was not made public until their work was well known.[6] Therefore it had no apparent causal role, and I shall simplify my story by omitting it.[7] Instead, I pick up the main route to the discovery of oxygen with the work of Bayen, who, sometime before March 1774, discovered that red precipitate of mercury (HgO) could, by heating, be made to yield a gas. That aeriform product Bayen identified as fixed air (CO_2), a substance made familiar to most pneumatic chemists by the earlier work of Joseph Black.[8] A variety of other substances were known to yield the same gas.

At the beginning of August 1774, a few months after Bayen's work had appeared, Joseph Priestley repeated the experiment, though probably independently. Priestley, however, observed that

discussion is included in J. B. Conant, *The Overthrow of the Phlogiston Theory: The Chemical Revolution of 1775–1789*, Harvard Case Histories in Experimental Science, case 2 (Cambridge: Harvard University Press, 1950). A recent and indispensable review, which includes an account of the development of the priority controversy, is M. Daumas, *Lavoisier, théoricien et expérimentateur* (Paris, 1955), chaps. 2 and 3. H. Guerlac has added much significant detail to our knowledge of the early relations between Priestley and Lavoisier in his "Joseph Priestley's First Papers on Gases and Their Reception in France," *Journal of the History of Medicine* 12 (1957): 1 and in his very recent monograph, *Lavoisier: The Crucial Year* (Ithaca: Cornell University Press, 1961). For Scheele see J. R. Partington, *A Short History of Chemistry*, 2d ed. (London, 1951), pp. 104–9.

6. For the dating of Scheele's work, see A. E. Nordenskjöld, *Carl Wilhelm Scheele, Nachgelassene Briefe und Aufzeichnungen* (Stockholm, 1892).

7. U. Bocklund ("A Lost Letter from Scheele to Lavoisier," *Lychnos*, 1957–58, pp. 39–62) argues that Scheele communicated his discovery of oxygen to Lavoisier in a letter of 30 Sept. 1774. Certainly the letter is important, and it clearly demonstrates that Scheele was ahead of both Priestley and Lavoisier at the time it was written. But I think the letter is not quite so candid as Bocklund supposes, and I fail to see how Lavoisier could have drawn the discovery of oxygen from it. Scheele describes a procedure for reconstituting common air, not for producing a new gas, and that, as we shall see, is almost the same information that Lavoisier received from Priestley at about the same time. In any case, there is no evidence that Lavoisier performed the sort of experiment that Scheele suggested.

8. P. Bayen, "Essai d'expériences chymiques, faites sur quelques précipités de mercure, dans la vue de découvrir leur nature, Seconde partie," *Observations sur la physique* 3 (1774): 280–95, particularly pp. 289–91.

the gaseous product would support combustion and therefore changed the identification. For him the gas obtained on heating red precipitate was nitrous air (N_2O), a substance that he had himself discovered more than two years before.[9] Later in the same month Priestley made a trip to Paris and there informed Lavoisier of the new reaction. The latter repeated the experiment once more, both in November 1774 and in February 1775. But, because he used tests somewhat more elaborate than Priestley's, Lavoisier again changed the identification. For him, as of May 1775, the gas released by red precipitate was neither fixed air nor nitrous air. Instead, it was "[atmospheric] air itself entire without alteration . . . even to the point that . . . it comes out more pure."[10] Meanwhile, however, Priestley had also been at work, and, before the beginning of March 1775, he, too, had concluded that the gas must be "common air." Until this point all of the men who had produced a gas from red precipitate of mercury had identified it with some previously known species.[11]

The remainder of this story of discovery is briefly told. During March 1775 Priestley discovered that his gas was in several respects very much "better" than common air, and he therefore reidentified the gas once more, this time calling it "dephlogisticated air," that is, atmospheric air deprived of its normal complement of phlogiston. This conclusion Priestley published in the *Philosophical Transactions*, and it was apparently that publication which led Lavoisier to reexamine his own results.[12] The reexamination began during February 1776 and within a year had led Lavoisier to the conclusion that the gas was actually a separable component of the atmospheric air which both he and Priestley had previously thought

9. J. B. Conant, *The Overthrow of the Phlogiston Theory*, pp. 34–40.

10. Ibid., p. 23. A useful translation of the full text is available in Conant.

11. For simplicity I use the term *red precipitate* throughout. Actually, Bayen used the precipitate; Priestley used both the precipitate and the oxide produced by direct calcination of mercury; and Lavoisier used only the latter. The difference is not without importance, for it was not unequivocally clear to chemists that the two substances were identical.

12. There has been some doubt about Priestley's having influenced Lavoisier's thinking at this point, but, when the latter returned to experimenting with the gas in February 1776, he recorded in his notebooks that he had obtained "l'air dephlogistique de M. Priestley" (M. Daumas, *Lavoisier*, p. 36).

of as homogeneous. With this point reached, with the gas recognized as an irreducibly distinct species, we may conclude that the discovery of oxygen had been completed.

But to return to my initial question, when shall we say that oxygen was discovered and what criteria shall we use in answering that question? If discovering oxygen is simply holding an impure sample in one's hands, then the gas had been "discovered" in antiquity by the first man who ever bottled atmospheric air. Undoubtedly, for an experimental criterion, we must at least require a relatively pure sample like that obtained by Priestley in August 1774. But during 1774 Priestley was unaware that he had discovered anything except a new way to produce a relatively familiar species. Throughout that year his "discovery" is scarcely distinguishable from the one made earlier by Bayen, and neither case is quite distinct from that of the Reverend Stephen Hales, who had obtained the same gas more than forty years before.[13] Apparently to discover something one must also be aware of the discovery and know as well what it is that one has discovered.

But, that being the case, how much must one know? Had Priestley come close enough when he identified the gas as nitrous air? If not, was either he or Lavoisier significantly closer when he changed the identification to common air? And what are we to say about Priestley's next identification, the one made in March 1775? Dephlogisticated air is still not oxygen or even, for the phlogistic chemist, a quite unexpected sort of gas. Rather it is a particularly pure atmospheric air. Presumably, then, we wait for Lavoisier's work in 1776 and 1777, work which led him not merely to isolate the gas but to see what it was. Yet even that decision can be questioned, for in 1777 and to the end of his life Lavoisier insisted that oxygen was an atomic "principle of acidity" and that oxygen *gas* was formed only when that "principle" united with caloric, the matter of heat.[14] Shall we therefore say that oxygen had not yet been discovered in 1777? Some may be tempted to do so. But the principle of acidity was not banished from chemistry until after 1810 and caloric lingered on until the 1860s. Oxygen had, however, become a standard chemical substance long before either of

13. J. R. Partington, *A Short History of Chemistry*, p. 91.

14. For the traditional elements in Lavoisier's interpretations of chemical reactions, see H. Metzger, *La philosophie de la matière chez Lavoisier* (Paris, 1935), and Daumas, *Lavoisier*, chap. 7.

those dates. Furthermore, what is perhaps the key point, it would probably have gained that status on the basis of Priestley's work alone without benefit of Lavoisier's still partial reinterpretation.

I conclude that we need a new vocabulary and new concepts for analyzing events like the discovery of oxygen. Though undoubtedly correct, the sentence "Oxygen was discovered" misleads by suggesting that discovering something is a single simple act unequivocally attributable, if only we knew enough, to an individual and an instant in time. When the discovery is unexpected, however, the latter attribution is always impossible and the former often is as well. Ignoring Scheele, we can, for example, safely say that oxygen had not been discovered before 1774; probably we would also insist that it had been discovered by 1777 or shortly thereafter. But within those limits any attempt to date the discovery or to attribute it to an individual must inevitably be arbitrary. Furthermore, it must be arbitrary just because discovering a new sort of phenomenon is necessarily a complex process which involves recognizing both *that* something is and *what* it is. Observation and conceptualization, fact and the assimilation of fact to theory, are inseparably linked in the discovery of scientific novelty. Inevitably, that process extends over time and may often involve a number of people. Only for discoveries in my second category—those whose nature is known in advance—can discovering *that* and discovering *what* occur together and in an instant.

Two last, simpler, and far briefer examples will simultaneously show how typical the case of oxygen is and also prepare the way for a somewhat more precise conclusion. On the night of 13 March 1781, the astronomer William Herschel made the following entry in his journal: "In the quartile near Zeta Tauri . . . is a curious either nebulous star or perhaps a comet."[15] That entry is generally said to record the discovery of the planet Uranus, but it cannot quite have done that. Between 1690 and Herschel's observation in 1781 the same object had been seen and recorded at least seventeen times by men who took it to be a star. Herschel differed from them only in supposing that, because in his telescope it appeared especially large, it might actually be a *comet*! Two additional observations on 17 and 19 March confirmed that suspicion by show-

15. P. Doig, *A Concise History of Astronomy* (London: Chapman, 1950), pp. 115–16.

ing that the object he had observed moved among the stars. As a result, astronomers throughout Europe were informed of the discovery, and the mathematicians among them began to compute the new comet's orbit. Only several months later, after all those attempts had repeatedly failed to square with observation, did the astronomer Lexell suggest that the object observed by Herschel might be a planet. And only when additional computations, using a planet's rather than a comet's orbit, proved reconcilable with observation was that suggestion generally accepted. At what point during 1781 do we want to say that the planet Uranus was discovered? And are we entirely and unequivocally clear that it was Herschel rather than Lexell who discovered it?

Or consider still more briefly the story of the discovery of X rays, a story which opens on the day in 1895 when the physicist Roentgen interrupted a well-precedented investigation of cathode rays because he noticed that a barium platinocyanide screen far from his shielded apparatus glowed when the discharge was in process.[16] Additional investigations—they required seven hectic weeks during which Roentgen rarely left the laboratory—indicated that the cause of the glow traveled in straight lines from the cathode ray tube, that the radiation cast shadows, that it could not be deflected by a magnet, and much else besides. Before announcing his discovery Roentgen had convinced himself that his effect was not due to cathode rays themselves but to a new form of radiation with at least some similarity to light. Once again the question suggests itself: When shall we say that X rays were actually discovered? Not, in any case, at the first instant, when all that had been noted was a glowing screen. At least one other investigator had seen that glow and, to his subsequent chagrin, discovered nothing at all. Nor, it is almost as clear, can the moment of discovery be pushed back to a point during the last week of investigation. By that time Roentgen was exploring the properties of the new radiation he had *already* discovered. We may have to settle for the remark that X rays emerged in Würzburg between 8 November and 28 December 1895.

The characteristics shared by these examples are, I think, common to all the episodes by which unanticipated novelties become

16. L. W. Taylor, *Physics, the Pioneer Science* (Boston: Houghton Mifflin Co., 1941), p. 790.

subjects for scientific attention. I therefore conclude these brief remarks by discussing three such common characteristics, ones which may help to provide a framework for the further study of the extended episodes we customarily call "discoveries."

In the first place, notice that all three of our discoveries—oxygen, Uranus, and X rays—began with the experimental or observational isolation of an anomaly, that is, with nature's failure to conform entirely to expectation. Notice, further, that the process by which that anomaly was educed displays simultaneously the apparently incompatible characteristics of the inevitable and the accidental. In the case of X rays, the anomalous glow which provided Roentgen's first clue was clearly the result of an accidental disposition of his apparatus. But by 1895 cathode rays were a normal subject for research all over Europe; that research quite regularly juxtaposed cathode-ray tubes with sensitive screens and films; as a result, Roentgen's accident was almost certain to occur elsewhere, as in fact it had. Those remarks, however, should make Roentgen's case look very much like those of Herschel and Priestley. Herschel first observed his oversized and thus anomalous star in the course of a prolonged survey of the northern heavens. That survey was, except for the magnification provided by Herschel's instruments, precisely of the sort that had repeatedly been carried through before and that had occasionally resulted in prior observations of Uranus. And Priestley, too—when he isolated the gas that behaved almost but not quite like nitrous air and then almost but not quite like common air—was seeing something unintended and wrong in the outcome of a sort of experiment for which there was much European precedent and which had more than once before led to the production of the new gas.

These features suggest the existence of two normal requisites for the beginning of an episode of discovery. The first, which throughout this paper I have largely taken for granted, is the individual skill, wit, or genius to recognize that something has gone wrong in ways that may prove consequential. Not any and every scientist would have noted that no unrecorded star should be so large, that the screen ought not to have glowed, that nitrous air should not have supported life. But that requisite presupposes another which is less frequently taken for granted. Whatever the level of genius available to observe them, anomalies do not emerge from the normal course of scientific research until both instruments and concepts have developed sufficiently to make their emergence likely and to make the anomaly which results recognizable as a violation of expectation.[17] To say that an unexpected discovery begins only when something goes wrong is to say that it begins only when scientists know well both how their instruments and how nature should behave. What distinguished Priestley, who saw an anomaly, from Hales, who did not, is largely the considerable articulation of pneumatic techniques and expectations that had come into being during the four decades which separate their two isolations of oxygen.[18] The very number of claimants indicates that after 1770 the discovery could not have been postponed for long.

The role of anomaly is the first of the characteristics shared by our three examples. A second can be considered more briefly, for it has provided the main theme for the body of my text. Though awareness of anomaly marks the beginning of a discovery, it marks only the beginning. What necessarily follows, if anything at all is to be discovered, is a more or less extended period during which the individual and often many members of his group struggle to make the anomaly lawlike. Invariably that period demands additional observation or experimentation as well as repeated cogitation. While it continues, scientists repeatedly revise their expectations, usually their instrumental standards, and sometimes their most fundamental theories as well. In this sense discoveries have a proper internal history as well as prehistory and a posthistory. Furthermore, within the rather vaguely delimited interval of internal history, there is no single moment or day which the historian, however complete his data, can identify as the point at which the discovery was made. Often, when several individuals are involved, it is even impossible unequivocally to identify any one of them as the discoverer.

Finally, turning to the third of these selected common characteristics, note briefly what happens as the period of discovery draws to a close. A full discussion of that question would require addi-

17. Though the point cannot be argued here, the conditions which make the emergence of anomaly likely and those which make anomaly recognizable are to a very great extent the same. That fact may help us understand the extraordinarily large amount of simultaneous discovery in the sciences.

18. A useful sketch of the development of pneumatic chemistry is included in Partington, *A Short History of Chemistry*, chap. 6.

tional evidence and a separate paper, for I have had little to say about the aftermath of discovery in the body of my text. Nevertheless, the topic must not be entirely neglected, for it is in part a corollary of what has already been said.

Discoveries are often described as mere additions or increments to the growing stockpile of scientific knowledge, and that description has helped make the unit discovery seem a significant measure of progress. I suggest, however, that it is fully appropriate only to those discoveries which, like the elements that filled missing places in the periodic table, were anticipated and sought in advance and which therefore demanded no adjustment, adaptation, and assimilation from the profession. Though the sorts of discoveries we have here been examining are undoubtedly additions to scientific knowledge, they are also something more. In a sense that I can now develop only in part, they also react back upon what has previously been known, providing a new view of some previously familiar objects and simultaneously changing the way in which even some traditional parts of science are practiced. Those in whose area of special competence the new phenomenon falls often see both the world and their work differently as they emerge from the extended struggle with anomaly which constitutes the discovery of that phenomenon.

William Herschel, for example, when he increased by one the time-honored number of planetary bodies, taught astronomers to see new things when they looked at the familiar heavens even with instruments more traditional than his own. That change in the vision of astronomers must be a principal reason why, in the half century after the discovery of Uranus, twenty additional circumsolar bodies were added to the traditional seven.[19] A similar trans-

19. R. Wolf, *Geschichte der Astronomie* (Munich, 1877), pp. 513–15, 683–93. The prephotographic discoveries of the asteroids is often seen as an effect of the invention of Bode's law. But that law cannot be the full explanation and may not even have played a large part. Piazzi's discovery of Ceres, in 1801, was made in ignorance of the current speculation about a missing planet in the "hole" between Mars and Jupiter. Instead, like Herschel, Piazzi was engaged on a star survey. More important, Bode's law was old by 1800 (ibid., p. 683), but only one man before that date seems to have thought it worthwhile to look for another planet. Finally, Bode's law, by itself, could only suggest the utility of looking for additional planets; it did not tell astronomers where to look. Clearly, however, the drive to look for additional planets dates from Herschel's work on Uranus.

formation is even clearer in the aftermath of Roentgen's work. In the first place, established techniques for cathode-ray research had to be changed, for scientists found they had failed to control a relevant variable. Those changes included both the redesign of old apparatus and revised ways of asking old questions. In addition, those scientists most concerned experienced the same transformation of vision that we have just noted in the aftermath of the discovery of Uranus. X rays were the first new sort of radiation discovered since infrared and ultraviolet at the beginning of the century. But within less than a decade after Roentgen's work, four more were disclosed by the new scientific sensitivity (for example, to fogged photographic plates) and by some of the new instrumental techniques that had resulted from Roentgen's work and its assimilation.[20]

Very often these transformations in the established techniques of scientific practice prove even more important than the incremental knowledge provided by the discovery itself. That could at least be argued in the cases of Uranus and of X rays; in the case of my third example, oxygen, it is categorically clear. Like the work of Herschel and Roentgen, that of Priestley and Lavoisier taught scientists to view old situations in new ways. Therefore, as we might anticipate, oxygen was not the only new chemical species to be identified in the aftermath of their work. But, in the case of oxygen, the readjustments demanded by assimilation were so profound that they played an integral and essential role—though they were not by themselves the cause—in the gigantic upheaval of chemical theory and practice which has since been known as the chemical revolution. I do not suggest that every unanticipated discovery has consequences for science so deep and so far-reaching as those which followed the discovery of oxygen. But I do suggest that every such discovery demands, from those most concerned, the sorts of readjustment that, when they are more obvious, we equate

20. For α-, β-, and γ-radiation, discovery of which dates from 1896, see Taylor, *Physics*, pp. 800–804. For the fourth new form of radiation, N rays, see D. J. S. Price, *Science Since Babylon* (New Haven: Yale University Press, 1961), pp. 84–89. That N rays were ultimately the source of a scientific scandal does not make them less revealing of the scientific community's state of mind.

with scientific revolution. It is, I believe, just because they demand readjustments like these that the process of discovery is necessarily and inevitably one that shows structure and that therefore extends in time.

8 The Function of Measurement in Modern Physical Science

Reprinted by permission from *Isis* 52 (1961): 161–93.

At the University of Chicago, the façade of the Social Science Research Building bears Lord Kelvin's famous dictum: "If you cannot measure, your knowledge is meager and unsatisfactory."[1] Would that statement be there if it had been written, not by a physicist, but by a sociologist, political scientist, or economist? Or again, would terms like "meter reading" and "yardstick" recur so frequently in contemporary discussions of epistemology and scientific method were it not for the prestige of modern physical science and the fact that measurement so obviously bulks large in its research? Suspecting that the answer to both these questions is no, I find my assigned role in this conference particularly challenging. Because physical science is so often seen as *the* paradigm of sound knowledge and because quantitative techniques seem to provide an essential clue to its success, the question how measurement has actually functioned for the past three centuries in physical science arouses more than its natural and intrinsic interest. Let me there-

1. For the façade see *Eleven Twenty-Six: A Decade of Social Science Research*, ed. Louis Wirth (Chicago, 1940), p. 169. The sentiment there inscribed recurs in Kelvin's writings, but I have found no formulation closer to the Chicago quotation than the following: "When you cannot express it in numbers, your knowledge is of a meagre and unsatisfactory kind." See Sir William Thomson, "Electrical Units of Measurement," *Popular Lectures and Addresses*, 3 vols. (London, 1889–91), 1:73.

fore make my general position clear at the start. Both as an ex-physicist and as an historian of physical science I feel sure that, for at least a century and a half, quantitative methods have indeed been central to the development of the fields I study. On the other hand, I feel equally convinced that our most prevalent notions both about the function of measurement and about the source of its special efficacy are derived largely from myth.

Partly because of this conviction and partly for more autobiographical reasons,[2] I shall employ in this paper an approach rather different from that of most other contributors to this conference. Until almost its close my essay will include no narrative of the increasing deployment of quantitative techniques in physical science since the close of the Middle Ages. Instead, the two central questions of this paper—how has measurement actually functioned in physical science, and what has been the source of its special efficacy—will be approached directly. For this purpose, and for it alone, history will truly be "philosophy teaching by example."

Before permitting history to function even as a source of examples, we must, however, grasp the full significance of allowing it any function at all. To that end my paper opens with a critical discussion of what I take to be the most prevalent image of scientific measurement, an image that gains much of its plausibility and force from the manner in which computation and measurement enter into a profoundly unhistorical source, the science text. That discussion, confined to the next section, will suggest that there is a textbook image or myth of science and that it may be systematically misleading. The actual function of measurement—either in the search for new theories or in the confirmation of those already at hand—must be sought in the journal literature, which displays not finished and accepted theories, but theories in the process of development. After that point in the discussion, history will necessarily become our guide, and the following two sections will present a more valid image of the most usual functions of measurement drawn from that source. The section that follows will employ the resulting description to ask why measurement should have

2. The central sections of this paper, which was added to the present program at a late date, are abstracted from my essay, "The Role of Measurement in the Development of Natural Science," a multilithed revision of a talk first given to the Social Sciences Colloquium of the University of California, Berkeley.

proved so extraordinarily effective in physical research. Only after that, in the concluding section, shall I attempt a synoptic view of the route by which measurement has come increasingly to dominate physical science during the past three hundred years.

One more caveat proves necessary before beginning. A few participants in this conference seem occasionally to mean by measurement any unambiguous scientific experiment or observation. Thus, Professor Boring supposes that Descartes was measuring when he demonstrated the inverted retinal image at the back of the eyeball; presumably he would say the same about Franklin's demonstration of the opposite polarity of the two coatings on a Leyden jar. Undoubtedly, experiments like these are among the most significant and fundamental that the physical sciences have known, but I see no virtue in describing their results as measurements. In any case, that terminology would obscure what are perhaps the most important points to be made in this paper. I shall therefore suppose that a measurement (or a fully quantified theory) always produces actual numbers. Experiments like Descartes' or Franklin's, above, will be classified as qualitative or as nonnumerical, without, I hope at all implying that they are therefore less important. Only with that distinction between qualitative and quantitative available can I hope to show that large amounts of qualitative work have usually been prerequisite to fruitful quantification in the physical sciences. And only if that point can be made shall we be in a position even to ask about the effects of introducing quantitative methods into sciences that had previously proceeded without major assistance from them.

Textbook Measurement

To a very much greater extent than we ordinarily realize, our image of physical science and of measurement is conditioned by science texts. In part that influence is direct: textbooks are the sole source of most people's firsthand acquaintance with the physical sciences. Their indirect influence is, however, undoubtedly larger and more pervasive. Textbooks or their equivalent are the unique repository of the finished achievements of modern physical scientists. It is with the analysis and propagation of these achievements that most writings on the philosophy of science and most interpretations of science for the nonscientist are concerned. As many autobiographies attest, even the research scientist does not always free himself from the textbook image gained during his first exposures to science.[3]

I shall shortly indicate why the textbook mode of presentation must inevitably be misleading, but let us first examine that presentation itself. Since most participants in this conference have already been exposed to at least one textbook of physical science, I restrict attention to the schematic tripartite summary in the following figure. It displays, in the upper left, a series of theoretical and "lawlike" statements, $(x)\phi_i(x)$, which together constitute the theory of the science being described.[4] The center of the diagram repre-

Theory

$(x)\phi_1(x)$

$(x)\phi_2(x)$

. . . . .

$(x)\phi_n(x)$

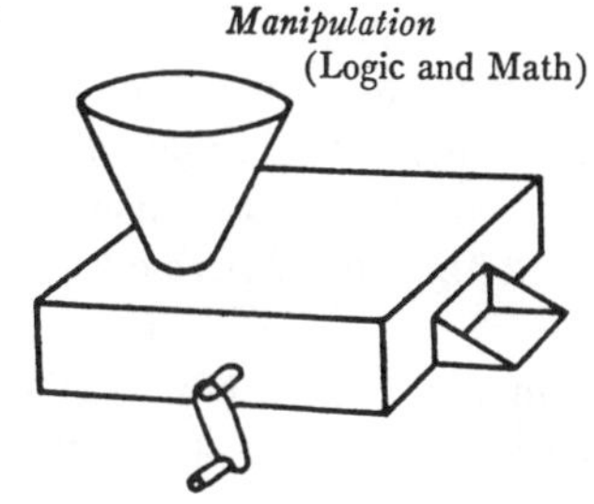

Results

Theory	Experiment
1.414	1.418
1.732	1.725
2.236	2.237

3. This phenomenon is examined in more detail in my monograph, *The Structure of Scientific Revolutions*, to appear when completed as vol. 2, no. 2, in the *International Encyclopedia of Unified Science*. Many other aspects of the textbook image of science, its sources and its strengths, are also examined in that place.

4. Obviously not all the statements required to constitute most theories are of this particular logical form, but the complexities have no relevance to

sents the logical and mathematical equipment employed in manipulating the theory. "Lawlike" statements from the upper left are to be imagined fed into the hopper at the top of the machine together with certain "initial conditions" specifying the situation to which the theory is being applied. The crank is then turned; logical and mathematical operations are internally performed; and numerical predictions for the application at hand emerge from the chute at the front of the machine. These predictions are entered in the left-hand column of the table that appears in the lower right of the figure. The right-hand column contains the numerical results of actual measurements, placed there so that they may be compared with the predictions derived from the theory. Most texts of physics, chemistry, astronomy, and the like contain many data of this sort, though they are not always presented in tabular form. Some of you will, for example, be more familiar with equivalent graphical presentations.

The table at the lower right is of particular concern, for it is there that the results of measurement appear explicitly. What may we take to be the significance of such a table and of the numbers it contains? I suppose that there are two usual answers: the first, immediate and almost universal; the other, perhaps more important, but very rarely explicit.

Most obviously the results in the table seem to function as a test of theory. If corresponding numbers in the two columns agree, the theory is acceptable; if they do not, the theory must be modified or rejected. This is the function of measurement as confirmation, here seen emerging, as it does for most readers, from the textbook formulation of a finished scientific theory. For the time being I shall assume that some such function is also regularly exemplified in normal scientific practice and can be isolated in writings whose purpose is not exclusively pedagogic. At this point we need only notice that on the question of practice, textbooks provide no evidence whatsoever. No textbook ever included a table that either intended or managed to infirm the theory the text was written to describe. Readers of current science texts accept the theories there expounded on the authority of the author and the

the points made here. R. B. Braithwaite, *Scientific Explanation* (Cambridge, 1953) includes a useful, though very general, description of the logical structure of scientific theories.

scientific community, not because of any tables that these texts contain. If the tables are read at all, as they often are, they are read for another reason.

I shall inquire for this other reason in a moment but must first remark on the second putative function of measurement, that of exploration. Numerical data like those collected in the right-hand column of our table can, it is often supposed, be useful in suggesting new scientific theories or laws. Some people seem to take for granted that numerical data are more likely to be productive of new generalizations than any other sort. It is that special productivity, rather than the function of measurement in confirmation, that probably accounts for Kelvin's dictum's being inscribed on the façade at the University of Chicago.[5]

It is by no means obvious that our ideas about this function of numbers are related to the textbook schema outlined in the diagram above, yet I see no other way to account for the special efficacy often attributed to the results of measurement. We are, I suspect, here confronted with a vestige of an admittedly outworn belief that laws and theories can be arrived at by some process like "running the machine backwards." Given the numerical data in the "Experiment" column of the table, logico-mathematical manipulation (aided, all would now insist, by "intuition") can proceed to the statement of the laws that underlie the numbers. If any process even remotely like this is involved in discovery—if, that is, laws and theories are forged directly from data by the mind—then the superiority of numerical to qualitative data is immediately apparent. The results of measurement are neutral and precise; they cannot mislead. Even more important, numbers are subject to mathematical manipulation; more than any other form of data, they can be assimilated to the semimechanical textbook schema.

I have already implied my skepticism about these two prevalent descriptions of the function of measurement. In the next two sections each of these functions will be further compared with ordinary scientific practice. But it will help first critically to pursue our examination of textbook tables. By doing so I would hope to suggest that our stereotypes about measurement do not even quite fit

5. Professor Frank Knight, for example, suggests that to social scientists the "practical meaning [of Kelvin's statement] tends to be: 'If you cannot measure, measure anyhow' " (*Eleven Twenty-Six*, p. 169).

the textbook schema from which they seem to derive. Though the numerical tables in a textbook do not there function either for exploration or confirmation, they are there for a reason. That reason we may perhaps discover by asking what the author of a text can mean when he says that the numbers in the "Theory" and "Experiment" column of a table "agree."

At best the criterion must be agreement within the limits of accuracy of the measuring instruments employed. Since computation from theory can usually be pushed to any desired number of decimal places, exact or numerical agreement is impossible in principle. But anyone who has examined the tables in which the results of theory and experiment are compared must recognize that agreement of this more modest sort is rather rare. Almost always the application of a physical theory involves some approximation (in fact, the plane is *not* "frictionless," the vacuum is *not* "perfect," the atoms are *not* "unaffected" by collisions), and the theory is not therefore expected to yield quite precise results. Or the construction of the instrument may involve approximations (e.g., the "linearity" of vacuum tube characteristics) that cast doubt upon the significance of the last decimal place that can be unambiguously read from their dial. Or it may simply be recognized that, for reasons not clearly understood, the theory whose results have been tabulated or the instrument used in measurement provides only estimates. For one of these reasons or another, physical scientists rarely expect agreement quite within instrumental limits. In fact, they often distrust it when they see it. At least on a student lab report overly close agreement is usually taken as presumptive evidence of data manipulation. That no experiment gives quite the expected numerical result is sometimes called the "fifth law of thermodynamics."[6] The fact that, unlike some other scientific laws, it has acknowledged exceptions does not diminish its utility as a guiding principle.

It follows that what scientists seek in numerical tables is not usually "agreement" at all, but what they often call "reasonable agreement." Furthermore, if we now ask for a criterion of "reasonable agreement," we are literally forced to look in the tables them-

6. The first three laws of thermodynamics are well known outside the trade. The "fourth law" states that no piece of experimental apparatus works the first time it is set up. We shall examine evidence for the fifth law below.

selves. Scientific practice exhibits no consistently applied or consistently applicable external criterion. "Reasonable agreement" varies from one part of science to another, and within any part of science it varies with time. What to Ptolemy and his immediate successors was reasonable agreement between astronomical theory and observation was to Copernicus incisive evidence that the Ptolemaic system must be wrong.[7] Between the times of Cavendish (1731–1810) and Ramsay (1852–1916), a similar change in accepted chemical criteria for "reasonable agreement" led to the study of the noble gases.[8] These divergences are typical and they are matched by those between contemporary branches of the scientific community. In parts of spectroscopy "reasonable agreement" means agreement in the first six or eight left-hand digits in the numbers of a table of wave lengths. In the theory of solids, by contrast, two-place agreement is often considered very good indeed. Yet there are parts of astronomy in which any search for even so limited an agreement must seem utopian. In the theoretical study of stellar magnitudes agreement to a multiplicative factor of ten is often taken to be "reasonable."

Notice that we have now inadvertently answered the question from which we began. We have, that is, said what "agreement" between theory and experiment must mean if that criterion is to be drawn from the tables of a science text. But in doing so we have gone full circle. I began by asking, at least by implication, what characteristic the numbers of the table must exhibit if they are to be said to "agree." I now conclude that the only possible criterion is the mere fact that they appear, together with the theory from which they are derived, in a professionally accepted text. When they appear in a text, tables of numbers drawn from theory and experiments cannot demonstrate anything but "reasonable agreement." And even that they demonstrate only by tautology, since they alone provide the definition of "reasonable agreement" that has been accepted by the profession. That, I think, is why the tables are there: they define "reasonable agreement." By studying them, the reader learns what can be expected of the theory. An acquaintance with the tables is part of an acquaintance with the

7. T. S. Kuhn, *The Copernican Revolution* (Cambridge, Mass., 1957), pp. 72–76, 135–43.

8. William Ramsay, *The Gases of the Atmosphere: The History of Their Discovery* (London, 1896), chaps. 4 and 5.

theory itself. Without the tables, the theory would be essentially incomplete. With respect to measurement, it would be not so much untested as untestable. Which brings us very close to the conclusion that, once it has been embodied in a text—which for present purposes means, once it has been adopted by the profession—no theory is recognized to be testable by any quantitative tests that it has not already passed.[9]

Perhaps these conclusions are not surprising. Certainly they should not be. Textbooks are, after all, written some time after the discoveries and confirmation procedures whose *outcomes* they record. Furthermore, they are written for purposes of pedagogy. The objective of a textbook is to provide the reader, in the most economical and easily assimilable form, with a statement of what the contemporary scientific community believes it knows and of the principal uses to which that knowledge can be put. Information about how that knowledge was acquired (discovery) and about why it was accepted by the profession (confirmation) would at best be excess baggage. Though including that information would almost certainly increase the "humanistic" values of the text and might conceivably breed more flexible and creative scientists, it would inevitably detract from the ease of learning the contemporary scientific language. To date only the last objective has been taken seriously by most writers of textbooks in the natural sciences. As a result, though texts may be the right place for philosophers to discover the logical structure of finished scientific theories, they are more likely to mislead than to help the unwary individual who asks about productive methods. One might equally

9. To pursue this point would carry us far beyond the subject of this paper, but it should be pursued because, if I am right, it relates to the important contemporary controversy over the distinction between analytic and synthetic truth. To the extent that a scientific theory must be accompanied by a statement of the evidence for it in order to have empirical meaning, the full theory (which includes the evidence) must be analytically true. For a statement of the philosophical problem of analyticity see W. V. Quine, "Two Dogmas of Empiricism" and other essays in *From a Logical Point of View* (Cambridge, Mass., 1953). For a stimulating, but loose, dis- n of the occasionally analytic status of scientific laws, see N. R. Han- *'atterns of Discovery* (Cambridge, 1958), pp. 93–118. A new discus- ıf the philosophical problem, including copious references to the con- sial literature, is Alan Pasch, *Experience and the Analytic: A Recon- tion of Empiricism* (Chicago, 1958).

appropriately go to a college language text for an authoritative characterization of the corresponding literature. Language texts, like science texts, teach how to *read* literature, not how to create or evaluate it. What signposts they supply to these latter points are most likely to point in the wrong direction.[10]

Motives for Normal Measurement

These considerations dictate our next step. We must ask how measurement comes to be juxtaposed with laws and theories in science texts. Furthermore, we must go for an answer to the journal literature, the medium through which natural scientists report their own original work and in which they evaluate that done by others.[11] Recourse to this body of literature immediately casts doubt upon one implication of the standard textbook schema. Only a miniscule fraction of even the best and most creative measurements undertaken by natural scientists are motivated by a desire to discover new quantitative regularities or to confirm old ones. Almost as small a fraction turn out to have had either of these effects. There are a few that did so, and I shall have something to say about them in the next two sections. But it will help first to discover just why these exploratory and confirmatory measurements are so rare. In

10. The monograph cited in note 3 will argue that the misdirection supplied by science texts is both systematic and functional. It is by no means clear that a more accurate image of the scientific processes would enhance the research efficiency of physical scientists.

11. It is, of course, somewhat anachronistic to apply the terms "journal literature" and "textbooks" in the whole of the period I have been asked to discuss. But I am concerned to emphasize a pattern of professional communication whose origins at least can be found in the seventeenth century and which has increased in rigor ever since. There was a time (different in different sciences) when the pattern of communication in a science was much the same as that still visible in the humanities and many of the social sciences, but in all the physical sciences this pattern is at least a century gone, and in many of them it disappeared even earlier than that. Now all publication of research results occurs in journals read only by the profession. Books are exclusively textbooks, compendia, popularizations, or philosophical reflections, and writing them is a somewhat suspect, because nonprofessional, activity. Needless to say this sharp and rigid separation between articles and books, research and nonresearch writings, greatly increases the strength of what I have called the textbook image.

this section and most of the next, I therefore restrict myself to measurement's most usual function in the normal practice of science.[12]

Probably the rarest and most profound sort of genius in physical science is that displayed by men who, like Newton, Lavoisier, or Einstein, enunciate a whole new theory that brings potential order to a vast number of natural phenomena. Yet radical reformulations of this sort are extremely rare, largely because the state of science very seldom provides occasion for them. Moreover, they are not the only truly essential and creative events in the development of scientific knowledge. The new order provided by a revolutionary new theory in the natural sciences is always overwhelmingly a *potential* order. Much work and skill, together with occasional genius, are required to make it *actual*. And actual it must be made, for only through the process of actualization can occasions for new theoretical reformulations be discovered. The bulk of scientific practice is thus a complex and consuming mopping-up operation that consolidates the ground made available by the most recent theoretical breakthrough and that provides essential preparation for the breakthrough to follow. In such mopping-up operations, measurement has its overwhelmingly most common scientific function.

Just how important and difficult these consolidating operations can be is indicated by the present state of Einstein's general theory of relativity. The equations embodying that theory have proved so difficult to apply that (excluding the limiting case in which the equations reduce to those of special relativity) they have so far yielded only three predictions that can be compared with observation.[13] Men of undoubted genius have totally failed to develop others, and the problem remains worth their attention. Until it is

12. Here and elsewhere in this paper I ignore the very large amount of measurement done simply to gather factual information. I think of such measurements as specific gravities, wave lengths, spring constants, boiling points, etc., undertaken in order to determine parameters that must be inserted into scientific theories but whose numerical outcome those theories do not (or did not in the relevant period) predict. This sort of measurement is not without interest, but I think it widely understood. In any case, considering it would too greatly extend the limits of this paper.

13. These are the deflection of light in the sun's gravitational field, the precession of the perihelion of Mercury, and the red shift of light from distant stars. Only the first two are actually quantitative predictions in the present state of the theory.

solved, Einstein's general theory remains a largely fruitless, because unexploitable, achievement.[14]

Undoubtedly the general theory of relativity is an extreme case, but the situation it illustrates is typical. Consider, for a somewhat more extended example, the problem that engaged much of the best eighteenth-century scientific thought, that of deriving testable numerical predictions from Newton's three laws of motion and from his principle of universal gravitation. When Newton's theory was first enunciated late in the seventeenth century, only his third law (equality of action and reaction) could be directly investigated by experiment, and the relevant experiments applied only to very special cases.[15] The first direct and unequivocal demonstrations of the second law awaited the development of the Atwood machine, a subtly conceived piece of laboratory apparatus that was not invented until almost a century after the appearance of the *Principia*.[16] Direct quantitative investigations of gravitational attraction proved even more difficult and were not presented in the scientific literature until 1798.[17] Newton's first law

14. The difficulties in producing concrete applications of the general theory of relativity need not prevent scientists from attempting to exploit the scientific viewpoint embodied in that theory. But, perhaps unfortunately, it seems to be doing so. Unlike the special theory, general relativity is today very little studied by students of physics. Within fifty years we may conceivably have totally lost sight of this aspect of Einstein's contribution.

15. The most relevant and widely employed experiments were performed with pendula. Determination of the recoil when two pendulum bobs collided seems to have been the main conceptual and experimental tool used in the seventeenth century to determine what dynamical "action" and "reaction" were. See A. Wolf, *A History of Science, Technology, and Philosophy in the Sixteenth and Seventeenth Centuries*, new ed. prepared by D. McKie (London, 1950), pp. 155, 231–35; and R. Dugas, *La mécanique au xvii*e *siècle* (Neuchâtel, 1954), pp. 283–98; and *Sir Isaac Newton's Mathematical Principles of Natural Philosophy and His System of the World*, ed. F. Cajori (Berkeley, 1934), pp. 21–28. Wolf (p. 155) describes the third law as "the only *physical* law of the three."

16. See the excellent description of this apparatus and the discussion of Atwood's reasons for building it in Hanson, *Patterns of Discovery*, pp. 100–102 and notes to these pages.

17. A. Wolf, *A History of Science, Technology, and Philosophy in the Eighteenth Century*, 2d ed. revised by D. McKie (London, 1952), pp. 111–13. There are some precursors of Cavendish's measurements of 1798, but it is only after Cavendish that measurement begins to yield consistent results.

cannot, to this day, be directly compared with the results of laboratory measurement, though developments in rocketry make it likely that we have not much longer to wait.

It is, of course, direct demonstrations, like those of Atwood, that figure most largely in natural science texts and in elementary laboratory exercises. Because they are simple and unequivocal, they have the greatest pedagogic value. That they were not and could scarcely have been available for more than a century after the publication of Newton's work makes no pedagogic difference. At most it only leads us to mistake the nature of scientific achievement.[18] But if Newton's contemporaries and successors had been forced to wait that long for quantitative evidence, apparatus capable of providing it would never have been designed. Fortunately there was another route, and much eighteenth-century scientific talent followed it. Complex mathematical manipulations, exploiting all the laws together, permitted a few other sorts of prediction capable of being compared with quantitative observation, particularly with laboratory observations of pendula and with astronomical observations of the motions of the moon and planets. But these predictions presented another and equally severe problem, that of essential approximations.[19] The suspensions of laboratory pendula are neither weightless nor perfectly elastic; air resistance damps the motion of the bob; besides, the bob itself is of finite size, and there is the question of which point of the bob should be used in computing the pendulum's length. If these three aspects of

18. Modern laboratory apparatus designed to help students study Galileo's law of free fall provides a classic, though perhaps quite necessary, example of the way pedagogy misdirects the historical imagination about the relation between creative science and measurement. None of the apparatus now used could possibly have been built in the seventeenth century. One of the best and most widely disseminated pieces of equipment, for example, allows a heavy bob to fall between a pair of parallel vertical rails. These rails are electrically charged every 1/100th of a second, and the spark that then passes through the bob from rail to rail records the bob's position on a chemically treated tape. Other pieces of apparatus involve electric timers, etc. For the historical difficulties of making measurements relevant to this law, see below.

19. All the applications of Newton's laws involve approximations of some sort, but in the following examples the approximations have a quantitative importance that they do not possess in those that precede.

the experimental situation are neglected, only the roughest sort of quantitative agreement between theory and observation can be expected. But determining how to reduce them (only the last is fully eliminable) and what allowance to make for the residue are themselves problems of the utmost difficulty. Since Newton's day much brilliant research has been devoted to their challenge.[20]

The problems encountered when applying Newton's laws to astronomical prediction are even more revealing. Since each of the bodies in the solar system attracts and is attracted by every other, precise prediction of celestial phenomena demanded, in Newton's day, the application of his laws to the simultaneous motions and interactions of eight celestial bodies. (These were the sun, moon, and six known planets. I ignore the other planetary satellites.) The result is a mathematical problem that has never been solved exactly. To get equations that could be solved, Newton was forced to the simplifying assumption that each of the planets was attracted only by the sun, and the moon only by the earth. With this assumption, he was able to derive Kepler's famous laws, a wonderfully convincing argument for his theory. But deviation of planets from the motions predicted by Kepler's laws is quite apparent to simple quantitative telescopic observation. To discover how to treat these deviations by Newtonian theory, it was necessary to devise mathematical estimates of the "perturbations" produced in a basically Keplerian orbit by the interplanetary forces neglected in the initial derivation of Kepler's laws. Newton's mathematical genius was displayed at its best when he produced a first crude estimate for the perturbation of the moon's motion caused by the sun. Improving his answer and developing similar approximate answers for the planets exercised the greatest mathematicians of the eighteenth and early nineteenth centuries, including Euler, Lagrange, Laplace, and Gauss.[21] Only as a result of their work was it possible to recognize the anomaly in Mercury's motion that was ultimately to be explained by Einstein's general theory. That anomaly had previously been hidden within the limits of "reasonable agreement."

20. Wolf (*Eighteenth Century*, pp. 75–81) provides a good preliminary description of this work.

21. Ibid., pp. 96–101. William Whewell, *History of the Inductive Sciences*, rev. ed., 3 vols. (London, 1847), 2: 213–71.

As far as it goes, the situation illustrated by quantitative application of Newton's laws is, I think, perfectly typical. Similar examples could be produced from the history of the corpuscular, the wave, or the quantum mechanical theory of light, from the history of electromagnetic theory, quantitative chemical analysis, or any other of the numerous natural scientific theories with quantitative implications. In each of these cases, it proved immensely difficult to find many problems that permitted quantitative comparison of theory and observation. Even when such problems were found, the highest scientific talents were often required to invent apparatus, reduce perturbing effects, and estimate the allowance to be made for those that remained. This is the sort of work that most physical scientists do most of the time *insofar as their work is quantitative.* Its objective is, on the one hand, to improve the measure of "reasonable agreement" characteristic of the theory in a given application and, on the other, to open up new areas of application and establish new measures of "reasonable agreement" applicable to them. For anyone who finds mathematical or manipulative puzzles challenging, this can be fascinating and intensely rewarding work. And there is always the remote possibility that it will pay an additional dividend: something may go wrong.

Yet unless something does go wrong—a situation to be explored in a later section—these finer and finer investigations of the quantitative match between theory and observation cannot be described as attempts at discovery or at confirmation. The man who is successful proves his talents, but he does so by getting a result that the entire scientific community had anticipated someone would someday achieve. His success lies only in the explicit demonstration of a *previously implicit* agreement between theory and the world. No novelty in nature has been revealed. Nor can the scientist who is successful in this sort of work quite be said to have "confirmed" the theory that guided his research. For if success in his venture "confirms" the theory, then failure ought certainly to "infirm" it, and nothing of the sort is true in this case. Failure to solve one of these puzzles counts only against the scientist; he has put in a great deal of time on a project whose outcome is not worth publication; the conclusion to be drawn, if any, is only that his talents were not adequate to it. If measurement ever leads to discovery or to confirmation, it does not do so in the most usual of all its applications.

The Effects of Normal Measurement

There is a second significant aspect of the normal problem of measurement in natural science. So far we have considered why scientists usually measure; now we must consider the results that they get when they do so. Immediately another stereotype enforced by textbooks is called in question. In textbooks the numbers that result from measurement usually appear as the archetypes of the "irreducible and stubborn facts" to which the scientist must, by struggle, make his theories conform. But in scientific practice, as seen through the journal literature, the scientist often seems rather to be struggling with facts, trying to force them into conformity with a theory he does not doubt. Quantitative facts cease to seem simply "the given." They must be fought for and with, and in this fight the theory with which they are to be compared proves the most potent weapon. Often scientists cannot get numbers that compare well with theory until they know what numbers they should be making nature yield.

Part of this problem is simply the difficulty in finding techniques and instruments that permit the comparison of theory with quantitative measurements. We have already seen that it took almost a century to invent a machine that could give a straightforward quantitative demonstration of Newton's second law. But the machine that Charles Atwood described in 1784 was not the first instrument to yield quantitative information relevant to that law. Attempts in this direction had been made ever since Galileo's description of his classic inclined plane experiment in 1638.[22] Galileo's brilliant intuition had seen in this laboratory device a way of investigating how a body moves when acted upon only by its own weight. After the experiment he announced that measurement of the distance covered in a measured time by a sphere rolling down the plane confirmed his prior thesis that the motion was uniformly accelerated. As reinterpreted by Newton, this result exemplified the second law for the special case of a uniform force. But Galileo did not report the numbers he had gotten, and a group of the best scientists in France announced their total failure to get compar-

22. For a modern English version of the original see Galileo Galilei, *Dialogues Concerning Two New Sciences*, trans. Henry Crew and A. De Salvio (Evanston and Chicago, 1946), pp. 171–72.

able results. In print they wondered whether Galileo could himself have tried the experiment.[23]

In fact, it is almost certain that Galileo did perform the experiment. If he did, he must surely have gotten quantitative results that seemed to him in *adequate* agreement with the law ($s = \frac{1}{2} at^2$) that he had shown to be a consequence of uniform acceleration. But anyone who has noted the stopwatches or electric timers, and the long planes or heavy flywheels needed to perform this experiment in modern elementary laboratories may legitimately suspect that Galileo's results were not in *unequivocal* agreement with his law. Quite possibly the French group looking even at the same data would have wondered how they could seem to exemplify uniform acceleration. This is, of course, largely speculation. But the speculative element casts no doubt upon my present point: whatever its source, disagreement between Galileo and those who tried to repeat his experiment was entirely natural. If Galileo's generalization had not sent men to the very border of existing instrumentation, an area in which experimental scatter and disagreement about interpretation were inevitable, then no genius would have been required to make it. His example typifies one important aspect of theoretical genius in the natural sciences—it is a genius that leaps ahead of the facts, leaving the rather different talent of the experimentalist and instrumentalist to catch up. In this case catching up took a long time. The Atwood machine was designed because, in the middle of the eighteenth century, some of the best Continental scientists still wondered whether acceleration provided the proper measure of force. Though their doubts derived from more than measurement, measurement was still sufficiently equivocal to fit a variety of different quantitative conclusions.[24]

The preceding example illustrates the difficulties and displays the role of theory in reducing scatter in the results of measurement. There is, however, more to the problem. When measurement is insecure, one of the tests for reliability of existing instruments and manipulative techniques must inevitably be their ability to give results that compare favorably with existing theory. In some parts of natural science, the adequacy of experimental technique can be

23. This whole story and more is brilliantly set forth in A. Koyré, "An Experiment in Measurement," *Proceedings of the American Philosophical Society* 97 (1953): 222–37.

24. Hanson, *Patterns of Discovery*, p. 101.

judged only in this way. When that occurs, one may not even speak of "insecure" instrumentation or technique, implying that these could be improved without recourse to an external theoretical standard.

For example, when John Dalton first conceived of using chemical measurements to elaborate an atomic theory that he had initially drawn from meteorological and physical observations, he began by searching the existing chemical literature for relevant data. Soon he realized that significant illumination could be obtained from those groups of reactions in which a single pair of elements, for example, nitrogen and oxygen, entered into more than one chemical combination. If his atomic theory were right, the constituent molecules of these compounds should differ only in the ratio of the number of whole atoms of each element that they contained. The three oxides of nitrogen might, for example, have molecules N_2O, NO, and NO_2, or they might have some other similarly simple arrangement.[25] But whatever the particular arrangements, if the weight of nitrogen were the same in the samples of the three oxides, then the weights of oxygen in the three samples should be related to each other by simple whole-number proportions. Generalization of this principle to all groups of compounds formed from the same group of elements produced Dalton's law of multiple proportions.

Needless to say, Dalton's search of the literature yielded some data that, in his view, sufficiently supported the law. But—and this is the point of the illustration—much of the then extant data did not support Dalton's law at all. For example, the measurements of the French chemist Proust on the two oxides of copper yielded, for a given weight of copper, a weight ratio for oxygen of 1.47:1. On Dalton's theory the ratio ought to have been 2:1, and Proust

25. This is not, of course, Dalton's original notation. In fact, I am somewhat modernizing and simplifying this whole account. It can be reconstructed more fully from A. N. Meldrum, "The Development of the Atomic Theory: (1) Berthollet's Doctrine of Variable Proportions," *Manchester Memoirs* 54 (1910): 1–16; and "(6) The Reception accorded to the Theory advocated by Dalton," ibid. 55 (1911): 1–10; L. K. Nash, *The Atomic Molecular Theory*, Harvard Case Histories in Experimental Science, case 4 (Cambridge, Mass., 1950); and "The Origins of Dalton's Chemical Atomic Theory," *Isis*, 47 (1956): 110–16. See also the useful discussions of atomic weight scattered through J. R. Partington, *A Short History of Chemistry*, 2d ed. (London, 1951).

is just the chemist who might have been expected to confirm the prediction. He was, in the first place, a fine experimentalist. Besides, he was then engaged in a major controversy involving the oxides of copper, a controversy in which he upheld a view very close to Dalton's. But, at the beginning of the nineteenth century, chemists did not know how to perform quantitative analyses that displayed multiple proportions. By 1850 they had learned, but only by letting Dalton's theory lead them. Knowing what results they should expect from chemical analyses, chemists were able to devise techniques that got them. As a result chemistry texts can now state that quantitative analysis confirms Dalton's atomism and forget that, historically, the relevant analytic techniques are based upon the very theory they are said to confirm. Before Dalton's theory was announced, measurement did not give the same results. There are self-fulfilling prophecies in the physical as well as in the social sciences.

That example seems to me quite typical of the way measurement responds to theory in many parts of the natural sciences. I am less sure that my next, and far stranger, example is equally typical, but colleagues in nuclear physics assure me that they repeatedly encounter similar irreversible shifts in the results of measurement.

Very early in the nineteenth century, P. S. de Laplace, perhaps the greatest and certainly the most famous physicist of his day, suggested that the recently observed heating of a gas when rapidly compressed might explain one of the outstanding numerical discrepancies of theoretical physics. This was the disagreement, approximately 20 per cent, between the predicted and measured values of the speed of sound in air—a discrepancy that had attracted the attention of all Europe's best mathematical physicists since Newton had first pointed it out. When Laplace's suggestion was made, it defied numerical confirmation (note the recurrence of this typical difficulty), because it demanded refined measurements of the thermal properties of gases, measurements that were beyond the capacity of apparatus designed for measurements on solids and liquids. But the French academy offered a prize for such measurements, and in 1813 the prize was won by two brilliant young experimentalists, Delaroche and Berard, men whose names are still cited in contemporary scientific literature. Laplace soon made use of these measurements in an indirect theoretical computation of the speed of sound in air, and the discrepancy between theory and measurement dropped from 20 per cent to 2.5 per cent, a recognized triumph in view of the state of measurement.[26]

But today no one can explain how this triumph can have occurred. Laplace's interpretation of Delaroche and Berard's figures made use of the caloric theory in a region where our own science is quite certain that that theory differs from directly relevant quantitative experiment by about 40 per cent. There is, however, also a 12 per cent discrepancy between the measurements of Delaroche and Berard and the results of equivalent experiments today. We are no longer able to get their quantitative result. Yet, in Laplace's perfectly straightforward and essential computation from the theory, these two discrepancies, experimental and theoretical, cancelled to give close final agreement between the predicted and measured speed of sound. We may not, I feel sure, dismiss this as the result of mere sloppiness. Both the theoretician and the experimentalists involved were men of the very highest caliber. Rather we must here see evidence of the way in which theory and experiment may guide each other in the exploration of areas new to both.

These examples may enforce the point drawn initially from the examples in the last section. Exploring the agreement between theory and experiment into new areas or to new limits of precision is a difficult, unremitting, and, for many, exciting job. Though its object is neither discovery nor confirmation, its appeal is quite sufficient to consume almost the entire time and attention of those physical scientists who do quantitative work. It demands the very best of their imagination, intuition, and vigilance. In addition—when combined with those of the last section—these examples may show something more. They may, that is, indicate why new laws of nature are so very seldom discovered simply by inspecting the results of measurements made without advance knowledge of those laws. Because most scientific laws have so few quantitative points of contact with nature, because investigations of those contact points usually demand such laborious instrumentation and approximation, and because nature itself needs to be forced to yield the appropriate results, the route from theory or law to measurement can almost never be travelled backward. Numbers gathered with-

26. T. S. Kuhn, "The Caloric Theory of Adiabatic Compression," *Isis* 49 (1958): 132–40.

out some knowledge of the regularity to be expected almost never speak for themselves. Almost certainly they remain just numbers.

This does not mean that no one has ever discovered a quantitative regularity merely by measuring. Boyle's law relating gas pressure with gas volume, Hooke's law relating spring distortion with applied force, and Joule's relationship between heat generated, electrical resistance, and electric current were all the direct results of measurement. There are other examples besides. But, partly just because they are so exceptional and partly because they never occur until the scientist measuring knows *everything but* the particular form of the quantitative result he will obtain, these exceptions show just how improbable quantitative discovery by quantitative measurement is. The cases of Galileo and Dalton—men who intuited a quantitative result as the simplest expression of a qualitative conclusion and then fought nature to confirm it—are very much the more typical scientific events. In fact, even Boyle did not find his law until both he and two of his readers had suggested that precisely that law (the simplest quantitative form that yielded the observed qualitative regularity) ought to result if the numerical results were recorded.[27] Here, too, the quantitative implications of a qualitative theory led the way.

One more example may make clear at least some of the prerequisites for this exceptional sort of discovery. The experimental search for a law or laws describing the variation with distance of the forces between magnetized and between electrically charged bodies began in the seventeenth century and was actively pursued through the eighteenth. Yet only in the decades immediately preceding Coulomb's classic investigations of 1785 did measurement yield even an approximately unequivocal answer to these questions. What made the difference between success and failure seems to have been the belated assimilation of a lesson learned from a part of Newtonian theory. Simple force laws, like the inverse square law for gravitational attraction, can generally be expected only between mathematical points or bodies that approximate them. The more complex laws of attraction between gross bodies can be derived from the simpler law governing the attraction of points by

27. Marie Boas, *Robert Boyle and Seventeenth-Century Chemistry* (Cambridge, 1958), p. 44.

summing all the forces between all the pairs of points in the two bodies. But these laws will seldom take a simple mathematical form unless the distance between the two bodies is large compared with the dimensions of the attracting bodies themselves. Under these circumstances the bodies will behave as points, and experiment may reveal the resulting simple regularity.

Consider only the historically simpler case of electrical attractions and repulsions.[28] During the first half of the eighteenth century—when electrical forces were explained as the results of effluvia emitted by the entire charged body—almost every experimental investigation of the force law involved placing a charged body a measured distance below one pan of a balance and then measuring the weight that had to be placed in the other pan to just overcome the attraction. With this arrangement of apparatus, the attraction varies in no simple way with distance. Furthermore, the complex way in which it does vary depends critically upon the size and material of the attracted pan. Many of the men who tried this technique therefore concluded by throwing up their hands; others suggested a variety of laws including both the inverse square and the inverse first power; measurement had proved totally equivocal. Yet it did not have to be so. What was needed and what was gradually acquired from more qualitative investigations during the middle decades of the century was a more "Newtonian" approach to the analysis of electrical and magnetic phenomena.[29] As this

28. Much relevant material will be found in Duane Roller and Duane H. D. Roller, *The Development of the Concept of Electric Charge: Electricity from the Greeks to Coulomb*, Harvard Case Histories in Experimental Science, case 8 (Cambridge, Mass., 1954), and in Wolf, *Eighteenth Century*, pp. 239–50, 268–71.

29. A fuller account would have to describe both the earlier and the later approaches as "Newtonian." The conception that electric force results from effluvia is partly Cartesian but in the eighteenth century its *locus-classicus* was the aether theory developed in Newtons *Opticks*. Coulomb's approach and that of several of his contemporaries depends far more directly on the mathematical theory in Newton's *Principia*. For the differences between these books, their influence in the eighteenth century, and their impact on the development of electrical theory, see I. B. Cohen, *Franklin and Newton: An Inquiry into Speculative Newtonian Experimental Science and Franklin's Work in Electricity as an Example Thereof* (Philadelphia, 1956).

evolved, experimentalists increasingly sought not the attraction between bodies but that between point poles and point charges. In that form the experimental problem was rapidly and unequivocally resolved.

This illustration shows once again how large an amount of theory is needed before the results of measurement can be expected to make sense. But, and this is perhaps the main point, when that much theory is available, the law is very likely to have been guessed without measurement. Coulomb's result, in particular, seems to have surprised few scientists. Though his measurements were necessary to produce a firm consensus about electrical and magnetic attractions—they had to be done; science cannot survive on guesses—many practitioners had already concluded that the law of attraction and repulsion must be inverse square. Some had done so by simple analagy to Newton's gravitational law; others by a more elaborate theoretical argument; still others from equivocal data. Coulomb's law was very much "in the air" before its discoverer turned to the problem. If it had not been, Coulomb might not have been able to make nature yield it.

Two possible misinterpretations of my argument must now be set aside. First, if what I have said is right, nature undoubtedly responds to the theoretical predispositions with which she is approached by the measuring scientist. But that is not to say either that nature will respond to any theory at all or that she will ever respond very much. Reexamine, for a historically typical example, the relationship between the caloric and dynamical theory of heat. In their abstract structures and in the conceptual entities they presuppose, these two theories are quite different and, in fact, incompatible. But, during the years when the two vied for the allegiance of the scientific community, the theoretical predictions that could be derived from them were very nearly the same.[30] If they had not been, the caloric theory would never have been a widely accepted tool of professional research nor would it have succeeded in disclosing the very problems that made transition to the dynamical theory possible. It follows that any measurement which, like that of Delaroche and Berard, "fit" one of these theories must have "very nearly fit" the other, and it is only within the experimental

30. Kuhn, "The Caloric Theory of Adiabatic Compression."

spread covered by the phrase "very nearly" that nature proved able to respond to the theoretical predisposition of the measurer.

That response could not have occurred with "any theory at all." There are logically possible theories of, say, heat that no sane scientist could ever have made nature fit, and there are problems, mostly philosophical, that make it worth inventing and examining theories of that sort. But those are not our problems, because those merely "conceivable" theories are not among the options open to the practicing scientist. His concern is with theories that seem to fit what is known about nature, and all these theories, however different their structure, will necessarily seem to yield very similar predictive results. If they can be distinguished at all by measurements, those measurements will usually strain the limits of existing experimental techniques. Furthermore, within the limits imposed by those techniques, the numerical differences at issue will very often prove to be quite small. Only under these conditions and within these limits can one expect nature to respond to preconception. On the other hand, these conditions and limits are just the ones typical in the historical situation.

If this much about my approach is clear, the second possible misunderstanding can be dealt with more easily. By insisting that a quite highly developed body of theory is ordinarily prerequisite to fruitful measurement in the physical sciences, I may seem to have implied that in these sciences theory must always lead experiment and that the latter has at best a decidedly secondary role. But that implication depends upon identifying "experiment" with "measurement," an identification I have already explicitly disavowed. It is only because significant quantitative comparison of theories with nature comes at such a late stage in the development of a science that theory has seemed to have so decisive a lead. If we had been discussing the *qualitative* experimentation that dominates the earlier developmental stages of a physical science and that continues to play a role later on, the balance would be quite different. Perhaps, even then, we would not wish to say that experiment is prior to theory (though experience surely is), but we would certainly find vastly more symmetry and continuity in the ongoing dialogue between the two. Only some of my conclusions about the role of measurement in physical science can be readily extrapolated to experimentation at large.

Extraordinary Measurement

To this point I have restricted attention to the role of measurement in the normal practice of natural science, the sort of practice in which all scientists are mostly, and most scientists are always, engaged. But natural science also displays abnormal situations—times when research projects go consistently astray and when no usual techniques seem quite to restore them—and it is through these rare situations that measurement shows its greatest strengths. In particular, it is through abnormal states of scientific research that measurement comes occasionally to play a major role in discovery and in confirmation.

Let me first try to clarify what I mean by an "abnormal situation" or by what I am elsewhere calling a "crisis state."[31] I have already indicated that it is a response by some part of the scientific community to its awareness of an anomaly in the ordinarily concordant relationship between theory and experiment. But it is not, let us be clear, a response called forth by any and every anomaly. As the preceding pages have shown, current scientific practice always embraces countless discrepancies between theory and experiment. During the course of his career, every natural scientist again and again notices *and passes by* qualitative and quantitative anomalies that just conceivably might, if pursued, have resulted in fundamental discovery. Isolated discrepancies with this potential occur so regularly that no scientist could bring his research problems to a conclusion if he paused for many of them. In any case, experience has repeatedly shown that, in overwhelming proportion, these discrepancies disappear upon closer scrutiny. They may prove to be instrumental effects, or they may result from previously unnoticed approximations in the theory, or they may, simply and mysteriously, cease to occur when the experiment is repeated under slightly different conditions. More often than not the efficient procedure is therefore to decide that the problem has "gone sour," that it presents hidden complexities, and that it is time to put it aside in favor of another. Fortunately or not, that is good scientific procedure.

But anomalies are not always dismissed, and of course they should not be. If the effect is particularly large when compared

31. See note 3.

with well-established measures of "reasonable agreement" applicable to similar problems, or if it seems to resemble other difficulties encountered repeatedly before, or if, for personal reasons, it intrigues the experimenter, then a special research project is likely to be dedicated to it.[32] At that point the discrepancy will probably vanish through an adjustment of theory or apparatus; as we have seen, few anomalies resist persistent effort for long. But it may resist, and, if it does, we may have the beginning of a "crisis" or "abnormal situation" affecting those in whose usual area of research the continuing discrepancy lies. They, at least, having exhausted all the usual recourses of approximation and instrumentation, may be forced to recognize that something has gone wrong, and their behavior as scientists will change accordingly. At this point, to a vastly greater extent than at any other, the scientist will start to search at random, trying anything at all which he thinks may conceivably illuminate the nature of his difficulty. If that difficulty endures long enough, he and his colleagues may even begin to wonder whether their entire approach to the now problematic range of natural phenomena is not somehow askew.

This is, of course, an immensely condensed and schematic description. Unfortunately, it will have to remain so, for the anatomy of the crisis state in natural science is beyond the scope of this paper. I shall remark only that these crises vary greatly in scope: they may emerge and be resolved within the work of an individual; more often they will involve most of those engaged in a particular scientific specialty; occasionally they will engross most of the members of an entire scientific profession. But, however widespread their impact, there are only a few ways in which they may be resolved. Sometimes, as has often happened in chemistry and astronomy, more refined experimental techniques or a finer scrutiny of the theoretical approximations will eliminate the discrepancy entirely. On other occasions, though I think not often, a discrepancy that has repeatedly defied analysis is simply left as a known anomaly, encysted within the body of more successful applications of the theory. Newton's theoretical value for the speed of sound and the

32. A recent example of the factors determining pursuit of an anomaly has been investigated by Bernard Barber and Renée C. Fox, "The Case of the Floppy-Eared Rabbits: An Instance of Serendipity Gained and Serendipity Lost," *American Sociological Review* 64 (1958): 128–36.

observed precession of the perihelion of Mercury provide obvious examples of effects which, though since explained, remained in the scientific literature as known anomalies for half a century or more. But there are still other modes of resolution, and they are the ones which give crises in science their fundamental importance. Often crises are resolved by the discovery of a new natural phenomenon; occasionally their resolution demands a fundamental revision of existing theory.

Obviously crisis is not a prerequisite for discovery in the natural sciences. We have already noticed that some discoveries, like Boyle's law and Coulomb's law, emerge naturally as a quantitative specification of what is qualitatively already known. Many other discoveries, more often qualitative than quantitative, result from preliminary exploration with a new instrument, for example, the telescope, battery, or cyclotron. In addition, there are the famous "accidental discoveries," Galvani and the twitching frog's legs, Roentgen and X rays, Becquerel and the fogged photographic plates. The last two categories of discovery are not, however, always independent of crises. It is probably the ability to recognize a significant anomaly against the background of current theory that most distinguishes the successful victim of an "accident" from those of his contemporaries who passed the same phenomenon by. (Is this not part of the sense of Pasteur's famous phrase, "In the fields of observation, chance favors only the prepared mind"?)[33] In addition, the new instrumental techniques that multiply discoveries are often themselves by-products of crises. Volta's invention of the battery was, for example, the outcome of a long attempt to assimilate Galvani's observations of frog's legs to existing electrical theory. And, over and above these somewhat questionable cases, there are a large number of discoveries that are quite clearly the outcome of prior crises. The discovery of the planet Neptune was the product of an effort to account for known anomalies in the orbit of Uranus.[34] The nature of both chlorine and carbon monoxide was discovered through attempts to reconcile Lavoisier's new

33. From Pasteur's inaugural address at Lille in 1854 as quoted in René Vallery-Radot, *La Vie de Pasteur* (Paris, 1903), p. 88.

34. Angus Armitage, *A Century of Astronomy* (London, 1950), pp. 111–15.

chemistry with observation.[35] The so-called noble gases were the products of a long series of investigations initiated by a small but persistent anomaly in the measured density of nitrogen.[36] The electron was posited to explain some anomalous properties of electrical conduction through gases, and its spin was suggested to account for other sorts of anomalies observed in atomic spectra.[37] The discovery of the neutrino presents still another example, and the list could be extended.[38]

I am not certain how large these discoveries-through-anomaly would rank in a statistical survey of discovery in the natural sciences.[39] They are, however, certainly important, and they require disproportionate emphasis in this paper. To the extent that measurement and quantitative technique play an especially significant role in scientific discovery, they do so precisely because, by displaying serious anomaly, they tell scientists when and where to look for a new qualitative phenomenon. To the nature of that phenomenon, they usually provide no clues. When measurement de-

35. For chlorine see Ernst von Meyer, *A History of Chemistry from the Earliest Times to the Present Day*, trans. G. M'Gowan (London, 1891), pp. 224–27. For carbon monoxide see J. R. Partington, *A Short History of Chemistry*, 2d ed., pp. 113–16, 140–41; and J. R. Partington and D. McKie, "Historical Studies of the Phlogiston Theory: IV. Last Phases of the Theory," *Annals of Science* 4 (1939): 365.

36. See note 7.

37. For useful surveys of the experiments that led to the discovery of the electron see T. W. Chalmers, *Historic Researches: Chapters in the History of Physical and Chemical Discovery* (London, 1949), pp. 187–217, and J. J. Thomson, *Recollections and Reflections* (New York, 1937), pp. 325–71. For electron spin see F. K. Richtmeyer, E. H. Kennard, and T. Lauritsen, *Introduction to Modern Physics*, 5th ed. (New York, 1955), p. 212.

38. Rogers D. Rusk, *Introduction to Atomic and Nuclear Physics* (New York, 1958), pp. 328–30. I know of no other elementary account recent enough to include a description of the physical detection of the neutrino.

39. Because scientific attention is often concentrated upon problems that seem to display anomaly, the prevalence of discovery-through-anomaly may be one reason for the prevalence of simultaneous discovery in the sciences. For evidence that it is not the only one see T. S. Kuhn, "Conservation of Energy as an Example of Simultaneous Discovery," *Critical Problems in the History of Science*, ed. Marshall Clagett (Madison, 1959), pp. 321–56 (pp. 66–104 above), but notice that much of what is there said about the emergence of "conversion processes" also describes the evolution of a crisis state.

parts from theory, it is likely to yield mere numbers, and their very neutrality makes them particularly sterile as a source of remedial suggestions. But numbers register the departure from theory with an authority and finesse that no qualitative technique can duplicate, and that departure often is enough to start a search. Neptune might, like Uranus, have been discovered through an accidental observation; it had, in fact, been noticed by a few earlier observers who had taken it for a previously unobserved star. What was needed to draw attention to it and to make its discovery as nearly inevitable as historical events can be was its involvement, as a source of trouble, in existing quantitative observation and existing theory. It is hard to see how either electron spin or the neutrino could have been discovered in any other way.

The case both for crises and for measurement becomes vastly stronger as soon as we turn from the discovery of new natural phenomena to the invention of fundamental new theories. Though the sources of individual theoretical inspiration may be inscrutable (certainly they must remain so for this paper), the conditions under which inspiration occurs are not. I know of no fundamental theoretical innovation in natural science whose enunciation has not been preceded by clear recognition, often common to most of the profession, that something was the matter with the theory then in vogue. The state of Ptolemaic astronomy was a scandal before Copernicus's announcement.[40] Both Galileo's and Newton's contributions to the study of motion were initially focused upon difficulties discovered in ancient and medieval theory.[41] Newton's new theory of light and color originated in the discovery that existing theory would not account for the length of the spectrum, and the wave theory that replaced Newton's was announced in the midst of growing concern about anomalies in the relation of diffraction and polarization to Newton's theory.[42] Lavoisier's new chemistry

40. Kuhn, *Copernican Revolution*, pp. 138–40, 270–71; A. R. Hall, *The Scientific Revolution, 1500–1800* (London, 1954), pp. 13–17. Note particularly the role of agitation for calendar reform in intensifying the crisis.

41. Kuhn, *Copernican Revolution*, pp. 237–60, and items in bibliography on pp. 290–91.

42. For Newton see T. S. Kuhn, "Newton's Optical Papers," in *Isaac Newton's Papers and Letters on Natural Philosophy*, ed. I. B. Cohen (Cambridge, Mass., 1958), pp. 27–45. For the wave theory see E. T. Whittaker,

was born after the observation of anomalous weight relations in combustion; thermodynamics from the collision of two existing nineteenth-century physical theories; quantum mechanics from a

History of the Theories of Aether and Electricity, vol. 1, *The Classical Theories*, 2d ed. (London, 1951), pp. 94–109, and Whewell, *Inductive Sciences*, 2:396–466. These references clearly delineate the crisis that characterized optics when Fresnel independently began to develop the wave theory after 1812. But they say too little about eighteenth-century developments to indicate a crisis prior to Young's earlier defense of the wave theory in and after 1801. In fact, it is not altogether clear that there was one, or at least that there was a new one. Newton's corpuscular theory of light had never been quite universally accepted, and Young's early opposition to it was based entirely upon anomalies that had been generally recognized and often exploited before. We may need to conclude that most of the eighteenth century was characterized by a low-level crisis in optics, for the dominant theory was never immune to fundamental criticism and attack.

That would be sufficient to make the point that is of concern here, but I suspect a careful study of the eighteenth-century optical literature will permit a still stronger conclusion. A cursory look at that body of literature suggests that the anomalies of Newtonian optics were far more apparent and pressing in the two decades before Young's work than they had been before. During the 1780s the availability of achromatic lenses and prisms led to numerous proposals for an astronomical determination of the relative motion of the sun and stars. (The references in Whittaker, *Aether and Electricity*, 1:109, lead directly to a far larger literature.) But these all depended upon light's moving more quickly in glass than in air and thus gave new relevance to an old controversy. L'Abbé Haüy demonstrated experimentally ("Sur la double réfraction du Spath d'Islande," *Memoires de l'Academie*, 1788, pp. 34–60) that Huyghens' wave-theoretical treatment of double refraction had yielded better results than Newton's corpuscular treatment. The resulting problem led to the prize offered by the French academy in 1808 and thus to Malus' discovery of polarization by reflection in the same year. Or again, the *Philosophical Transactions* for 1796, 1797, and 1798 contain a series of articles, two by Brougham and one by Prevost, which show still other difficulties in Newton's theory. According to Prevost, in particular, the sorts of forces which must be exerted on light at an interface in order to explain reflection and refraction are not compatible with the sorts of forces needed to explain inflection (*Philosophical Transactions* 84 [1798]: 325–28. Biographers of Young might pay more attention than they have to the two Brougham papers in the preceding volumes. These display an intellectual commitment that goes a long way to explain Brougham's subsequent vitriolic attack upon Young in the pages of the *Edinburgh Review*.)

variety of difficulties surrounding black-body radiation, specific heat, and the photoelectric effect.[43] Furthermore, though this is not the place to show it, each of these difficulties, except the optical one observed by Newton, had been a source of concern before (but usually not long before) the theory that resolved it was announced.

I suggest, therefore, that though a crisis or an "abnormal situation" is only one of the routes to *discovery* in the natural sciences, it is prerequisite to *fundamental inventions of theory*. Furthermore, I suspect that in the creation of the particularly deep crisis that usually precedes theoretical innovation, measurement makes one of its two most significant contributions to scientific advance. Most of the anomalies isolated in the preceding paragraph were quantitative or had a significant quantitative component, and, though the subject again carries us beyond the bounds of this essay, there is excellent reason why this should have been the case.

Unlike discoveries of new natural phenomena, innovations in scientific theory are not simply additions to the sum of what is already known. Almost always (always, in the mature sciences) the acceptance of a new theory demands the rejection of an older one. In the realm of theory, innovation is thus necessarily destructive as well as constructive. But, as the preceding pages have repeatedly indicated, theories are, even more than laboratory instruments, the essential tools of the scientist's trade. Without their constant assistance, even the observations and measurements made by the scientist would scarcely be scientific. A threat to theory is therefore a threat to the scientific life, and, though the scientific enterprise progresses through such threats, the individual scientist ignores them while he can. Particularly, he ignores them if his own prior practice has already committed him to the use of the threatened theory.[44] It follows that new theoretical suggestions, destructive of

43. Richtmeyer, Kennard, and Lauritsen, *Modern Physics*, pp. 89–94, 124–32, and 409–14. A more elementary account of the black-body problem and of the photoelectric effect is included in Gerald Holton, *Introduction to Concepts and Theories in Physical Science* (Cambridge, Mass., 1953), pp. 528–45.

44. Evidence for this effect of prior experience with a theory is provided by the well-known, but inadequately investigated, youthfulness of famous innovators as well as by the way in which younger men tend to cluster to the newer theory. Planck's statement about the latter phenomenon needs no citation. An earlier and particularly moving version of the same sentiment

old practices, rarely if ever emerge in the absence of a crisis that can no longer be suppressed.

No crisis is, however, so hard to suppress as one that derives from a quantitative anomaly that has resisted all the usual efforts at reconciliation. Once the relevant measurements have been stabilized and the theoretical approximations fully investigated, a quantitative discrepancy proves persistently obtrusive to a degree that few qualitative anomalies can match. By their very nature, qualitative anomalies usually suggest *ad hoc* modifications of theory that will disguise them, and once these modifications have been suggested there is little way of telling whether they are "good enough." An established quantitative anomaly, in contrast, usually suggests nothing except trouble, but at its best it provides a razor-sharp instrument for judging the adequacy of proposed solutions. Kepler provides a brilliant case in point. After prolonged struggle to rid astronomy of pronounced quantitative anomalies in the motion of Mars, he invented a theory accurate to 8′ of arc, a measure of agreement that would have astounded and delighted any astronomer who did not have access to the brilliant observations of Tycho Brahe. But from long experience Kepler knew Brahe's observations to be accurate to 4′ of arc. To us, he said, Divine goodness has given a most diligent observer in Tycho Brahe, and it is therefore right that we should with a grateful mind make use of this gift to find the true celestial motions. Kepler next attempted computations with noncircular figures. The outcome of those trials was his first two laws of planetary motion, the laws that for the first time made the Copernican system work.[45]

Two brief examples should make clear the differential effectiveness of qualitative and quantitative anomalies. Newton was apparently led to his new theory of light and color by observing the surprising elongation of the solar spectrum. Opponents of his new theory quickly pointed out that the existence of elongation had been known before and that it could be treated by existing theory. Qualitatively they were quite right. But utilizing Snell's quantitative law of refraction (a law that had been available to scientists

is provided by Darwin in the last chapter of *The Origin of Species* (see the 6th ed. [New York, 1889], 2:295–96).

45. J. L. E. Dreyer, *A History of Astronomy from Thales to Kepler*, 2d ed. (New York, 1953), pp. 385–93.

for less than three decades), Newton was able to show that the elongation predicted by existing theory was quantitatively far smaller than the one observed. On this quantitative discrepancy, all previous qualitative explanations of elongation broke down. Given the quantitative law of refraction, Newton's ultimate, and in this case quite rapid, victory was assured.[46] The development of chemistry provides a second striking illustration. It was well known, long before Lavoisier, that some metals gain weight when they are calcined (i.e., roasted). Furthermore, by the middle of the eighteenth century this qualitative observation was recognized to be incompatible with at least the simplest versions of the phlogiston theory, a theory that said phlogiston *escaped* from the metal during calcination. But so long as the discrepancy remained qualitative, it could be disposed of in numerous ways: perhaps phlogiston had negative weight, or perhaps fire particles lodged in the roasted metal. There were other suggestions besides, and together they served to reduce the urgency of the qualitative problem. The development of pneumatic techniques, however, transformed the qualitative anomaly into a quantitative one. In the hands of Lavoisier, they showed how much weight was gained and where it came from. These were data with which the earlier qualitative theories could not deal. Though phlogiston's adherents gave vehement and skillful battle, and though their qualitative arguments were fairly persuasive, the quantitative arguments for Lavoisier's theory proved overwhelming.[47]

These examples were introduced to illustrate how difficult it is

46. Kuhn, "Newton's Optical Papers," pp. 31–36.

47. This is a slight oversimplification, since the battle between Lavoisier's new chemistry and its opponents really implicated more than combustion processes, and the full range of relevant evidence cannot be treated in terms of combustion alone. Useful elementary accounts of Lavoisier's contributions can be found in J. B. Conant, *The Overthrow of the Phlogiston Theory*, Harvard Case Histories in Experimental Science, case 2 (Cambridge, Mass., 1950), and D. McKie, *Antoine Lavoisier: Scientist, Economist, Social Reformer* (New York, 1952). Maurice Daumas, *Lavoisier, théoricien et expérimentateur* (Paris, 1955) is the best recent scholarly review. J. H. White, *The Phlogiston Theory* (London, 1932) and especially J. R. Partington and D. McKie, "Historical Studies of the Phlogiston Theory: IV. Last Phases of the Theory," *Annals of Science* 4 (1939): 113–49, give most detail about the conflict between the new theory and the old.

to explain away established quantitative anomalies, and to show how much more effective these are than qualitative anomalies in establishing unevadable scientific crises. But the examples also show something more. They indicate that measurement can be an immensely powerful weapon in the battle between two theories, and that, I think, is its second particularly significant function. Furthermore, it is for this function—aid in the choice between theories—and for it alone, that we must reserve the word "confirmation." We must, that is, if "confirmation" is intended to denote a procedure anything like what scientists ever do. The measurements that display an anomaly and thus create crisis may tempt the scientist to leave science or to transfer his attention to some other part of the field. But, if he stays where he is, anomalous observations, quantitative or qualitative, cannot tempt him to abandon his theory *until another one is suggested to replace it.* Just as a carpenter, while he retains his craft, cannot discard his toolbox because it contains no hammer fit to drive a particular nail, so the practitioner of science cannot discard established theory because of a felt inadequacy. At least he cannot do so until shown some other way to do his job. In scientific practice the real confirmation questions always involve the comparison of two theories with each other and with the world, not the comparison of a single theory with the world. In these three-way comparisons, measurement has a particular advantage.

To see where the advantage of measurement resides, I must once more step briefly, and hence dogmatically, beyond the bounds of this essay. In the transition from an earlier to a later theory, there is very often a loss as well as a gain of explanatory power.[48] Newton's theory of planetary and projectile motion was fought vehemently for more than a generation because, unlike its main competitors, it demanded the introduction of an inexplicable force that acted directly upon bodies at a distance. Cartesian theory, for example, had attempted to explain gravity in terms of the direct collisions between elementary particles. To accept Newton meant to

48. This point is central to the reference cited in note 3. In fact, it is largely the necessity of balancing gains and losses and the controversies that so often result from disagreements about an appropriate balance that make it appropriate to describe changes of theory as "revolutions."

abandon the possibility of any such explanation, or so it seemed to most of Newton's immediate successors.[49] Similarly, though the historical detail is more equivocal, Lavoisier's new chemical theory was opposed by a number of men who felt that it deprived chemistry of one principal traditional function—the explanation of the qualitative properties of bodies in terms of the particular combination of chemical "principles" that composed them.[50] In each case the new theory was victorious, but the price of victory was the abandonment of an old and partly achieved goal. For eighteenth-century Newtonians it gradually became "unscientific" to ask for the cause of gravity; nineteenth-century chemists increasingly ceased to ask for the causes of particular qualities. Yet subsequent experience has shown that there was nothing *intrinsically* "unscientific" about these questions. General relativity does explain gravitational attraction and quantum mechanics does explain many of the qualitative characteristics of bodies. We now know what makes some bodies yellow and others transparent. But in gaining this immensely important understanding, we have had to regress, in certain respects, to an older set of notions about the bounds of scientific inquiry. Problems and solutions that had to be abandoned in embracing classic theories of modern science are again very much with us.

The study of the confirmation procedures as they are practiced in the sciences is therefore often the study of what scientists will and will not give up in order to gain other particular advantages. That problem has scarcely even been stated before, and I can therefore scarcely guess what its fuller investigation would reveal. But impressionistic study strongly suggests one significant conclusion. I know of no case in the development of science which ex-

49. Cohen, *Franklin and Newton*, chap. 4; Pierre Brunet, *L'introduction des théories de Newton en France au xviiie siècle* (Paris, 1931).

50. On this traditional task of chemistry see E. Meyerson, *Identity and Reality*, trans. K. Lowenberg (London, 1930), chap. 10, particularly pp. 331–36. Much essential material is also scattered through Hélène Metzger, *Les doctrines chimiques en France du début du xviie à la fin du xviiie siècle*, vol. 1 (Paris, 1923), and *Newton, Stahl, Boerhaave, et la doctrine chimique* (Paris, 1930). Notice particularly that the phlogistonists, who looked upon ores as elementary bodies from which the metals were compounded by addition of phlogiston, could explain why the metals were so much more like each other than were the ores from which they were compounded. All metals had a principle, phlogiston, in common. No such explanation was possible with Lavoisier's theory.

hibits a loss of quantitative accuracy as a consequence of the transition from an earlier to a later theory. Nor can I imagine a debate between scientists in which, however hot the emotions, the search for greater numerical accuracy in a previously quantified field would be called "unscientific." Probably for the same reasons that make them particularly effective in creating scientific crises, the comparison of numerical predictions, *where they have been available*, has proved particularly successful in bringing scientific controversies to a close. Whatever the price in redefinitions of science, its methods, and its goals, scientists have shown themselves consistently unwilling to compromise the numerical success of their theories. Presumably there are other such desiderata as well, but one suspects that, in case of conflict, measurement would be the consistent victor.

Measurement in the Development of Physical Science

To this point we have taken for granted that measurement did play a central role in physical science and have asked about the nature of that role and the reasons for its peculiar efficacy. Now we must ask, though too late to anticipate a comparably full response, about the way in which physical science came to make use of quantitative techniques at all. To make that large and factual question manageable, I select for discussion only those parts of an answer which relate particularly closely to what has already been said.

One recurrent implication of the preceding discussion is that much qualitative research, both empirical and theoretical, is normally prerequisite to fruitful quantification of a given research field. In the absence of such prior work, the methodological directive, "Go ye forth and measure," may well prove only an invitation to waste time. If doubts about this point remain, they should be quickly resolved by a brief review of the role played by quantitative techniques in the emergence of the various physical sciences. Let me begin by asking what role such techniques had in the Scientific Revolution that centered in the seventeenth century.

Since any answer must now be schematic, I begin by dividing the fields of physical science studied during the seventeenth century into two groups. The first, to be labeled the traditional sciences, consists of astronomy, optics, and mechanics, all of them fields that had received considerable qualitative and quantitative development in antiquity and during the Middle Ages. These fields are to be contrasted with what I shall call the Baconian sciences, a new cluster of research areas that owed their status *as sciences* to the characteristic insistence of seventeenth-century natural philosophers upon experimentation and the compilation of natural histories, including histories of the crafts. To this second group belong particularly the study of heat, of electricity, of magnetism, and of chemistry. Only chemistry had been much explored before the Scientific Revolution, and the men who explored it had almost all been either craftsmen or alchemists. If we except a few of the Islamic practitioners of the art, the emergence of a rational and systematic chemical tradition cannot be dated earlier than the late sixteenth century.[51] Magnetism, heat, and electricity emerged still more slowly as independent subjects for learned study. Even more clearly than chemistry, they are novel by-products of the Baconian elements in the "new philosophy."[52]

The separation of traditional from Baconian sciences provides an important analytic tool, because the man who looks to the Scientific Revolution for examples of productive measurement in physical science will find them only in the sciences of the first group. Further, and perhaps more revealing, even in these traditional sciences measurement was most often effective just when it could be performed with well-known instruments and applied to very nearly traditional concepts. In astronomy, for example, it was Tycho Brahe's enlarged and better-calibrated version of medieval instruments that made the decisive quantitative contribution. The telescope, a characteristic novelty of the seventeenth century, was scarcely used quantitatively until the last third of the century, and that quantitative use had no effect on astronomical theory until Bradley's discovery of aberration in 1729. Even that discovery was

51. Boas, *Robert Boyle*, pp. 48–66.

52. For electricity see, Roller and Roller, *Concept of Electric Charge*, and Edgar Zilsel, "The Origins of William Gilbert's Scientific Method," *Journal of the History of Ideas* 2 (1941): 1–32. I agree with those who feel Zilsel exaggerates the importance of a single factor in the genesis of electrical science and, by implication, of Baconianism, but the craft influences he describes cannot conceivably be dismissed. There is no equally satisfactory discussion of the development of thermal science before the eighteenth century, but Wolf (*Sixteenth and Seventeenth Centuries*, pp. 82–92 and 275–81) will illustrate the transformation produced by Baconianism.

isolated. Only during the second half of the eighteenth century did astronomy begin to experience the full effects of the immense improvements in quantitative observation that the telescope permitted.[53] Or again, as previously indicated, the novel inclined plane experiments of the seventeenth century were not nearly accurate enough to have alone been the source of the law of uniform acceleration. What is important about them—and they are critically important—is the conception that such measurements could have relevance to the problems of free fall and of projectile motion. That conception implies a fundamental shift in both the idea of motion and the techniques relevant to its analysis. But clearly no such conception could have evolved as it did if many of the subsidiary concepts needed for its exploitation had not existed, at least as developed embryos, in the works of Archimedes and of the scholastic analysts of motion.[54] Here again the effectiveness of quantitative work depended upon a long-standing prior tradition.

Perhaps the best test case is provided by optics, the third of my traditional sciences. In this field during the seventeenth century, real quantitative work was done with both new and old instruments, and the work done with old instruments on well-known phenomena proved the more important. The reformulation of optical theory during the Scientific Revolution turned upon Newton's prism experiments, and for these there was much qualitative precedent. Newton's innovation was the quantitative analysis of a well-known qualitative effect, and that analysis was possible only because of the discovery, a few decades before Newton's work, of Snell's law of refraction. That law is the vital quantitative novelty in the optics of the seventeenth century. It was, however, a law that had been sought by a series of brilliant investigators since the time of Ptolemy, and all had used apparatus quite similar to that which Snell employed. In short, the research which led to Newton's new

53. Wolf, *Eighteenth Century*, pp. 102–45, and Whewell, *Inductive Sciences*, 2:213–371. Particularly in the latter, notice the difficulty in separating advances due to improved instrumentation from those due to improved theory. This difficulty is not due primarily to Whewell's mode of presentation.

54. For pre-Galilean work, see Marshall Clagett, *The Science of Mechanics in the Middle Ages* (Madison, Wis., 1959), particularly parts 2 and 3. For Galileo's use of this work, see Alexander Koyré, *Etudes galiléennes*, 3 vols. (Paris, 1939), particularly vols. 1 and 2.

theory of light and color was of an essentially traditional nature.[55]

Much in seventeenth-century optics was, however, by no means traditional. Interference, diffraction, and double refraction were all first discovered in the half-century before Newton's *Opticks* appeared; all were totally unexpected phenomena; and all were known to Newton.[56] On two of them Newton conducted careful quantitative investigations. Yet the real impact of these novel phenomena upon optical theory was scarcely felt until the work of Young and Fresnel a century later. Though Newton was able to develop a brilliant preliminary theory for interference effects, neither he nor his immediate successors even noted that that theory agreed with quantitative experiment only for the limited case of perpendicular incidence. Newton's measurements of diffraction produced only the most qualitative theory, and on double refraction he seems not even to have attempted quantitative work of his own. Both Newton and Huyghens announced mathematical laws governing the refraction of the extraordinary ray, and the latter showed how to account for this behavior by considering the expansion of a spheroidal wave front. But both mathematical discussions involved large extrapolations from scattered quantitative data of doubtful accuracy. And almost a hundred years elapsed before quantitative experiments proved able to distinguish between these two quite different mathematical formulations.[57] As with the other optical phenomena discovered during the Scientific Revolution, most of the eighteenth century was needed for the additional exploration and instrumentation prerequisite to quantitative exploitation.

Turning now to the Baconian sciences, which throughout the Scientific Revolution possessed few old instruments and even fewer well-wrought concepts, we find quantification proceeding even

55. A. C. Crombie, *Augustine to Galileo* (London, 1952), pp. 70–82, and Wolf, *Sixteenth and Seventeenth Centuries*, pp. 244–54.

56. Wolf, *Sixteenth and Seventeenth Centuries*, pp. 254–64.

57. For the seventeenth-century work (including Huyghen's geometric construction) see ibid. The eighteenth-century investigations of these phenomena have scarcely been studied, but for what is known, see Joseph Priestley, *History . . . of Discoveries relating to Vision, Light, and Colours* (London, 1772), pp. 279–316, 498–520, 548–62. The earliest examples I know of more precise work on double refraction are R. J. Haüy, "Sur la double réfraction du Spath d'Islande" (see n. 42), and W. H. Wollaston, "On the Oblique Refraction of Iceland Crystal," *Philosophical Transactions* 92 (1802): 381–86.

more slowly. Though the seventeenth century saw many new instruments, of which a number were quantitative and others potentially so, only the new barometer disclosed significant quantitative regularities when applied to new fields of study. And even the barometer is only an apparent exception, for pneumatics, the field of its application, was able to borrow *en bloc* the concepts of a far older field, hydrostatics. As Torricelli put it, the barometer measured pressure "at the bottom of an ocean of the element air."[58] In the field of magnetism the only significant seventeenth-century measurements, those of declination and dip, were made with one or another modified version of the traditional compass, and these measurements did little to improve the understanding of magnetic phenomena. For a more fundamental quantification, magnetism, like electricity, awaited the work of Coulomb, Gauss, Poisson, and others in the late eighteenth and early nineteenth centuries. Before that work could be done, a better qualitative understanding of attraction, repulsion, conduction, and other such phenomena was needed. The instruments that produced a lasting quantification had then to be designed with these initially qualitative conceptions in mind.[59] Furthermore, the decades in which success was at last achieved are almost the same ones that produced the first effective contacts between measurement and theory in the study of chemistry and of heat.[60] Successful quantification of the Baconian sci-

58. See I. H. B. and A. G. H. Spiers, *The Physical Treatises of Pascal* (New York, 1937), p. 164. This whole volume displays the way in which seventeenth-century pneumatics took over concepts from hydrostatics.

59. For the quantification and early mathematization of electrical science, see Roller and Roller, *Concept of Electric Charge*, pp. 66–80; Whittaker, *Aether and Electricity*, 1:53–66; and W. C. Walker, "The Detection and Estimation of Electric Charge in the Eighteenth Century," *Annals of Science* 1 (1936): 66–100.

60. For heat see, Douglas McKie and N. H. de V. Heathcote, *The Discovery of Specific and Latent Heats* (London, 1935). In chemistry it may well be impossible to fix any date for the "first effective contacts between measurement and theory." Volumetric or gravimetric measures were always an ingredient of chemical recipes and assays. By the seventeenth century, for example in the work of Boyle, weight gain or loss was often a clue to the theoretical analysis of particular reactions. But until the middle of the eighteenth century, the significance of chemical measurement seems always to have been either descriptive (as in recipes) or qualitative (as in demonstrating a weight gain without significant reference to its magnitude). Only

ences had scarcely begun before the last third of the eighteenth century and only realized its full potential in the nineteenth. That realization—exemplified in the work of Fourier, Clausius, Kelvin, and Maxwell—is one facet of a second scientific revolution no less consequential than the seventeenth-century revolution. Only in the nineteenth century did the Baconian physical sciences undergo the transformation which the group of traditional sciences had experienced two or more centuries before.

Since Professor Guerlac's paper is devoted to chemistry and since I have already sketched some of the bars to quantification of electrical and magnetic phenomena, I take my single more extended illustration from the study of heat. Unfortunately, much of the research upon which such a sketch should be based remains to be done. What follows is necessarily more tentative than what has gone before.

Many of the early experiments involving thermometers read like investigations *of* that new instrument rather than investigations *with* it. How could anything else have been the case during a period when it was totally unclear what the thermometer measured? Its readings obviously depended upon the "degree of heat" but apparently in immensely complex ways. "Degree of heat" had for a long time been defined by the senses, and the senses responded quite differently to bodies which produced the same thermometric readings. Before the thermometer could become unequivocally a laboratory instrument rather than an experimental subject, thermometric reading had to be seen as the direct measure of "degree of heat," and sensation had simultaneously to be viewed as a complex and equivocal phenomenon dependent upon a number of different parameters.[61]

in the work of Black, Lavoisier, and Richter does measurement begin to play a fully quantitative role in the development of chemical laws and theories. For an introduction to these men and their work, see J. R. Partington, *A Short History of Chemistry*, 2d ed., pp. 93–97, 122–28, and 161–63.

61. Maurice Daumas (*Les instruments scientifiques aux xvii*ᵉ *et xviii*ᵉ *siècles* [Paris, 1953], pp. 78–80) provides an excellent brief account of the slow stages in the deployment of the thermometer as a scientific instrument. Robert Boyle's *New Experiments and Observations Touching Cold* illustrates the need during the seventeenth century to demonstrate that properly constructed thermometers must replace the senses in thermal measurements even though the two give divergent results. See *Works of the Honourable Robert Boyle*, ed. T. Birch, 5 vols. (London, 1744), 2:240–43.

That conceptual reorientation seems to have been completed in at least a few scientific circles before the end of the seventeenth century, but no rapid discovery of quantitative regularities followed. First scientists had to be forced to a bifurcation of "degree of heat" into "quantity of heat," on the one hand, and "temperature," on the other. In addition they had to select for close scrutiny, from the immense multitude of available thermal phenomena, the ones that could most readily be made to reveal quantitative law. These proved to be two in number: mixing two components of a single fluid initially at different temperatures and radiant heating of two different fluids in identical vessels. Even when attention was focused upon these phenomena, however, scientists still did not get unequivocal or uniform results. As Heathcote and McKie have brilliantly shown, the last stages in the development of the concepts of specific and latent heat display intuited hypotheses constantly interacting with stubborn measurement, each forcing the other into line.[62] Still other sorts of work were required before the contributions of Laplace, Poisson, and Fourier could transform the study of thermal phenomena into a branch of mathematical physics.[63]

This sort of pattern, reiterated both in the other Baconian sciences and in the extension of traditional sciences to new instruments and new phenomena thus provides one additional illustration of this paper's most persistent thesis. *The road from scientific law to scientific measurement can rarely be traveled in the reverse direction.* To discover quantitative regularity one must normally know what regularity one is seeking and one's instruments must be designed accordingly; even then nature may not yield consistent or generalizable results without a struggle. So much for my major thesis. The preceding remarks about the way in which quantification entered the modern physical sciences should, however, also recall this paper's minor thesis, for they redirect attention to the immense efficacy of quantitative experimentation undertaken within

62. For the elaboration of calorimetric concepts, see E. Mach, *Die Principien der Wärmelehre* (Leipzig, 1919), pp. 153–81, and McKie and Heathcote, *Specific and Latent Heats.* The discussion of Krafft's work in the latter (pp. 59–63) provides a particularly striking example of the problems in making measurement work.

63. Gaston Bachelard, *Etude sur l'évolution d'un problème de physique* (Paris, 1928), and Kuhn, "Caloric Theory of Adiabatic Compression."

the context of a fully mathematized theory. Sometime between 1800 and 1850 there was an important change in the character of research in many of the physical sciences, particularly in the cluster of research fields known as physics. That change is what makes me call the mathematization of Baconian physical science one facet of a second scientific revolution.

It would be absurd to pretend that mathematization was more than a facet. The first half of the nineteenth century also witnessed a vast increase in the scale of the scientific enterprise, major changes in patterns of scientific organization, and a total reconstruction of scientific education.[64] But these changes affected all the sciences in much the same way. They ought not to explain the characteristics that differentiate the newly mathematized sciences of the nineteenth century from other sciences of the same period. Though my sources are now impressionistic, I feel quite sure that there are such characteristics. Let me hazard the following prediction. Analytic, and in part statistical, research would show that physicists, as a group, have displayed since about 1840 a greater ability to concentrate their attention on a few key areas of research than have their colleagues in less completely quantified fields. In the same period, if I am right, physicists would prove to have been more successful than most other scientists in decreasing the length of controversies about scientific theories and in increasing the strength of the consensus that emerged from such controversies. In short, I believe that the nineteenth-century mathematization of physical science produced vastly refined professional criteria for problem selection and that it simultaneously very much increased the effectiveness of professional verification procedures.[65] These are, of course, just the changes that the discussion in the preceding section would lead us to expect. A critical and compara-

64. S. F. Mason (*Main Currents of Scientific Thought* [New York, 1956], pp. 352–63) provides an excellent brief sketch of these institutional changes. Much additional material is scattered through J. T. Merz, *History of European Thought in the Nineteenth Century*, vol. 1 (London, 1923).

65. For an example of effective problem selection, note the esoteric quantitative discrepancies which isolated the three problems—photoelectric effect, k-body radiation, and specific heats—that gave rise to quantum meics. For the new effectiveness of verification procedures, note the speed which this radical new theory was adopted by the profession.

tive analysis of the development of physics during the past century and a quarter should provide an acid test of those conclusions.

Pending that test, can we conclude anything at all? I venture the following paradox: The full and intimate quantification of any science is a consummation devoutly to be wished. Nevertheless, it is not a consummation that can effectively be sought by measuring. As in individual development, so in the scientific group, maturity comes most surely to those who know how to wait.

Appendix

Reflecting on the other papers and on the discussion that continued throughout the conference, two additional points that had reference to my own paper seem worth recording. Undoubtedly there were others as well, but my memory has proved more than usually unreliable. Professor Price raised the first point, which gave rise to considerable discussion. The second followed from an aside by Professor Spengler, and I shall consider its consequences first.

Professor Spengler expressed great interest in my concept of "crises" in the development of a science or of a scientific specialty, but added that he had had difficulty discovering more than one such episode in the development of economics. This raised for me the perennial, but perhaps not very important question about whether or not the social sciences are really sciences at all. Though I shall not even attempt to answer it in that form, a few further remarks about the possible absence of crises in the development of a social science may illuminate some part of what is at issue.

As developed in the section Extraordinary Measurement, above, the concept of a crisis implies a prior unanimity of the group that experiences one. Anomalies, by definition, exist only with respect to firmly established expectations. Experiments can create a crisis by consistently going wrong only for a group that has previously experienced everything's seeming to go right. Now, as my earlier sections should indicate quite fully, in the mature physical sciences most things generally do go right. The entire professional community can therefore ordinarily agree about the fundamental concepts, tools, and problems of its science. Without that professional consensus, there would be no basis for the sort of puzzle-solving activity in which, as I have already urged, most physical scientists are normally engaged. In the physical sciences disagreement about fundamentals is, like the search for basic innovations, reserved for periods of crisis.[66] It is, however, by no means equally clear that a consensus of anything like similar strength and scope ordinarily characterizes the social sciences. Experience with my university colleagues and a fortunate year spent at the Center for Advanced Study in the Behavioral Sciences suggest that the fundamental agreement which physicists, say, can normally take for granted has only recently begun to emerge in a few areas of social science research. Most other areas are still characterized by fundamental disagreements about the definition of the field, its paradigm achievements, and its problems. While that situation obtains (as it did also in earlier periods of the development of the various physical sciences), either there can be no crises or there can never be anything else.

Professor Price's point was very different and far more historical. He suggested, and I think quite rightly, that my historical epilogue failed to call attention to a very important change in the attitude of physical scientists toward measurement that occurred during the Scientific Revolution. In commenting on Dr. Crombie's paper, Price had pointed out that not until the late sixteenth century did astronomers begin to record continuous series of observations of planetary position. (Previously they had restricted themselves to occasional quantitative observations of special phenomena.) Only in that same late period, he continued, did astronomers begin to be critical of their quantitative data, recognizing, for example, that a recorded celestial position is a clue to an astronomical fact rather than the fact itself. When discussing my paper, Professor Price pointed to still other signs of a change in the attitude toward measurement during the Scientific Revolution. For one thing, he emphasized, many more numbers were recorded. More important, perhaps, people like Boyle, when announcing laws derived from measurement, began for the first time to record their quantitative

66. I have developed some other significant concomitants of this professional consensus in my paper, "The Essential Tension: Tradition and Innovation in Scientific Research," in Calvin W. Taylor, ed., *The Third (1959) University of Utah Research Conference on the Identification of Creative Scientific Talent* (Salt Lake City, 1959), pp. 162–77 (pp. 225–39 below).

data, *whether or not they perfectly fit the law*, rather than simply stating the law itself.

I am somewhat doubtful that this transition in attitude toward numbers proceeded quite so far in the seventeenth century as Professor Price seemed occasionally to imply. Hooke, for one example, did not report the numbers from which he derived his law of elasticity; no concept of "significant figures" seems to have emerged in the experimental physical sciences before the *nineteenth* century. But I cannot doubt that the change was in process and that it is very important. At least in another sort of paper, it deserves detailed examination which I very much hope it will get. Pending that examination, however, let me simply point out how very closely the development of the phenomena emphasized by Professor Price fits the pattern I have already sketched in describing the effects of seventeenth-century Baconianism.

In the first place, except perhaps in astronomy, the seventeenth-century change in attitude toward measurement looks very much like a response to the novelties of the methodological program of the "new philosophy." Those novelties were not, as has so often been supposed, consequences of the belief that observation and experiment were basic to science. As Crombie has brilliantly shown, that belief and an accompanying methodological philosophy were highly developed during the Middle Ages.[67] Instead, the novelties of method in the "new philosophy" included a belief that lots and lots of experiments would be necessary (the plea for natural histories) and an insistence that all experiments and observations be reported in full and naturalistic detail, preferably accompanied by the names and credentials of witnesses. Both the increased frequency with which numbers were recorded and the decreased tendency to round them off are precisely congruent with those more general Baconian changes in the attitude toward experimentation at large.

Furthermore, whether or not its source lies in Baconianism, the effectiveness in the seventeenth century of the new attitude toward numbers developed in very much the same way as the effectiveness of the other Baconian novelties discussed in my concluding section. In dynamics, as Professor Koyré has repeatedly shown, the new

67. See particularly his *Robert Grosseteste and the Origins of Experimental Science, 1100–1700* (Oxford, 1953).

attitude had almost no effect before the later eighteenth century. The other two traditional sciences, astronomy and optics, were affected sooner by the change, but only in their most nearly traditional parts. And the Baconian sciences, heat, electricity, chemistry, and the like, scarcely begin to profit from the new attitude until after 1750. Again it is in the work of Black, Lavoisier, Coulomb, and their contemporaries that the first truly significant effects of the change are seen. And the full transformation of physical science due to that change is scarcely visible before the work of Ampère, Fourier, Ohm, and Kelvin. Professor Price has, I think, isolated another very significant seventeenth-century novelty. But like so many of the other novel attitudes displayed by the "new philosophy," the significant effects of this new attitude toward measurement were scarcely manifested in the seventeenth century at all.

9

The Essential Tension: Tradition and Innovation in Scientific Research

Reprinted by permission from *The Third (1959) University of Utah Research Conference on the Identification of Scientific Talent,* ed. C. W. Taylor (Salt Lake City: University of Utah Press, 1959), pp. 162–74.

I am grateful for the invitation to participate in this important conference, and I interpret it as evidence that students of creativity themselves possess the sensitivity to divergent approaches that they seek to identify in others. But I am not altogether sanguine about the outcome of your experiment with me. As most of you already know, I am no psychologist, but rather an ex-physicist now working in the history of science. Probably my concern is no less with creativity than your own, but my goals, my techniques, and my sources of evidence are so very different from yours that I am far from sure how much we do, or even *should*, have to say to each other. These reservations imply no apology: rather they hint at my central thesis. In the sciences, as I shall suggest below, it is often better to do one's best with the tools at hand than to pause for contemplation of divergent approaches.

If a person of my background and interests has anything relevant to suggest to this conference, it will not be about your central concerns, the creative personality and its early identification. But implicit in the numerous working papers distributed to participants in this conference is an image of the scientific process and of the scientist; that image almost certainly conditions many of the experiments you try as well as the conclusions you draw; and about it the physicist-historian may well have something to say. I shall restrict my attention to one aspect of this image—an aspect epitomized as follows in one of the working papers: The basic scientist "must lack prejudice to a degree where he can look at the most 'self-evident' facts or concepts without necessarily accepting them, and, conversely, allow his imagination to play with the most unlikely possibilities" (Selye, 1959). In the more technical language supplied by other working papers (Getzels and Jackson), this aspect of the image recurs as an emphasis upon "divergent thinking, . . . the freedom to go off in different directions, . . . rejecting the old solution and striking out in some new direction."

I do not at all doubt that this description of "divergent thinking" and the concomitant search for those able to do it are entirely proper. Some divergence characterizes all scientific work, and gigantic divergences lie at the core of the most significant episodes in scientific development. But both my own experience in scientific research and my reading of the history of sciences lead me to wonder whether flexibility and open-mindedness have not been too exclusively emphasized as the characteristics requisite for basic research. I shall therefore suggest below that something like "convergent thinking" is just as essential to scientific advance as is divergent. Since these two modes of thought are inevitably in conflict, it will follow that the ability to support a tension that can occasionally become almost unbearable is one of the prime requisites for the very best sort of scientific research.

I am elsewhere studying these points more historically, with emphasis on the importance to scientific development of "revolutions."[1] These are episodes—exemplified in their most extreme and readily recognized form by the advent of Copernicanism, Darwinism, or Einsteinianism—in which a scientific community abandons one time-honored way of regarding the world and of pursuing science in favor of some other, usually incompatible, approach to its discipline. I have argued in the draft that the historian constantly encounters many far smaller but structurally similar revolutionary episodes and that they are central to scientific advance. Contrary to a prevalent impression, most new discoveries and theories in the sciences are not merely additions to the existing stockpile of scientific knowledge. To assimilate them the scientist must usually rearrange the intellectual and manipulative equipment he has previously relied upon, discarding some elements of his prior belief

1. *The Structure of Scientific Revolutions* (Chicago, 1962).

and practice while finding new significances in and new relationships between many others. Because the old must be revalued and reordered when assimilating the new, discovery and invention in the sciences are usually intrinsically revolutionary. Therefore, they do demand just that flexibility and open-mindedness that characterize, or indeed define, the divergent thinker. Let us henceforth take for granted the need for these characteristics. Unless many scientists possessed them to a marked degree, there would be no scientific revolutions and very little scientific advance.

Yet flexibility is not enough, and what remains is not obviously compatible with it. Drawing from various fragments of a project still in progress, I must now emphasize that revolutions are but one of two complementary aspects of scientific advance. Almost none of the research undertaken by even the greatest scientists is designed to be revolutionary, and very little of it has any such effect. On the contrary, normal research, even the best of it, is a highly convergent activity based firmly upon a settled consensus acquired from scientific education and reinforced by subsequent life in the profession. Typically, to be sure, this convergent or consensus-bound research ultimately results in revolution. Then, traditional techniques and beliefs are abandoned and replaced by new ones. But revolutionary shifts of a scientific tradition are relatively rare, and extended periods of convergent research are the necessary preliminary to them. As I shall indicate below, only investigations firmly rooted in the contemporary scientific tradition are likely to break that tradition and give rise to a new one. That is why I speak of an "essential tension" implicit in scientific research. To do his job the scientist must undertake a complex set of intellectual and manipulative commitments. Yet his claim to fame, if he has the talent and good luck to gain one, may finally rest upon his ability to abandon this net of commitments in favor of another of his own invention. Very often the successful scientist must simultaneously display the characteristics of the traditionalist and of the iconoclast.[2]

2. Strictly speaking, it is the professional group rather than the individual scientist that must display both these characteristics simultaneously. In a fuller account of the ground covered in this paper that distinction between individual and group characteristics would be basic. Here I can only note that, though recognition of the distinction weakens the conflict or tension referred to above, it does not eliminate it. Within the group some individ-

The multiple historical examples upon which any full documentation of these points must depend are prohibited by the time limitations of the conference. But another approach will introduce you to at least part of what I have in mind—an examination of the nature of education in the natural sciences. One of the working papers for this conference (Getzels and Jackson) quotes Guilford's very apt description of scientific education as follows: "[It] has emphasized abilities in the areas of convergent thinking and evaluation, often at the expense of development in the area of divergent thinking. We have attempted to teach students how to arrive at 'correct' answers that our civilization has taught us are correct. . . . Outside the arts [and I should include most of the social sciences] we have generally discouraged the development of divergent-thinking abilities, unintentionally." That characterization seems to me eminently just, but I wonder whether it is equally just to deplore the product that results. Without defending plain bad teaching, and granting that in this country the trend to convergent thinking in all education may have proceeded entirely too far, we may nevertheless recognize that a rigorous training in convergent thought has been intrinsic to the sciences almost from their origin. I suggest that they could not have achieved their present state or status without it.

Let me try briefly to epitomize the nature of education in the natural sciences, ignoring the many significant yet minor differences between the various sciences and between the approaches of different educational institutions. The single most striking feature of this education is that, to an extent totally unknown in other creative fields, it is conducted entirely through textbooks. Typically, undergraduate *and* graduate students of chemistry, physics, astronomy, geology, or biology acquire the substance of their fields from books written especially for students. Until they are ready, or very nearly ready, to commence work on their own dissertations, they are neither asked to attempt trial research projects nor exposed to the immediate products of research done by others, that is, to the professional communications that scientists write for each other. There are no collections of "readings" in the natural sciences. Nor

uals may be more traditionalistic, others more iconoclastic, and their contributions may differ accordingly. Yet education, institutional norms, and the nature of the job to be done will inevitably combine to insure that all group members will, to a greater or lesser extent, be pulled in both directions.

are science students encouraged to read the historical classics of their fields—works in which they might discover other ways of regarding the problems discussed in their textbooks, but in which they would also meet problems, concepts, and standards of solution that their future professions have long since discarded and replaced.

In contrast, the various textbooks that the student does encounter display different subject matters, rather than, as in many of the social sciences, exemplifying different approaches to a single problem field. Even books that compete for adoption in a single course differ mainly in level and in pedagogic detail, not in substance or conceptual structure. Last, but most important of all, is the characteristic technique of textbook presentation. Except in their occasional introductions, science textbooks do not describe the sorts of problems that the professional may be asked to solve and the variety of techniques available for their solution. Rather, these books exhibit concrete problem solutions that the profession has come to accept as paradigms, and they then ask the student, either with a pencil and paper or in the laboratory, to solve for himself problems very closely related in both method and substance to those through which the textbook or the accompanying lecture has led him. Nothing could be better calculated to produce "mental sets" or *Einstellungen*. Only in their most elementary courses do other academic fields offer as much as a partial parallel.

Even the most faintly liberal educational theory must view this pedagogic technique as anathema. Students, we would all agree, must begin by learning a good deal of what is already known, but we also insist that education give them vastly more. They must, we say, learn to recognize and evaluate problems to which no unequivocal solution has yet been given; they must be supplied with an arsenal of techniques for approaching these future problems; and they must learn to judge the relevance of these techniques and to evaluate the possibly partial solutions which they can provide. In many respects these attitudes toward education seem to me entirely right, and yet we must recognize two things about them. First, education in the natural sciences seems to have been totally unaffected by their existence. It remains a dogmatic initiation in a pre-established tradition that the student is not equipped to evaluate. Second, at least in the period when it was followed by a term in an apprenticeship relation, this technique of exclusive exposure to a rigid tradition has been immensely productive of the most consequential sorts of innovations.

I shall shortly inquire about the pattern of scientific practice that grows out of this educational initiation and will then attempt to say why that pattern proves quite so successful. But first, an historical excursion will reinforce what has just been said and prepare the way for what is to follow. I should like to suggest that the various fields of natural science have not always been characterized by rigid education in exclusive paradigms, but that each of them acquired something like that technique at precisely the point when the field began to make rapid and systematic progress. If one asks about the origin of our contemporary knowledge of chemical composition, of earthquakes, of biological reproduction, of motion through space, or of any other subject matter known to the natural sciences, one immediately encounters a characteristic pattern that I shall here illustrate with a single example.

Today, physics textbooks tell us that light exhibits some properties of a wave and some of a particle: both textbook problems and research problems are designed accordingly. But both this view and these textbooks are products of an early twentieth-century revolution. (One characteristic of scientific revolutions is that they call for the rewriting of science textbooks.) For more than half a century before 1900, the books employed in scientific education had been equally unequivocal in stating that light was wave motion. Under those circumstances scientists worked on somewhat different problems and often embraced rather different sorts of solutions to them. The nineteenth-century textbook tradition does not, however, mark the beginning of our subject matter. Throughout the eighteenth century and into the early nineteenth, Newton's *Opticks* and the other books from which men learned science taught almost all students that light was particles, and research guided by this tradition was again different from that which succeeded it. Ignoring a variety of subsidiary changes within these three successive traditions, we may therefore say that our views derive historically from Newton's views by way of two revolutions in optical thought, each of which replaced one tradition of convergent research with another. If we make appropriate allowances for changes in the locus and materials of scientific education, we may say that each of these three traditions was embodied in the sort of education by exposure to unequivocal paradigms that I briefly epitomized above. Since Newton, education and research in physical optics have normally been highly convergent.

The history of theories of light does not, however, begin with Newton. If we ask about knowledge in the field before his time, we encounter a significantly different pattern—a pattern still familiar in the arts and in some social sciences, but one which has largely disappeared in the natural sciences. From remote antiquity until the end of the seventeenth century there was no single set of paradigms for the study of physical optics. Instead, many men advanced a large number of different views about the nature of light. Some of these views found few adherents, but a number of them gave rise to continuing schools of optical thought. Although the historian can note the emergence of new points of view as well as changes in the relative popularity of older ones, there was never anything resembling consensus. As a result, a new man entering the field was inevitably exposed to a variety of conflicting viewpoints; he was forced to examine the evidence for each, and there always was good evidence. The fact that he made a choice and conducted himself accordingly could not entirely prevent his awareness of other possibilities. This earlier mode of education was obviously more suited to produce a scientist without prejudice, alert to novel phenomena, and flexible in his approach to his field. On the other hand, one can scarcely escape the impression that, during the period characterized by this more liberal educational practice, physical optics made very little progress.[3]

The preconsensus (we might here call it the divergent) phase in the development of physical optics is, I believe, duplicated in the history of all other scientific specialties, excepting only those that were born by the subdivision and recombination of pre-existing disciplines. In some fields, like mathematics and astronomy, the first firm consensus is prehistoric. In others, like dynamics, geo-

3. The history of physical optics before Newton has recently been well described by Vasco Ronchi in *Histoire de la lumière*, trans. J. Taton (Paris, 1956). His account does justice to the element I elaborate too little above. Many fundamental contributions to physical optics were made in the two millennia before Newton's work. Consensus is not prerequisite to a sort of progress in the natural sciences, any more than it is to progress in the social sciences or the arts. It is, however, prerequisite to the sort of progress that we now generally refer to when distinguishing the natural sciences from the arts and from most social sciences.

metric optics, and parts of physiology, the paradigms that produced a first consensus date from classical antiquity. Most other natural sciences, though their problems were often discussed in antiquity, did not achieve a first consensus until after the Renaissance. In physical optics, as we have seen, the first firm consensus dates only from the end of the seventeenth century; in electricity, chemistry, and the study of heat, it dates from the eighteenth; while in geology and the nontaxonomic parts of biology no very real consensus developed until after the first third of the nineteenth century. This century appears to be characterized by the emergence of a first consensus in parts of a few of the social sciences.

In all the fields named above, important work was done before the achievement of the maturity produced by consensus. Neither the nature nor the timing of the first consensus in these fields can be understood without a careful examination of both the intellectual and the manipulative techniques developed before the existence of unique paradigms. But the transition to maturity is not less significant because individuals practiced science before it occurred. On the contrary, history strongly suggests that, though one can practice science—as one does philosophy or art or political science—without a firm consensus, this more flexible practice will not produce the pattern of rapid consequential scientific advance to which recent centuries have accustomed us. In that pattern, development occurs from one consensus to another, and alternate approaches are not ordinarily in competition. Except under quite special conditions, the practitioner of a mature science does not pause to examine divergent modes of explanation or experimentation.

I shall shortly ask how this can be so—how a firm orientation toward an apparently unique tradition can be compatible with the practice of the disciplines most noted for the persistent production of novel ideas and techniques. But it will help first to ask what the education that so successfully transmits such a tradition leaves to be done. What can a scientist working within a deeply rooted tradition and little trained in the perception of significant alternatives hope to do in his professional career? Once again limits of time force me to drastic simplification, but the following remarks will at least suggest a position that I am sure can be documented in detail.

In pure or basic science—that somewhat ephemeral category of research undertaken by men whose most immediate goal is to increase understanding rather than control of nature—the characteristic problems are almost always repetitions, with minor modifications, of problems that have been undertaken and partially resolved before. For example, much of the research undertaken within a scientific tradition is an attempt to adjust existing theory or existing observation in order to bring the two into closer and closer agreement. The constant examination of atomic and molecular spectra during the years since the birth of wave mechanics, together with the design of theoretical approximations for the prediction of complex spectra, provides one important instance of this typical sort of work. Another was provided by the remarks about the eighteenth-century development of Newtonian dynamics in the paper on measurement supplied to you in advance of the conference.[4] The attempt to make existing theory and observation conform more closely is not, of course, the only standard sort of research problem in the basic sciences. The development of chemical thermodynamics or the continuing attempts to unravel organic structure illustrate another type—the extension of existing theory to areas that it is expected to cover but in which it has never before been tried. In addition, to mention a third common sort of research problem, many scientists constantly collect the concrete data (e.g., atomic weights, nuclear moments) required for the application and extension of existing theory.

These are normal research projects in the basic sciences, and they illustrate the sorts of work on which all scientists, even the greatest, spend most of their professional lives and on which many spend all. Clearly their pursuit is neither intended nor likely to produce fundamental discoveries or revolutionary changes in scientific theory. Only if the validity of the contemporary scientific tradition is assumed do these problems make much theoretical or any practical sense. The man who suspected the existence of a totally new type of phenomenon or who had basic doubts about the validity of existing theory would not think problems so closely modeled on textbook paradigms worth undertaking. It follows that the man who does undertake a problem of this sort—and that

4. A revised version appeared in *Isis* 52 (1961): 161–93 (pp. 178–224 above).

means all scientists at most times—aims to elucidate the scientific tradition in which he was raised rather than to change it. Furthermore, the fascination of his work lies in the difficulties of elucidation rather than in any surprises that the work is likely to produce. Under normal conditions the research scientist is not an innovator but a solver of puzzles, and the puzzles upon which he concentrates are just those which he believes can be both stated and solved within the existing scientific tradition.

Yet—and this is the point—the ultimate effect of this tradition-bound work has invariably been to change the tradition. Again and again the continuing attempt to elucidate a currently received tradition has at last produced one of those shifts in fundamental theory, in problem field, and in scientific standards to which I previously referred as scientific revolutions. At least for the scientific community as a whole, work within a well-defined and deeply ingrained tradition seems more productive of tradition-shattering novelties than work in which no similarly convergent standards are involved. How can this be so? I think it is because no other sort of work is nearly so well suited to isolate for continuing and concentrated attention those loci of trouble or causes of crisis upon whose recognition the most fundamental advances in basic science depend.

As I have indicated in the first of my working papers, new theories and, to an increasing extent, novel discoveries in the mature sciences are not born *de novo*. On the contrary, they emerge from old theories and within a matrix of old beliefs about the phenomena that the world does *and does not* contain. Ordinarily such novelties are far too esoteric and recondite to be noted by the man without a great deal of scientific training. And even the man with considerable training can seldom afford simply to go out and look for them, let us say by exploring those areas in which existing data and theory have failed to produce understanding. Even in a mature science there are always far too many such areas, areas in which no existing paradigms seem obviously to apply and for whose exploration few tools and standards are available. More likely than not the scientist who ventured into them, relying merely upon his receptivity to new phenomena and his flexibility to new patterns of organization, would get nowhere at all. He would rather return his science to its preconsensus or natural history phase.

Instead, the practitioner of a mature science, from the beginning of his doctoral research, continues to work in the regions for which the paradigms derived from his education and from the research of his contemporaries seem adequate. He tries, that is, to elucidate topographical detail on a map whose main outlines are available in advance, and he hopes—if he is wise enough to recognize the nature of his field—that he will some day undertake a problem in which the anticipated does *not* occur, a problem that goes wrong in ways suggestive of a fundamental weakness in the paradigm itself. In the mature sciences the prelude to much discovery and to all novel theory is not ignorance, but the recognition that something has gone wrong with existing knowledge and beliefs.

What I have said so far may indicate that it is sufficient for the productive scientist to adopt existing theory as a lightly held tentative hypothesis, employ it *faute de mieux* in order to get a start in his research, and then abandon it as soon as it leads him to a trouble spot, a point at which something has gone wrong. But though the ability to recognize trouble when confronted by it is surely a requisite for scientific advance, trouble must not be too easily recognized. The scientist requires a thoroughgoing commitment to the tradition with which, if he is fully successful, he will break. In part this commitment is demanded by the nature of the problems the scientist normally undertakes. These, as we have seen, are usually esoteric puzzles whose challenge lies less in the information disclosed by their solutions (all but its details are often known in advance) than in the difficulties of technique to be surmounted in providing any solution at all. Problems of this sort are undertaken only by men assured that there is a solution which ingenuity can disclose, and only current theory could possibly provide assurance of that sort. That theory alone gives meaning to most of the problems of normal research. To doubt it is often to doubt that the complex technical puzzles which constitute normal research have any solutions at all. Who, for example, would have developed the elaborate mathematical techniques required for the study of the effects of interplanetary attractions upon basic Keplerian orbits if he had not assumed that Newtonian dynamics, applied to the planets then known, would explain the last details of astronomical observation? But without that assurance, how would Neptune have been discovered and the list of planets changed?

In addition, there are pressing practical reasons for commitment. Every research problem confronts the scientist with anomalies whose sources he cannot quite identify. His theories and observations never quite agree; successive observations never yield quite the same results; his experiments have both theoretical and phenomenological by-products which it would take another research project to unravel. Each of these anomalies or incompletely understood phenomena could conceivably be the clue to a fundamental innovation in scientific theory or technique, but the man who pauses to examine them one by one never completes his first project. Reports of effective research repeatedly imply that all but the most striking and central discrepancies could be taken care of by current theory if only there were time to take them on. The men who make these reports find most discrepancies trivial or uninteresting, an evaluation that they can ordinarily base only upon their faith in current theory. Without that faith their work would be wasteful of time and talent.

Besides, lack of commitment too often results in the scientist's undertaking problems that he has little chance of solving. Pursuit of an anomaly is fruitful only if the anomaly is more than nontrivial. Having discovered it, the scientist's first efforts and those of his profession are to do what nuclear physicists are now doing. They strive to generalize the anomaly, to discover other and more revealing manifestations of the same effect, to give it structure by examining its complex interrelationships with phenomena they still feel they understand. Very few anomalies are susceptible to this sort of treatment. To be so they must be in explicit and unequivocal conflict with some structurally central tenet of current scientific belief. Therefore, their recognition and evaluation once again depend upon a firm commitment to the contemporary scientific tradition.

This central role of an elaborate and often esoteric tradition is what I have principally had in mind when speaking of the essential tension in scientific research. I do not doubt that the scientist must be, at least potentially, an innovator, that he must possess mental flexibility, and that he must be prepared to recognize troubles where they exist. That much of the popular stereotype is surely correct, and it is important accordingly to search for indices of the corresponding personality characteristics. But what is no part of our stereotype and what appears to need careful integration with it is the other face of this same coin. We are, I think, more likely fully to exploit our potential scientific talent if we recognize the extent to which the basic scientist must also be a firm traditionalist, or, if I am using your vocabulary at all correctly, a convergent thinker. Most important of all, we must seek to understand how these two superficially discordant modes of problem solving can be reconciled both within the individual and within the group.

Everything said above needs both elaboration and documentation. Very likely some of it will change in the process. This paper is a report on work in progress. But, though I insist that much of it is tentative and all of it incomplete, I still hope that the paper has indicated why an educational system best described as an initiation into an unequivocal tradition should be thoroughly compatible with successful scientific work. And I hope, in addition, to have made plausible the historical thesis that no part of science has progressed very far or very rapidly before this convergent education and correspondingly convergent normal practice became possible. Finally, though it is beyond my competence to derive personality correlates from this view of scientific development, I hope to have made meaningful the view that the productive scientist must be a traditionalist who enjoys playing intricate games by preestablished rules in order to be a successful innovator who discovers new rules and new pieces with which to play them.

As first planned, my paper was to have ended at this point. But work on it, against the background supplied by the working papers distributed to conference participants, has suggested the need for a postscript. Let me therefore briefly try to eliminate a likely ground of misunderstanding and simultaneously suggest a problem that urgently needs a great deal of investigation.

Everything said above was intended to apply strictly only to basic science, an enterprise whose practitioners have ordinarily been relatively free to choose their own problems. Characteristically, as I have indicated, these problems have been selected in areas where paradigms were clearly applicable but where exciting puzzles remained about how to apply them and how to make nature conform to the results of the application. Clearly the inventor and applied scientist are not generally free to choose puzzles of this sort. The problems among which they may choose are likely to be largely determined by social, economic, or military circumstances external to the sciences. Often the decision to seek a cure for a virulent disease, a new source of household illumination, or an alloy able to withstand the intense heat of rocket engines must be made with little reference to the state of the relevant science. It is, I think, by no means clear that the personality characteristics requisite for pre-eminence in this more immediately practical sort of work are altogether the same as those required for a great achievement in basic science. History indicates that only a few individuals, most of whom worked in readily demarcated areas, have achieved eminence in both.

I am by no means clear where this suggestion leads us. The troublesome distinctions between basic research, applied research, and invention need far more investigation. Nevertheless, it seems likely, for example, that the applied scientist, to whose problems no scientific paradigm need be fully relevant, may profit by a far broader and less rigid education than that to which the pure scientist has characteristically been exposed. Certainly there are many episodes in the history of technology in which lack of more than the most rudimentary scientific education has proved to be an immense help. This group scarcely needs to be reminded that Edison's electric light was produced in the face of unanimous scientific opinion that the arc light could not be "subdivided," and there are many other episodes of this sort.

This must not suggest, however, that mere differences in education will transform the applied scientist into a basic scientist or vice versa. One could at least argue that Edison's personality, ideal for the inventor and perhaps also for the "oddball" in applied science, barred him from fundamental achievements in the basic sciences. He himself expressed great scorn for scientists and thought of them as wooly-headed people to be hired when needed. But this did not prevent his occasionally arriving at the most sweeping and irresponsible scientific theories of his own. (The pattern recurs in the early history of electrical technology: both Tesla and Gramme advanced absurd cosmic schemes that they thought deserved to replace the current scientific knowledge of their day.) Episodes like this reinforce an impression that the personality requisites of the pure scientist and of the inventor may be quite different, perhaps with those of the applied scientist lying somewhere between.[5]

Is there a further conclusion to be drawn from all this? One speculative thought forces itself upon me. If I read the working papers correctly, they suggest that most of you are really in search of the *inventive* personality, a sort of person who does emphasize divergent thinking but whom the United States has already produced in abundance. In the process you may be ignoring certain of the essential requisites of the basic scientist, a rather different sort of person, to whose ranks America's contributions have as yet been notoriously sparse. Since most of you are, in fact, Americans, this correlation may not be entirely coincidental.

5. For the attitude of scientists toward the technical possibility of the incandescent light see Francis A. Jones, *Thomas Alva Edison* (New York, 1908), pp. 99–100, and Harold C. Passer, *The Electrical Manufacturers, 1875–1900* (Cambridge, Mass., 1953), pp. 82–83. For Edison's attitude toward scientists see Passer, ibid., pp. 180–81. For a sample of Edison's theorizing in realms otherwise subject to scientific treatments see Dagobert D. Runes, ed., *The Diary and Sundry Observations of Thomas Alva Edison* (New York, 1948), pp. 205–44, passim.

10 A Function for Thought Experiments

Reprinted by permission from *L'aventure de la science, Mélanges Alexandre Koyré* (Paris: Hermann, 1964), 2:307–34.

Thought experiments have more than once played a critically important role in the development of physical science. The historian, at least, must recognize them as an occasionally potent tool for increasing man's understanding of nature. Nevertheless, it is far from clear how they can ever have had very significant effects. Often, as in the case of Einstein's train struck by lightning at both ends, they deal with situations that have not been examined in the laboratory.[1] Sometimes, as in the case of the Bohr-Heisenberg microscope, they posit situations that could not be fully examined and that need not occur in nature at all.[2] That state of affairs gives rise to a series of

1. The famous train experiment first appears in Einstein's popularization of relativity theory, *Ueber die spezielle und allgemeine Relativitätstheorie (Gemeinverständlich)* (Braunschweig, 1916). In the fifth edition (1920), which I have consulted, the experiment is described on pp. 14–19. Notice that this thought experiment is only a simplified version of the one employed in Einstein's first paper on relativity, "Zur Elektrodynamik bewegter Körper," *Annalen der Physik* 17 (1905): 891–921. In that original thought experiment only one light signal is used, mirror reflection taking the place of the other.

2. W. Heisenberg, "Ueber den anschaulichen Inhalt der quantentheoretischen Kinematik und Mechanik," *Zeitschrift für Physik* 43 (1927): 172–98. N. Bohr, "The Quantum Postulate and the Recent Development of Atomic Theory," *Atti del Congresso Internazionale dei Fisici, 11–20 Settembre 1927*, vol. 2 (Bologna, 1928), pp. 565–88. The argument begins by

perplexities, three of which will be examined in this paper through the extended analysis of a single example. No single thought experiment can, of course, stand for all of those which have been historically significant. The category "thought experiment" is in any case too broad and too vague for epitome. Many thought experiments differ from the one examined here. But this particular example, being drawn from the work of Galileo, has an interest all its own, and that interest is increased by its obvious resemblance to certain of the thought experiments which proved effective in the twentieth-century reformulation of physics. Though I shall not argue the point, I suggest the example is typical of an important class.

The main problems generated by the study of thought experiments can be formulated as a series of questions. First, since the situation imagined in a thought experiment clearly may not be arbitrary, to what conditions of verisimilitude is it subject? In what sense and to what extent must the situation be one that nature could present or has in fact presented? That perplexity, in turn, points to a second. Granting that every successful thought experiment embodies in its design some prior information about the world, that information is not itself at issue in the experiment. On the contrary, if we have to do with a real thought experiment, the empirical data upon which it rests must have been both well-known and generally accepted before the experiment was even conceived. How, then, relying exclusively upon familiar data, can a thought experiment lead to new knowledge or to new understanding of nature? Finally, to put the third question most briefly of all, what sort of new knowledge or understanding can be so produced? What, if anything, can scientists hope to learn from thought experiments?

There is one rather easy set of answers to these questions, and I shall elaborate it, with illustrations drawn from both history and psychology, in the two sections immediately to follow. Those an-

treating the electron as a classical particle and discusses its trajectory both before and after its collision with the photon that is used to determine its position or velocity. The outcome is to show that these measurements cannot be carried through classically and that the initial description has therefore assumed more than quantum mechanics allows. That violation of quantum mechanical principles does not, however, diminish the import of the thought experiment.

swers—which are clearly important but, I think, not quite right—suggest that the new understanding produced by thought experiments is not an understanding of *nature* but rather of the scientist's *conceptual apparatus*. On this analysis, the function of the thought experiment is to assist in the elimination of prior confusion by forcing the scientist to recognize contradictions that had been inherent in his way of thinking from the start. Unlike the discovery of new knowledge, the elimination of existing confusion does not seem to demand additional empirical data. Nor need the imagined situation be one that actually exists in nature. On the contrary, the thought experiment whose sole aim is to eliminate confusion is subject to only one condition of verisimilitude. The imagined situation must be one to which the scientist can apply his concepts in the way he has normally employed them before.

Because they are immensely plausible and because they relate closely to philosophical tradition, these answers require detailed and respectful examination. In addition, a look at them will supply us with essential analytic tools. Nevertheless, they miss important features of the historical situation in which thought experiments function, and the last two sections of this paper will therefore seek answers of a somewhat different sort. The third section, in particular, will suggest that it is significantly misleading to describe as "self-contradictory" or "confused" the situation of the scientist prior to the performance of the relevant thought experiment. We come closer if we say that thought experiments assist scientists in arriving at laws and theories different from the ones they had held before. In that case, prior knowledge can have been "confused" and "contradictory" only in the rather special and quite unhistorical sense which would attribute confusion and contradiction to all the laws and theories that scientific progress has forced the profession to discard. Inevitably, however, that description suggests that the effects of thought experimentation, even though it presents no new data, are much closer to those of actual experimentation than has usually been supposed. The last section will attempt to suggest how this could be the case.

The historical context within which actual thought experiments assist in the reformulation or readjustment of existing concepts is inevitably extraordinarily complex. I therefore begin with a simpler, because nonhistorical, example, choosing for the purpose a conceptual transposition induced in the laboratory by the brilliant Swiss child psychologist Jean Piaget. Justification for this apparent departure from our topic will appear as we proceed. Piaget dealt with children, exposing them to an actual laboratory situation and then asking them questions about it. In slightly more mature subjects, however, the same effect might have been produced by questions alone in the absence of any physical exhibit. If those same questions had been self-generated, we would be confronted with the pure thought-experimental situation to be exhibited in the next section from the work of Galileo. Since, in addition, the particular transposition induced by Galileo's experiment is very nearly the same as the one produced by Piaget in the laboratory, we may learn a good deal by beginning with the more elementary case.

Piaget's laboratory situation presented children with two toy autos of different colors, one red and the other blue.[3] During each experimental exposure both cars were moved uniformly in a straight line. On some occasions both would cover the same distance but in different intervals of time. In other exposures the times required were the same, but one car would cover a greater distance. Finally, there were a few experiments during which neither the distances nor the times were quite the same. After each run Piaget asked his subjects which car had moved faster and how the child could tell.

In considering how the children responded to the questions, I restrict attention to an intermediate group, old enough to learn something from the experiments and young enough so that its responses were not yet those of an adult. On most occasions the children in this group would describe as "faster" the auto that reached the goal first or that had led during most of the motion. Furthermore, they would continue to apply the term in this way even when they recognized that the "slower" car had covered more ground than the "faster" during the same amount of time. Examine, for example, an exposure in which both cars departed from the same line but in which the red started later and then caught the blue at the goal. The following dialogue, with the child's contribution in italics, is then typical. "Did they leave at the same

3. J. Piaget, *Les notions de mouvement et de vitesse chez l'enfant* (Paris, 1946), particularly chap. 6 and 7. The experiments described below are in the latter chapter.

time?"—"*No, the blue left first.*"—"Did they arrive together?"—"*Yes.*"—"Was one of the two faster, or were they the same?"—"*The blue went more quickly.*"[4] Those responses manifest what for simplicity I shall call the "goal-reaching" criterion for the application of "faster."

If goal reaching were the only criterion employed by Piaget's children, there would be nothing that the experiments alone could teach them. We would conclude that their concept of "faster" was different from an adult's but that, since they employed it consistently, only the intervention of parental or pedagogic authority would be likely to induce change. Other experiments, however, reveal the existence of a second criterion, and even the experiment just described can be made to do so. Almost immediately after the exposure recorded above, the apparatus was readjusted so that the red car started very late and had to move especially rapidly to catch the blue at the goal. In this case, the dialogue with the same child went as follows. "Did one go more quickly than the other?"—"*The red.*"—"How did you find that out?"—"I WATCHED IT."[5] Apparently, when motions are sufficiently rapid, they can be perceived directly and as such by children. (Compare the way adults "see" the motion of the second hand on a clock with the way they observe the minute hand's change of position.) Sometimes children employ that direct perception of motion in identifying the faster car. For lack of a better word I shall call the corresponding criterion "perceptual blurriness."

It is the coexistence of these two criteria, goal-reaching and perceptual blurriness, that makes it possible for the children to learn in Piaget's laboratory. Even without the laboratory, nature would sooner or later teach the same lesson as it has to the older children in Piaget's group. Not very often (or the children could not have preserved the concept for so long) but occasionally nature will present a situation in which a body whose directly perceived speed is lower nevertheless reaches the goal first. In this case the two clues conflict; the child may be led to say that both bodies are

4. Ibid., p. 160, my translation.

5. Ibid., p. 161, my emphasis. In this passage I have rendered "plus fort" as more quickly; in the previous passage the French was "plus vite." The experiments themselves indicate, however, that in this context though perhaps not in all, the answers to the questions "plus fort?" and "plus vite?" are the same.

"faster" or both "slower" or that the same body is both "faster" and "slower." That experience of paradox is the one generated by Piaget in the laboratory with occasionally striking results. Exposed to a single paradoxical experiment, children will first say one body was "faster" and then immediately apply the same label to the other. Their answers become critically dependent upon minor differences in the experimental arrangement and in the wording of the questions. Finally, as they become aware of the apparently arbitrary oscillation of their responses, those children who are either cleverest or best prepared will discover or invent the adult conception of "faster." With a bit more practice some of them will thereafter employ it consistently. Those are the children who have learned from their exposure to Piaget's laboratory.

But, to return to the set of questions which motivate this inquiry, what shall we say they have learned and from what have they learned it? For the moment I restrict myself to a minimal and quite traditional series of answers which will provide the point of departure for a later section. Because it included two independent criteria for applying the conceptual relation "faster," the mental apparatus which Piaget's children brought to his laboratory contained an implicit contradiction. In the laboratory the impact of a novel situation, including both exposures and interrogation, forced the children to an awareness of that contradiction. As a result, some of them changed their concept of "faster," perhaps by bifurcating it. The original concept was split into something like the adult's notion of "faster" and a separate concept of "reaching-goal-first." The children's conceptual apparatus was then probably richer and certainly more adequate. They had learned to avoid a significant conceptual error and thus to think more clearly.

Those answers, in turn, supply another, for they point to the single condition that Piaget's experimental situations must satisfy in order to achieve a pedagogic goal. Clearly those situations may not be arbitrary. A psychologist might, for quite different reasons, ask a child whether a tree or a cabbage were faster; furthermore, he would probably get an answer;[6] but the child would not learn to think more clearly. If he is to do that, the situation presented to

6. Questions just like this one have been used by Charles E. Osgood to obtain what he calls the "semantic profile" of various words. See his recent book, *The Measurement of Meaning* (Urbana, Ill., 1957).

him must, at the very least, be relevant. It must, that is, exhibit the cues which he customarily employs when he makes judgments of relative speed. On the other hand, though the cues must be normal, the full situation need not be. Presented with an animated cartoon showing the paradoxical motions, the child would reach the same conclusions about his concepts, even though nature itself were governed by the law that faster bodies always reach the goal first. There is, then, no condition of physical verisimilitude. The experimenter may imagine any situation he pleases so long as it permits the application of normal cues.

Turn now to an historical, but otherwise similar, case of concept revision, this one again promoted by the close analysis of an imagined situation. Like the children in Piaget's laboratory, Aristotle's *Physics* and the tradition that descends from it give evidence of two disparate criteria used in discussions of speed. The general point is well known but must be isolated for emphasis here. On most occasions Aristotle regards motion or change (the two terms are usually interchangeable in his physics) as a change of state. Thus, "every change is *from* something to something—as the word itself *metabole* indicates."[7] Aristotle's reiteration of statements like this indicates that he normally views any noncelestial motion as a finite completed act to be grasped as a whole. Correspondingly, he measures the amount and speed of a motion in terms of the parameters which describe its end points, the *termini a quo* and *ad quem* of medieval physics.

The consequences for Aristotle's notion of speed are both immediate and obvious. As he puts it himself, "The quicker of two things traverses a greater magnitude in an equal time, an equal magnitude in less time, and a greater magnitude in less time."[8] Or elsewhere, "There is equal velocity where *the same* change is accomplished in an equal time."[9] In these passages, as in many other parts of Aristotle's writings, the implicit notion of speed is very like what we should call "average speed," a quantity we equate with the ratio of total distance to total elapsed time. Like the child's goal-reaching criterion, this way of judging speed differs from our own.

7. Aristotle, *Physica*, trans. R. P. Hardie and R. K. Gaye, in *The Works of Aristotle*, vol. 2 (Oxford, 1930), 224^{b}35–225^{a}1.

8. Ibid., 232^{a}25–27.

9. Ibid., 249^{b}4–5.

But, again, the difference can do no harm so long as the average-velocity criterion is itself consistently employed.

Yet, again like Piaget's children, Aristotle is not, from a modern viewpoint, everywhere entirely consistent. He, too, seems to possess a criterion like the child's perceptual blurriness for judging speed. In particular, he does occasionally discriminate between the speed of a body near the beginning and near the end of its motion. For example, in distinguishing natural or unforced motions, which terminate in rest, from violent motions, which require an external mover, he says, "But whereas the velocity of that which comes to a standstill seems always to increase, the velocity of that which is carried violently seems always to decrease."[10] Here, as in a few similar passages, there is no mention of endpoints, of distance covered, or of time elapsed. Instead, Aristotle is grasping directly, and perhaps perceptually, an aspect of motion which we should describe as "instantaneous velocity" and which has properties quite different from average velocity. Aristotle, however, makes no such distinction. In fact, as we shall see below, important substantive aspects of his physics are conditioned by his failure to discriminate. As a result, those who use the Aristotelian concept of speed can be confronted with paradoxes quite like those with which Piaget confronted his children.

We shall examine in a moment the thought experiment which Galileo employed to make these paradoxes apparent, but must first note that by Galileo's time the concept of speed was no longer quite as Aristotle had left it. The well-known analytic techniques developed during the fourteenth century to treat latitude of forms had enriched the conceptual apparatus available to students of motion. In particular, it had introduced a distinction between the total velocity of a motion, on the one hand, and the intensity of velocity at each point of the motion, on the other. The second of these concepts was very close to the modern notion of instantaneous velocity; the first, though only after some important revisions by Galileo, was a long step toward the contemporary concept of average velocity.[11] Part of the paradox implicit in Aristotle's concept

10. Ibid., 230^{b}23–25.

11. For a detailed discussion of the entire question of the latitude of forms, see Marshall Clagett, *The Science of Mechanics in the Middle Ages* (Madison, Wis., 1959), part 2.

of speed was eliminated during the Middle Ages, two centuries and a half before Galileo wrote.

That medieval transformation of concepts was, however, incomplete in one important respect. Latitude of forms could be used for the comparison of two different motions only if they both had the same "extension," covered the same distance, that is, or consumed the same time. Richard Swineshead's statement of the Mertonian rule should serve to make apparent this too often neglected limitation: If an increment of velocity were uniformly acquired, then "just as much space would be traversed by means of that increment . . . as by means of the mean degree [or intensity of velocity] of that increment, assuming something were to be moved with that mean degree [of velocity] throughout the whole time."[12] Here the elapsed time must be the same for both motions, or the technique for comparison breaks down. If the elapsed times could be different, then a uniform motion of low intensity but long duration could have a greater total velocity than a more intense motion (i.e., one with greater instantaneous velocity) that lasted only a short time. In general, the medieval analysts of motion avoided this potential difficulty by restricting their attention to comparisons which their techniques could handle. Galileo, however, required a more general technique and in developing it (or at least in teaching it to others) he employed a thought experiment that brought the full Aristotelian paradox to the fore. We have two assurances that the difficulty was still very real during the first third of the seventeenth century. Galileo's pedagogic acuteness is one—his text was directed to real problems. More impressive, perhaps, is the fact that Galileo did not always succeed in evading the difficulty himself.[13]

12. Ibid., p. 290.

13. The most significant lapse of this sort occurs in "The Second Day" of Galileo's *Dialogue concerning the Two Chief World Systems* (see the translation by Stillman Drake [Berkeley, 1953], pp. 199–201). Galileo there argues that no material body, however light, will be thrown from a rotating earth even if the earth rotates far faster than it does. That result (which Galileo's system requires—his lapse, though surely not deliberate, is not unmotivated) is gained by treating the terminal velocity of a uniformly accelerated motion as though it were proportional to the distance covered by the motion. The proportion is, of course, a straightforward consequence of the Mertonian rule, but it is applicable only to motions that require the same time. Drake's notes to this passage should also be examined since they supply a somewhat different interpretation.

The relevant experiment is produced almost at the start of "The First Day" in Galileo's *Dialogue concerning the Two Chief World Systems*.[14] Salviati, who speaks for Galileo, asks his two interlocutors to imagine two planes, CB vertical and CA inclined, erected the same vertical distance over a horizontal plane, AB. To aid the imagination Salviati includes a sketch like the one below. Along

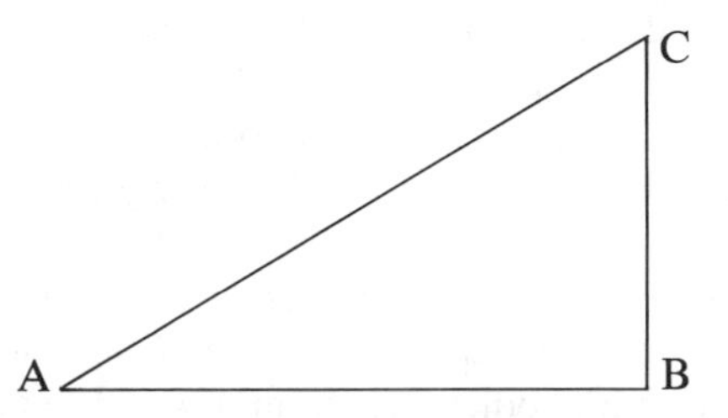

these two planes, two bodies are to be imagined sliding or rolling without friction from a common starting point at C. Finally, Salviati asks his interlocutors to concede that, when the sliding bodies reach A and B, respectively, they will have acquired the same impetus or speed, the speed necessary, that is, to carry them again to the vertical height from which they started.[15] That request also is granted, and Salviati proceeds to ask the participants in the dialogue which of the two bodies moves faster. His object is to make them realize that, using the concept of speed then current, they can be forced to admit that motion along the perpendicular is simultaneously faster than, equal in speed to, and slower than the motion along the incline. His further object is, by the impact of this paradox, to make his interlocutors and readers realize that speed ought not be attributed to the whole of a motion, but rather to its parts. In short, the thought experiment is, as Galileo himself points out, a propaedeutic to the full discussion of uniform and accelerated motion that occurs in "The Third Discourse" of his *Two New Sciences*. The argument itself I shall considerably condense and systematize since the detailed give and take of the dialogue

14. Ibid., pp. 22–27.

15. Galileo makes somewhat less use of this concession than I shall below. Strictly speaking, his argument does not depend upon it if the plane CA can be extended beyond A and if the body rolling along the extended plane continues to gain speed. For simplicity I shall restrict my systematized recapitulation to the unextended plane, following the lead supplied by Galileo in the first part of his text.

need not concern us. When first asked which body is faster, the interlocutors give the response we are all drawn to though the physicists among us should know better. The motion along the perpendicular, they say, is obviously the faster.[16] Here, two of the three criteria we have already encountered combine. While both bodies are in motion, the one moving along the perpendicular is the "more blurred." In addition, the perpendicular motion is the one that reaches its goal first.

This obvious and immensely appealing answer immediately, however, raises difficulties which are first recognized by the cleverer of the interlocutors, Sagredo. He points out (or very nearly—I am making this part of the argument slightly more binding than it is in the original) that the answer is incompatible with the initial concession. Since both bodies start from rest and since both acquire the same terminal velocity, they must have the same mean speed. How then can one be faster than the other? At this point Salviati reenters the discussion, reminding his listeners that the faster of two motions is usually defined as the one that covers the same distance in a lesser time. Part of the difficulty, he suggests, arises from the attempt to compare two motions that cover different distances. Instead, he urges, the participants in the dialogue should compare the times required by the two bodies in moving over a common standard distance. As a standard he selects the length of the vertical plane CB.

This, however, only makes the problem worse. CA is longer than CB, and the answer to the question, which body moves faster, turns out to depend critically upon where, along the incline CA, the standard length CB is measured. If it is measured down from the top of the incline, then the body moving on the perpendicular will complete its motion in less time than the body on the incline requires to move through a distance equal to CB. Motion along the perpendicular is therefore faster. On the other hand, if the standard distance is measured up from the bottom of the incline, the body moving on the perpendicular will need more time to complete its motion than the body on the incline will need to move through the same standard distance. Motion along the perpendicular is there-

16. Anyone who doubts that this is a very tempting and natural answer should try Galileo's question, as I have, on graduate students of physics. Unless previously told what will be involved, many of them give the same answer as Salviati's interlocutors.

fore slower. Finally, Salviati argues, if the distance CB is laid out along some appropriate internal part of the incline, then the times required by the two bodies to traverse the two standard segments will be the same. Motion on the perpendicular has the same speed as that on the incline. At this point the dialogue has provided three answers, each incompatible with both the others, to a single question about a single situation.

The result, of course, is paradox, and that is the way, or one of them, in which Galileo prepared his contemporaries for a change in the concepts employed when discussing, analyzing, or experimenting upon motion. Though the new concepts were not fully developed for the public until the appearance of the *Two New Sciences*, the *Dialogue* already shows where the argument is headed. "Faster" and "speed" must not be used in the traditional way. One may say that at a particular instant one body has a faster instantaneous speed than another body has at that same time or at another specified instant. Or one may say that a particular body traverses a particular distance more quickly than another traverses the same or some other distance. But the two sorts of statements do not describe the same characteristics of the motion. "Faster" means something different when applied, on the one hand, to the comparison of instantaneous rates of motion at particular instants, and, on the other, to the comparison of the times required for the completion of the whole of two specified motions. A body may be "faster" in one sense and not in the other.

That conceptual reform is what Galileo's thought experiment helped to teach, and we can therefore ask our old questions about it. Clearly, the minimal answers are the same ones supplied when considering the outcome of Piaget's experiments. The concepts that Aristotle applied to the study of motion were, in some part, self-contradictory, and the contradiction was not entirely eliminated during the Middle Ages. Galileo's thought experiment brought the difficulty to the fore by confronting readers with the paradox implicit in their mode of thought. As a result, it helped them to modify their conceptual apparatus.

If that much is right, then we can also see the criterion of verisimilitude to which the thought experiment had necessarily to conform. It makes no difference to Galileo's argument whether or not bodies actually execute uniformly accelerated motion when moving down inclined and vertical planes. It does not even matter whether, when the heights of these planes are the same, the two bodies actually reach equal instantaneous velocities at the bottom. Galileo does not bother to argue either of the points. For his purpose in this part of the *Dialogue*, it is quite sufficient that we may suppose these things to be the case. On the other hand, it does not follow that Galileo's choice of the experimental situation could be arbitrary. He could not, for example, usefully have suggested that we consider a situation in which the body vanished at the start of its motion from C and then reappeared shortly afterwards at A without having traversed the intervening distance. That experiment would illustrate limitations in the applicability of "faster," but, at least until the recognition of quantum jumps, those limitations would not have been informative. From them, neither we nor Galileo's readers could learn anything about the concepts traditionally employed. Those concepts were never intended to apply in such a case. In short, if this sort of thought experiment is to be effective, it must allow those who perform or study it to employ concepts in the same ways they have been employed before. Only if that condition is met can the thought experiment confront its audience with unanticipated consequences of their normal conceptual operations.

To this point, essential parts of my argument have been conditioned by what I take to be a philosophical position traditional in the analysis of scientific thought since at least the seventeenth century. If a thought experiment is to be effective, it must, as we have already seen, present a normal situation, that is, a situation which the man who analyzes the experiment feels well equipped by prior experience to handle. Nothing about the imagined situation may be entirely unfamiliar or strange. Therefore, if the experiment depends, as it must, upon prior experience of nature, that experience must have been generally familiar before the experiment was undertaken. This aspect of the thought-experimental situation has seemed to dictate one of the conclusions that I have so far consistently drawn. Because it embodies no new information about the world, a thought experiment can teach nothing that was not known before. Or, rather, it can teach nothing about the world. Instead, it teaches the scientist about his mental apparatus. Its function is limited to the correction of previous conceptual mistakes.

I suspect, however, that some historians of science may be uneasy about this conclusion, and I suggest that others should be. Somehow, it is too reminiscent of the familiar position which regards the Ptolemaic theory, the phlogiston theory, or the caloric theory as mere errors, confusions, or dogmatisms which a more liberal or intelligent science would have avoided from the start. In the climate of contemporary historiography, evaluations like these have come to seem less and less plausible, and that same air of implausibility infects the conclusion I have so far drawn in this paper. Aristotle, if no experimental physicist, was a brilliant logician. Would he, in a matter so fundamental to his physics, have committed an error so elementary as the one we have attributed to him? Or if he had, would his successors, for almost two millennia, have continued to make the same elementary mistake? Can a logical confusion be all that is involved, and can the function of thought experiments be so trivial as this entire point of view implies? I believe that the answer to all of these questions is no, and that the root of the difficulty is our assumption that, because they rely exclusively upon well-known data, thought experiments can teach nothing about the world. Though the contemporary epistemological vocabulary supplies no truly useful locutions, I want now to argue that from thought experiments most people learn about their concepts and the world together. In learning about the concept of speed Galileo's readers also learn something about how bodies move. What happens to them is very similar to what happens to a man, like Lavoisier, who must assimilate the result of a new unexpected experimental discovery.[17]

In approaching this series of central points, I first ask what can have been meant when we described the child's concept of faster and the Aristotelian concept of speed as "self-contradictory" or "confused." "Self-contradictory," at least, suggests that these concepts are like the logician's famous example, square-circle, but that cannot be quite right. Square-circle is self-contradictory in the sense that it could not be exemplified in any possible world. One cannot even imagine an object which would display the requisite qualities. Neither the child's concept nor Aristotle's, however, are

17. That remark presumes an analysis of the manner in which new discoveries emerge, for which see my paper, "The Historical Structure of Scientific Discovery," *Science* 136 (1962): 760–64 (pp. 165–77 above).

contradictory in that sense. The child's concept of faster is repeatedly exemplified in our own world; contradiction arises only when the child is confronted with that relatively rare sort of motion in which the perceptually *more* blurred object *lags* in reaching the goal. Similarly, Aristotle's concept of speed, with its two simultaneous criteria, can be applied without difficulty to most of the motions we see about us. Problems arise only for that class of motions, again rather rare, in which the criterion of instantaneous velocity and the criterion of average velocity lead to contradictory responses in qualitative applications. In both these cases the concepts are contradictory only in the sense that the individual who employs them *runs the risk* of self-contradiction. He may, that is, find himself in a situation where he can be forced to give incompatible answers to one and the same question.

That, of course, is not what is usually meant when the term "self-contradictory" is applied to a concept. It may well, however, be what we had in mind when we described the concepts examined above as "confused" or "inadequate to clear thought." Certainly those terms fit the situation better. They do, however, imply a standard for clarity and adequacy that we may have no right to apply. Ought we demand of our concepts, as we do not and could not of our laws and theories, that they be applicable to any and every situation that might conceivably arise in any possible world? Is it not sufficient to demand of a concept, as we do of a law or theory, that it be unequivocally applicable in every situation which we expect ever to encounter?

To see the relevance of those questions, imagine a world in which all motions occur at uniform speed. (That condition is more stringent than necessary, but it will make the argument clearer. The requisite weaker condition is that no body which is "slower" by either criterion shall ever overtake a "faster" body. I shall call motions which satisfy this weaker condition "quasiuniform.") In a world of that sort the Aristotelian concept of speed could never be jeopardized by an actual physical situation, for the instantaneous and average speed of any motion would always be the same.[18]

18. One can also imagine a world in which the two criteria employed by Piaget's children would never lead to contradiction, but it is more complex, and I shall therefore make no use of it in the argument that follows. Let me, however, hazard one testable guess about the nature of motion in that world. Unless copying their elders, children who view motion in the way

What, then, would we say if we found a scientist in this imaginary world consistently employing the Aristotelian concept of speed? Not, I think, that he was confused. Nothing could go wrong with his science or logic because of his application of the concept. Instead, given our own broader experience and our correspondingly richer conceptual apparatus, we would likely say that, consciously or unconsciously, he had embodied in his concept of speed his expectation that only uniform motions would occur in his world. We would, that is, conclude that his concept functioned in part as a law of nature, a law that was regularly satisfied in his world but that would only occasionally be satisfied in our own.

In Aristotle's case, of course, we cannot say quite this much. He did know, and occasionally admits, that falling bodies, for example, increase their speeds as they move. On the other hand, there is ample evidence that Aristotle kept this information at the very periphery of his scientific consciousness. Whenever he could, which was frequently, he regarded motions as uniform or as possessing the properties of uniform motion, and the results were consequential for much of his physics. In the preceding section, for example, we examined a passage from the *Physics*, which can pass for a definition of "faster motion": "The quicker of two things traverses a greater magnitude in an equal time, an equal magnitude in less time and a greater magnitude in less time." Compare this with the passage that follows it immediately: "Suppose that A is quicker than B. Now since of two things that which changes sooner is quicker, in the time FG, in which A has changed from C to D, B will not yet have arrived at D but will be short of it."[19] This statement is no longer quite a definition. Instead it is about the physical behavior of "quicker" bodies, and, as such, it holds only for bodies that are in uniform or quasi-uniform motion.[20] The whole burden of Galileo's thought experiment is to show that this statement and others like it—statements that seem to follow inevitably from the

described above should be relatively insensitive to the importance of a handicap to the winning of a race. Instead, everything should seem to depend upon the violence with which arms and legs are moved.

19. Aristotle, *Works* 2:232a28–31.

20. Actually, of course, the first passage cannot be a definition. Any one of the three conditions there stated could have that function, but taking the three to be equivalent, as Aristotle does, has the same physical implications which I here illustrate from the second passage.

only definition the traditional concept "faster" will support—do not hold in the world as we know it and that the concept therefore requires modification. Aristotle nevertheless proceeds to build his view of motion as quasi-uniform deeply into the fabric of his system. For example, in the paragraph just after that from which the preceding statements are taken, he employs those statements to show that space must be continuous if time is. His argument depends upon the assumption, implicit above, that, if a body B lags behind another body A at the end of a motion, it will have lagged at all intermediate points. In that case, B can be used to divide the space and A to divide the time. If one is continuous, the other must be too.[21] Unfortunately, however, the assumption need not hold if, for example, the slower motion is decelerating and the faster accelerating, yet Aristotle sees no need to bar motions of that sort. Here again his argument depends upon his attributing to all movements the qualitative properties of uniform change.

The same view of motion underlies the arguments by which Aristotle develops his so-called quantitative laws of motion.[22] For illustration, consider only the dependence of distance covered on the size of the body and upon elapsed time: "If, then, A the movent have moved B a distance C in a time D, then in the same time the same force A will move ½ B twice the distance C, and in ½ D it will move ½ B the whole distance C: for thus the rules of proportion will be observed."[23] With both force and medium given, that is, the distance covered varies directly with time and inversely with body size.

21. Aristotle, *Works* 2:232^{b}21–233^{a}13.

22. These laws are always described as "quantitative," and I follow that usage. But it is hard to believe they were meant to be quantitative in the sense of that term current in the study of motion since Galileo. Both in antiquity and the Middle Ages men who regularly thought measurement relevant to astronomy and who occasionally employed it in optics discussed these laws of motion without even a veiled reference to any sort of quantitative observation. Furthermore, the laws are never applied to nature except in arguments which rely on *reductio ad absurdum*. To me their intent seems qualitative—they are a statement, using the vocabulary of proportions, of several correctly observed qualitative regularities. This view may appear more plausible if we remember that after Eudoxus even geometric proportion was regularly interpreted as nonnumerical.

23. Aristotle, *Works* 2:249^{b}30–250^{a}4.

To modern ears this is inevitably a strange law, though perhaps not so strange as it has usually seemed.[24] But given the Aristotelian concept of speed—a concept that raises no problems in most of its applications—it is readily seen to be the only simple law available. If motion is such that average and instantaneous speed are identical, then, *ceteris paribus*, distance covered must be proportional to time. If, in addition, we assume with Aristotle (and Newton) that "two forces each of which separately moves one of two weights a given distance in a given time . . . will move the combined weights an equal distance in an equal time," then speed must be some function of the ratio of force to body size.[25] Aristotle's law follows directly by assuming the function to be the simplest one available, the ratio itself. Perhaps this does not seem a legitimate way to arrive at laws of motion, but Galileo's procedures were very often identical.[26] In this particular respect what principally differentiated Galileo from Aristotle is that the former started with a different conception of speed. Since he did not see all motions as quasi-uniform, speed was not the only measure of motion that could change with applied force, body size, and so on. Galileo could consider variations of acceleration as well.

These examples could be considerably multiplied, but my point may already be clear. Aristotle's concept of speed, in which something like the separate modern concepts of average and instantaneous speed were merged, was an integral part of his entire theory of motion and had implications for the whole of his physics. That role it could play because it was not simply a definition, confused or otherwise. Instead, it had physical implications and acted

24. For cogent criticism of those who find the law merely silly, see Stephen Toulmin, "Criticism in the History of Science: Newton on Absolute Space, Time and Motion, I," *Philosophical Review* 68 (1959): 1–29, particulary footnote 1.

25. Aristotle, *Works* 2: 250^{a}25–28.

26. For example, "When, therefore, I observe a stone initially at rest falling from an elevated position and continually acquiring new increments of speed, why should I not believe that such increases take place in a manner which is exceedingly simple and rather obvious to everybody? If now we examine the matter carefully we find no addition or increment more simple than that which repeats itself always in the same manner." Cf. Galileo Galilei, *Dialogues Concerning Two New Sciences*, trans. H. Crew and A. de Salvio (Evanston and Chicago, 1946), pp. 154–55. Galileo, however, did proceed to an experimental check.

in part as a law of nature. Those implications could never have been challenged by observation or by logic in a world where all motions were uniform or quasi-uniform, and Aristotle acted as though he lived in a world of that sort. Actually, of course, his world was different, but his concept nevertheless functioned so successfully that potential conflicts with observation went entirely unnoticed. And while they did so—until, that is, the potential difficulties in applying the concept began to become actual—we may not properly speak of the Aristotelian concept of speed as confused. We may, of course, say that it was "wrong" or "false" in the same sense that we apply those terms to outmoded laws and theories. In addition, we may say that, because the concept was false, the men who employed it were *liable to become confused*, as Salviati's interlocutors did. But we cannot, I think, find any intrinsic defect in the concept by itself. Its defects lay not in its logical consistency but in its failure to fit the full fine structure of the world to which it was expected to apply. That is why learning to recognize its defects was necessarily learning about the world as well as about the concept.

If the legislative content of individual concepts seems an unfamiliar notion, that is probably because of the context within which I have approached it here. To linguists the point has long been familiar, if controversial, through the writings of B. L. Whorf.[27] Braithwaite, following Ramsey, has developed a similar thesis by using logical models to demonstrate the inextricable mixture of law and definition which must characterize the function of even relatively elementary scientific concepts.[28] Still more to the point are the several recent logical discussions of the use of "reduction sentences" in forming scientific concepts. These are sentences which specify (in a logical form that need not here concern us) the observational or test conditions under which a given concept may be applied. In practice, they closely parallel the contexts in which most scientific concepts are actually acquired, and that makes their two most salient characteristics particularly significant. First, several reduction sentences—sometimes a great many—are

27. B. L. Whorf, *Language, Thought, and Reality: Selected Writings*, ed. John B. Carroll (Cambridge, Mass., 1956).

28. R. B. Braithwaite, *Scientific Explanation* (Cambridge, 1953), pp. 50–87. And see also W. V. O. Quine, "Two Dogmas of Empiricism," in *From a Logical Point of View* (Cambridge, Mass., 1953), pp. 20–46.

required to supply a given concept with the range of application required by its use in scientific theory. Second, as soon as more than one reduction sentence is used to introduce a single concept, those sentences turn out to imply "certain statements which have the character of empirical laws. . . . Sets of reduction sentences combine in a peculiar way the functions of concept and of theory formation."[29] That quotation, with the sentence that precedes it, very nearly describes the situation we have just been examining.

We need not, however, make the full transition to logic and philosophy of science in order to recognize the legislative function of scientific concepts. In another guise it is already familiar to every historian who has studied closely the evolution of concepts like element, species, mass, force, space, caloric, or energy.[30] These and many other scientific concepts are invariably encountered within a matrix of law, theory, and expectation from which they cannot be altogether extricated for the sake of definition. To discover what they mean the historian must examine both what is said about them and also the way in which they are used. In the process he regularly discovers a number of different criteria which govern their use and whose coexistence can be understood only by reference to many of the other scientific (and sometimes extrascientific) beliefs which guide the men who use them. It follows that those concepts

29. C. G. Hempel, *Fundamentals of Concept Formation in Empirical Science*, vol. 2, no. 7, in the *International Encyclopedia of Unified Science* (Chicago, 1952). The fundamental discussion of reduction sentences is in Rudolph Carnap, "Testability and Meaning," *Philosophy of Science* 3 (1936): 420–71, and 4 (1937): 2–40.

30. The cases of caloric and of mass are particularly instructive, the first because it parallels the case discussed above, the second because it reverses the line of development. It has often been pointed out that Sadi Carnot derived good experimental results from the caloric theory because his concept of heat combined characteristics that later had to be distributed between heat and entropy. (See my exchange with V. K. La Mer, *American Journal of Physics* 22 [1954]: 20–27; 23 [1955]: 91–102 and 387–89. The last of these items formulates the point in the way required here.) Mass, on the other hand, displays an opposite line of development. In Newtonian theory inertial mass and gravitational mass are separate concepts, measured by distinct means. An experimentally tested law of nature is needed to say that the two sorts of measurements will always, within instrumental limits, give the same results. According to general relativity, however, no separate experimental law is required. The two measurements *must* yield the same result because they measure the same quantity.

were not intended for application to any possible world, but only to the world as the scientist saw it. Their use is one index of his commitment to a larger body of law and theory. Conversely, the legislative content of that larger body of belief is in part carried by the concepts themselves. That is why, though many of them have histories coextensive with the histories of the sciences in which they function, their meaning and their criteria for use have so often and so drastically changed in the course of scientific development.

Finally, returning to the concept of speed, notice that Galileo's reformulation did not make it once and for all logically pure. No more than its Aristotelian predecessor was it free from implications about the way nature must behave. As a result, again like Aristotle's concept of speed, it could be called in question by accumulated experience, and that is what occurred at the end of the last century and the beginning of this one. The episode is too well known to require extended discussion. When applied to accelerated motions, the Galilean concept of speed implies the existence of a set of physically unaccelerated spatial reference systems. That is the lesson of Newton's bucket experiment, a lesson which none of the relativists of the seventeenth and eighteenth centuries were able to explain away. In addition, when applied to linear motions, the revised concept of speed used in this paper implies the validity of the so-called Galilean transformation equations, and these specify physical properties, for example the additivity of the velocity of matter or of light. Without benefit of any superstructure of laws and theories like Newton's, they provided immensely significant information about what the world is like.

Or, rather, they used to do so. One of the first great triumphs of twentieth-century physics was the recognition that that information could be questioned and the consequent recasting of the concepts of speed, space, and time. Furthermore, in that reconceptualization thought experiments again played a vital role. The historical process we examined above through the work of Galileo has since been repeated with respect to the same constellation of concepts. Perfectly possibly it may occur again, for it is one of the basic processes through which the sciences advance.

My argument is now very nearly complete. To discover the element still missing, let me briefly recapitulate the main points discussed so far. I began by suggesting that an important class of thought experiments functions by confronting the scientist with a contradiction or conflict implicit in his mode of thought. Recognizing the contradiction then appeared an essential propaedeutic to its elimination. As a result of the thought experiment, clear concepts were developed to replace the confused ones that had been in use before. Closer examination, however, disclosed an essential difficulty in that analysis. The concepts "corrected" in the aftermath of thought experiments displayed no *intrinsic* confusion. If their use raised problems for the scientist, those problems were like the ones to which the use of any experimentally based law or theory would expose him. They arose, that is, not from his mental equipment alone but from difficulties discovered in the attempt to fit that equipment to previously unassimilated experience. Nature rather than logic alone was responsible for the apparent confusion. This situation led me to suggest that from the sort of thought experiment here examined the scientist learns about the world as well as about his concepts. Historically their role is very close to the double one played by actual laboratory experiments and observations. First, thought experiments can disclose nature's failure to conform to a previously held set of expectations. In addition, they can suggest particular ways in which both expectation and theory must henceforth be revised.

But how—to raise the remaining problem—can they do so? Laboratory experiments play these roles because they supply the scientist with new and unexpected information. Thought experiments, on the contrary, must rest entirely on information already at hand. If the two can have such similar roles, that must be because, on occasions, thought experiments give the scientist access to information which is simultaneously at hand and yet somehow inaccessible to him. Let me now try to indicate, though necessarily briefly and incompletely, how this could be the case.

I have elsewhere pointed out that the development of a mature scientific specialty is normally determined largely by the closely integrated body of concepts, laws, theories, and instrumental techniques which the individual practitioner acquires from professional education.[31] That time-tested fabric of belief and expectation tells him what the world is like and simultaneously defines the prob-

31. For incomplete discussions of this and the following points see my papers, "The Function of Measurement in Modern Physical Science," *Isis* 52 (1961): 161–93 (pp. 178–224 above), and "The Function of Dogma in Scientific Research," in *Scientific Change*, A. C. Crombie, ed. (New York,

lems which still demand professional attention. Those problems are the ones which, when solved, will extend the precision and scope of the fit between existing belief, on the one hand, and observation of nature, on the other. When problems are selected in this way, past success ordinarily ensures future success as well. One reason why scientific research seems to advance steadily from solved problem to solved problem is that professionals restrict their attention to problems defined by the conceptual and instrumental techniques already at hand.

That mode of problem selection, however, though it makes short-term success particularly likely, also guarantees long-run failures that prove even more consequential to scientific advance. Even the data that this restricted pattern of research presents to the scientist never entirely or precisely fit his theory-induced expectations. Some of those failures to fit provide his current research problems; but others are pushed to the periphery of consciousness and some are suppressed entirely. Usually that inability to recognize and confront anomaly is justified in the event. More often than not minor instrumental adjustments or small articulations of existing theory ultimately reduce the apparent anomaly to law. Pausing over anomalies when they are first confronted is to invite continual distraction.[32] But not all anomalies do respond to minor adjustments of the existing conceptual and instrumental fabric. Among those that do not are some which, either because they are particularly striking or because they are educed repeatedly in many different laboratories, cannot be indefinitely ignored. Though they remain unassimilated, they impinge with gradually increasing force upon the consciousness of the scientific community.

As this process continues, the pattern of the community's research gradually changes. At first, reports of unassimilated observations appear more and more frequently in the pages of laboratory notebooks or as asides in published reports. Then more and more research is turned to the anomaly itself. Those who are attempting to make it lawlike will increasingly quarrel over the meaning of the concepts and theories which they have long held in

1963), pp. 347–69. The whole subject is treated more fully and with many additional examples in my essay *The Structure of Scientific Revolutions* (Chicago, 1962).

32. Much evidence on this point is to be found in Michael Polanyi, *Personal Knowledge* (Chicago, 1958), particularly chap. 9.

common without awareness of ambiguity. A few of them will begin critically to analyze the fabric of belief that has brought the community to its present impasse. On occasions even philosophy will become a legitimate scientific tool, which it ordinarily is not. Some or all of these symptoms of community crisis are, I think, the invariable prelude to the fundamental reconceptualization that the removal of an obdurate anomaly almost always demands. Typically, that crisis ends only when some particularly imaginative individual, or a group of them, weaves a new fabric of laws, theories, and concepts, one which can assimilate the previously incongruous experience and most or all of the previous assimilated experience as well.

This process of reconceptualization I have elsewhere labeled scientific revolution. Such revolutions need not be nearly so total as the preceding sketch implies, but they all share with it one essential characteristic. The data requisite for revolution have existed before at the fringe of scientific consciousness; the emergence of crisis brings them to the center of attention; and the revolutionary reconceptualization permits them to be seen in a new way.[33] What was vaguely known in spite of the community's mental equipment before the revolution is afterwards precisely known because of its mental equipment.

That conclusion, or constellation of conclusions, is, of course, both too grandiose and too obscure for general documentation here. I suggest, however, that in one limited application a number of its essential elements have been documented already. A crisis induced by the failure of expectation and followed by revolution is at the heart of the thought-experimental situations we have been examining. Conversely, thought experiment is one of the essential analytic tools which are deployed during crisis and which then help to promote basic conceptual reform. The outcome of thought experiments can be the same as that of scientific revolutions: they can enable the scientist to use as an integral part of his knowledge what that knowledge had previously made inaccessible to him. That is the sense in which they change his knowledge of the world. And

33. The phrase "permits them to be seen in a new way" must here remain a metaphor though I intend it quite literally. N. R. Hanson (*Patterns of Discovery* [Cambridge, 1958], pp. 4–30) has already argued that what scientists see depends upon their prior beliefs and training, and much evidence on this point will be found in the last reference cited in note 31, above.

it is because they can have that effect that they cluster so notably in the works of men like Aristotle, Galileo, Descartes, Einstein, and Bohr, the great weavers of new conceptual fabrics.

Return now briefly and for the last time to our own experiments, both Piaget's and Galileo's. What troubled us about them was, I think, that we found implicit in the preexperimental mentality laws of nature which conflicted with information we felt sure our subjects already possessed. Indeed, it was only because they possessed the information that they could learn from the experimental situation at all. Under those circumstances we were puzzled by their inability to see the conflict; we were unsure what they had still to learn; and we were therefore impelled to regard them as confused. That way of describing the situation was not, I think, altogether wrong, but it was misleading. Though my own concluding substitute must remain partly metaphor, I urge the following description instead.

For some time before we encountered them, our subjects had, in their transactions with nature, successfully employed a conceptual fabric different from the one we use ourselves. That fabric was time-tested; it had not yet confronted them with difficulties. Nevertheless, as of the time we encountered them, they had at last acquired a variety of experience which could not be assimilated by their traditional mode of dealing with the world. At this point they had at hand all the experience requisite to a fundamental recasting of their concepts, but there was something about that experience which they had not yet seen. Because they had not, they were subject to confusion and were perhaps already uneasy.[34] Full confusion, however, came only in the thought-experimental situation, and then it came as a prelude to its cure. By transforming felt anomaly to concrete contradiction, the thought experiment informed our subjects what was wrong. That first clear view of the misfit between experience and implicit expectation provided the clues necessary to set the situation right.

What characteristics must a thought experiment possess if it is to be capable of these effects? One part of my previous answer can still stand. If it is to disclose a misfit between traditional conceptual

34. Piaget's children were, of course, not uneasy (at least not for relevant reasons) until his experiments were exhibited to them. In the historical situation, however, thought experiments are generally called forth by a growing awareness that something somewhere is the matter.

apparatus and nature, the imagined situation must allow the scientist to employ his usual concepts in the way he has employed them before. It must not, that is, strain normal usage. On the other hand, the part of my previous answer which dealt with physical verisimilitude now needs revision. It presumed that thought experiments were directed to purely logical contradictions or confusions; any situation capable of displaying such contradictions would therefore suffice; there was then no condition of physical verisimilitude at all. If, however, we suppose that nature and conceptual apparatus are jointly implicated in the contradiction posed by thought experiments, a stronger condition is required. Though the imagined situation need not be even potentially realizable in nature, the conflict deduced from it must be one that nature itself could present. Indeed, even that condition is not quite strong enough. The conflict that confronts the scientist in the experimental situation must be one that, however unclearly seen, has confronted him before. Unless he has already had that much experience, he is not yet prepared to learn from thought experiments alone.

11 Logic of Discovery or Psychology of Research?

Reprinted by permission from *Criticism and the Growth of Knowledge*, ed. I. Lakatos and A. Musgrave (Cambridge: Cambridge University Press, 1970), pp. 1–22.

My object in these pages is to juxtapose the view of scientific development outlined in my book, *The Structure of Scientific Revolutions*, with the better known views of our chairman, Sir Karl Popper.[1] Ordinarily I should decline such an undertaking, for I am not so sanguine as Sir Karl about the utility of confrontations. Besides, I have admired his work for too long to turn critic easily at this date. Nevertheless, I am persuaded that for this occasion the attempt must be made. Even before my book was published two and a half years ago, I had begun to discover special and often puzzling characteristics of the relation between my views and his. That relation and the divergent reactions I have encountered to it

This paper was initially prepared at the invitation of P. A. Schilpp for his volume *The Philosophy of Karl R. Popper* (La Salle, Ill.: Open Court Publishing Co., 1974), pp. 798–819. I am most grateful to both Professor Schilpp and the publishers for permission to print it as part of the proceedings of this symposium before its appearance in the volume for which it was first solicited.

1. For purposes of the following discussion I have reviewed Sir Karl Popper's *Logic of Scientific Discovery* (1959), *Conjectures and Refutations* (1963), and *The Poverty of Historicism* (1957). I have also occasionally referred to his original *Logik der Forschung* (1935) and his *Open Society and Its Enemies* (1945). My own *Structure of Scientific Revolutions* (1962) provides a more extended account of many of the issues discussed below.

suggest that a disciplined comparison of the two may produce peculiar enlightenment. Let me say why I think this could occur.

On almost all the occasions when we turn explicitly to the same problems, Sir Karl's view of science and my own are very nearly identical.[2] We are both concerned with the dynamic process by which scientific knowledge is acquired rather than with the logical structure of the products of scientific research. Given that concern, both of us emphasize, as legitimate data, the facts and also the spirit of actual scientific life, and both of us turn often to history to find them. From this pool of shared data, we draw many of the same conclusions. Both of us reject the view that science progresses by accretion; both emphasize instead the revolutionary process by which an older theory is rejected and replaced by an incompatible new one;[3] and both deeply underscore the role played in this process by the occasional failure of the older theory to meet challenges posed by logic, experiment, or observation. Finally, Sir Karl and I are united in opposition to a number of the most characteristic theses of classical positivism. We both emphasize, for example, the intimate and inevitable entanglement of scientific observation with scientific theory; we are correspondingly skeptical of efforts to produce any neutral observation language; and we both insist that scientists may properly aim to invent theories that *explain* observed phenomena and that do so in terms of *real* objects, whatever the latter phrase may mean.

That list, though it by no means exhausts the issues about which Sir Karl and I agree,[4] is already extensive enough to place us in the

2. More than coincidence is presumably responsible for this extensive overlap. Though I had read none of Sir Karl's work before the appearance in 1959 of the English translation of his *Logik der Forschung* (by which time my book was in draft), I had repeatedly heard a number of his main ideas discussed. In particular, I had heard him discuss some of them when he was William James Lecturer at Harvard in the spring of 1950. These circumstances do not permit me to detail an intellectual debt to Sir Karl, but there must be one.

3. Elsewhere I use the term "paradigm" rather than "theory" to denote what is rejected and replaced during scientific revolutions. Some reasons for the change of term will emerge below.

4. Underlining one additional area of agreement about which there has been much misunderstanding may further highlight what I take to be the real differences between Sir Karl's views and mine. We both insist that

same minority among contemporary philosophers of science. Presumably that is why Sir Karl's followers have with some regularity provided my most sympathetic philosophical audience, one for which I continue to be grateful. But my gratitude is not unmixed. The same agreement that evokes the sympathy of this group too often misdirects its interest. Apparently Sir Karl's followers can often read much of my book as chapters from a late (and, for some, a drastic) revision of his classic, *The Logic of Scientific Discovery.* One of them asks whether the view of science outlined in my *Scientific Revolutions* has not long been common knowledge. A second, more charitably, isolates my originality as the demonstration that discoveries of fact have a life cycle very like that displayed by innovations of theory. Still others express general pleasure in the book but will discuss only the two comparatively secondary issues about which my disagreement with Sir Karl is most nearly explicit: my emphasis on the importance of deep commitment to tradition and my discontent with the implications of the term "falsification." All these men, in short, read my book through a quite special pair of spectacles, and there is another way to read it. The view through those spectacles is not wrong—my agreement with Sir Karl is real and substantial. Yet readers outside of the Popperian circle almost invariably fail even to notice that the agreement exists, and it is these readers who most often recognize (not necessarily with sympathy) what seem to me the central issues. I conclude that a gestalt switch divides readers of my book into two or more groups. What one of these sees as striking parallelism is virtually invisible to the others. The desire to understand how this can be so motivates the present comparison of my view with Sir Karl's.

The comparison must not, however, be a mere point-by-point juxtaposition. What demands attention is not so much the peripheral area in which our occasional secondary disagreements are to be isolated but the central region in which we appear to agree.

adherence to a tradition has an essential role in scientific development. He has written, for example, "Quantitatively and qualitatively by far the most important source of our knowledge—apart from inborn knowledge—is tradition" (Popper, *Conjectures and Refutations*, p. 27). Even more to the point, as early as 1948 Sir Karl wrote, "I do not think that we could ever free ourselves entirely from the bonds of tradition. The so-called freeing is really only a change from one tradition to another" (ibid., p. 122).

Sir Karl and I do appeal to the same data; to an uncommon extent we are seeing the same lines on the same paper; asked about those lines and those data, we often give virtually identical responses, or at least responses that inevitably seem identical in the isolation enforced by the question-and-answer mode. Nevertheless, experiences like those mentioned above convince me that our intentions are often quite different when we say the same things. Though the lines are the same, the figures which emerge from them are not. That is why I call what separates us a gestalt switch rather than a disagreement and also why I am at once perplexed and intrigued about how best to explore the separation. How am I to persuade Sir Karl, who knows everything I know about scientific development and who has somewhere or other said it, that what he calls a duck can be seen as a rabbit? How am I to show him what it would be like to wear my spectacles when he has already learned to look at everything I can point to through his own?

In this situation a change in strategy is called for, and the following suggests itself. Reading over once more a number of Sir Karl's principal books and essays, I encounter again a series of recurrent phrases which, though I understand them and do not quite disagree, are locutions that *I* could never have used in the same places. Undoubtedly they are most often intended as metaphors applied rhetorically to situations for which Sir Karl has elsewhere provided unexceptionable descriptions. Nevertheless, for present purposes these metaphors, which strike me as patently inappropriate, may prove more useful than straightforward descriptions. They may, that is, be symptomatic of contextual differences that a careful literal expression hides. If that is so, then these locutions may function not as the lines-on-paper but as the rabbit-ear, the shawl, or the ribbon-at-the-throat which one isolates when teaching a friend to transform his way of seeing a gestalt diagram. That, at least, is my hope for them. I have four such differences of locutions in mind and shall treat them *seriatim*.

Among the most fundamental issues on which Sir Karl and I agree is our insistence that an analysis of the development of scientific knowledge must take account of the way science has actually been practiced. That being so, a few of his recurrent generalizations startle me. One of these provides the opening sentences of the first chapter of the *Logic of Scientific Discovery:* "A scientist," writes Sir Karl, "whether theorist or experimenter, puts forward statements, or systems of statements, and tests them step by step. In the field of the empirical sciences, more particularly, he constructs hypotheses, or systems of theories, and tests them against experience by observation and experiment."[5] The statement is virtually a cliché, yet in application it presents three problems. It is ambiguous in its failure to specify which of two sorts of "statements" or "theories" are being tested. That ambiguity can, it is true, be eliminated by reference to other passages in Sir Karl's writings, but the generalization that results is historically mistaken. Furthermore, the mistake proves important, for the unambiguous form of the description misses just that characteristic of scientific practice which most nearly distinguishes the sciences from other creative pursuits.

There is one sort of "statement" or "hypothesis" that scientists do repeatedly subject to systematic test. I have in mind statements of an individual's best guesses about the proper way to connect his own research problem with the corpus of accepted scientific knowledge. He may, for example, conjecture that a given chemical unknown contains the salt of a rare earth, that the obesity of his experimental rats is due to a specified component in their diet, or that a newly discovered spectral pattern is to be understood as an effect of nuclear spin. In each case, the next steps in his research are intended to try out or test the conjecture or hypothesis. If it passes enough or stringent enough tests, the scientist has made a discovery or has at least resolved the puzzle he had been set. If not, he must either abandon the puzzle entirely or attempt to solve it with the aid of some other hypothesis. Many research problems, though by no means all, take this form. Tests of this sort are a standard component of what I have elsewhere labelled "normal science" or "normal research," an enterprise which accounts for the overwhelming majority of the work done in basic science. In no usual sense, however, are such tests directed to current theory. On the contrary, when engaged with a normal research problem, the scientist must *premise* current theory as the rules of his game. His object is to solve a puzzle, preferably one at which others have failed, and current theory is required to define that puzzle and to

5. Popper, *Logic of Scientific Discovery*, p. 27.

guarantee that, given sufficient brilliance, it can be solved.[6] Of course the practitioner of such an enterprise must often test the conjectural puzzle solution that his ingenuity suggests. But only his personal conjecture is tested. If it fails the test, only his own ability, not the corpus of current science, is impugned. In short, though tests occur frequently in normal science, these tests are of a peculiar sort, for in the final analysis it is the individual scientist rather than current theory which is tested.

This is not, however, the sort of test Sir Karl has in mind. He is above all concerned with the procedures through which science grows, and he is convinced that "growth" occurs not primarily by accretion but by the revolutionary overthrow of an accepted theory and its replacement by a better one.[7] (The subsumption under "growth" of "repeated overthrow" is itself a linguistic oddity whose *raison d'être* may become more visible as we proceed.) Taking this view, Sir Karl emphasizes tests performed to explore the limitations of accepted theory or to subject a current theory to maximum strain. Among his favorite examples, all of them startling and destructive in their outcome, are Lavoisier's experiments on calcination, the eclipse expedition of 1919, and the recent experiments on parity conservation.[8] All, of course, are classic tests, but in using them to characterize scientific activity Sir Karl misses something terribly important about them. Episodes like these are very rare in the development of science. When they occur, they are

6. For an extended discussion of normal science, the activity which practitioners are trained to carry on, see *The Structure of Scientific Revolutions*, pp. 23–42 and 135–42. It is important to notice that when I describe the scientist as a puzzle solver and Sir Karl describes him as a problem solver (e.g., in his *Conjectures and Refutations*, pp. 67, 222), the similarity of our terms disguises a fundamental divergence. Sir Karl writes (the italics are his), "Admittedly, our expectations, and thus our theories, may precede, historically, even our problems. *Yet science starts only with problems*. Problems crop up especially when we are disappointed in our expectations, or when our theories involve us in difficulties, in contradictions." I use the term "puzzle" in order to emphasize that the difficulties which *ordinarily* confront even the very best scientists are, like crossword puzzles or chess puzzles, challenges only to his ingenuity. *He* is in difficulty, not current theory. My point is almost the converse of Sir Karl's.

7. See Popper, *Conjectures and Refutations*, pp. 129, 215, and 221, for particularly forceful statements of this position.

8. For example, ibid., p. 220.

generally called forth either by a prior crisis in the relevant field (Lavoisier's experiments or Lee and Yang's[9]) or by the existence of a theory which competes with the existing canons of research (Einstein's general relativity). These are, however, aspects of or occasions for what I have elsewhere called "extraordinary research," an enterprise in which scientists do display very many of the characteristics Sir Karl emphasizes, but one which, at least in the past, has arisen only intermittently and under quite special circumstances in any scientific specialty.[10]

I suggest then that Sir Karl has characterized the entire scientific enterprise in terms that apply only to its occasional revolutionary parts. His emphasis is natural and common: the exploits of a Copernicus or Einstein make better reading than those of a Brahe or Lorentz; Sir Karl would not be the first if he mistook what I call normal science for an intrinsically uninteresting enterprise. Nevertheless, neither science nor the development of knowledge is likely to be understood if research is viewed exclusively through the revolutions it occasionally produces. For example, though testing of basic commitments occurs only in extraordinary science, it is normal science that discloses both the points to test and the manner of testing. Or again, it is for the normal, not the extraordinary practice of science that professionals are trained; if they are nevertheless eminently successful in displacing and replacing the theories on which normal practice depends, that is an oddity which must be explained. Finally, and this is for now my main point, a careful look at the scientific enterprise suggests that it is normal science, in which Sir Karl's sort of testing does not occur, rather than extraordinary science which most nearly distinguishes science from other enterprises. If a demarcation criterion exists (we must not, I think, seek a sharp or decisive one), it may lie just in that part of science which Sir Karl ignores.

In one of his most evocative essays, Sir Karl traces the origin of "the tradition of critical discussion [which] represents the only practicable way of expanding our knowledge" to the Greek philos-

9. For the work on calcination, see Guerlac, *Lavoisier: The Crucial Year* (1961). For the background of the parity experiments see Hafner and Presswood, "Strong Interference and Weak Interactions," *Science* 149 (1965): 503–10.

10. The point is argued at length in my *Structure of Scientific Revolutions*, pp. 52–97.

ophers between Thales and Plato, the men who, as he sees it, encouraged critical discussion both between schools and within individual schools.[11] The accompanying description of pre-Socratic discourse is most apt, but what is described does not at all resemble science. Rather it is the tradition of claims, counterclaims, and debates over fundamentals which, except perhaps during the Middle Ages, have characterized philosophy and much of social science ever since. Already by the Hellenistic period mathematics, astronomy, statics, and the geometric parts of optics had abandoned this mode of discourse in favor of puzzle solving. Other sciences, in increasing numbers, have undergone the same transition since. In a sense, to turn Sir Karl's view on its head, it is precisely the abandonment of critical discourse that marks the transition to a science. Once a field has made that transition, critical discourse recurs only at moments of crisis when the bases of the field are again in jeopardy.[12] Only when they must choose between competing theories do scientists behave like philosophers. That, I think, is why Sir Karl's brilliant description of the reasons for the choice between metaphysical systems so closely resembles my description of the reasons for choosing between scientific theories.[13] In neither choice, as I shall shortly try to show, can testing play a quite decisive role.

There is, however, good reason why testing has seemed to do so, and in exploring it Sir Karl's duck may at last become my rabbit. No puzzle-solving enterprise can exist unless its practitioners share criteria which, for that group and for that time, determine when a particular puzzle has been solved. The same criteria necessarily determine failure to achieve a solution, and anyone who chooses may view that failure as the failure of a theory to pass a test. Normally, as I have already insisted, it is not viewed that way. Only the practitioner is blamed, not his tools. But under the special circumstances which induce a crisis in the profession (e.g., gross failure, or repeated failure by the most brilliant professionals) the group's opinion may change. A failure that had previously been personal may then come to seem the failure of a theory under test.

11. Popper, *Conjectures and Refutations*, chap. 5, esp. pp. 148–52.

12. Though I was not then seeking a demarcation criterion, just these points are argued at length in my *Structure of Scientific Revolutions*, pp. 10–22 and 87–90.

13. Compare Popper, *Conjectures and Refutations*, pp. 192–200, with my *Structure of Scientific Revolutions*, pp. 143–58.

Thereafter, because the test arose from a puzzle and thus carried settled criteria of solution, it proves both more severe and harder to evade than the tests available within a tradition whose normal mode is critical discourse rather than puzzle solving.

In a sense, therefore, severity of test criteria is simply one side of the coin whose other face is a puzzle-solving tradition. That is why Sir Karl's line of demarcation and my own so frequently coincide. That coincidence is, however, only in their *outcome*; the *process* of applying them is very different, and it isolates distinct aspects of the activity about which the decision—science or non-science—is to be made. Examining the vexing cases, for example, psychoanalysis or Marxist historiography, for which Sir Karl tells us his criterion was initially designed,[14] I concur that they cannot now properly be labelled "science." But I reach that conclusion by a route far surer and more direct than his. One brief example may suggest that of the two criteria, testing and puzzle solving, the latter is at once the less equivocal and the more fundamental.

To avoid irrelevant contemporary controversies, I consider astrology rather than, say, psychoanalysis. Astrology is Sir Karl's most frequently cited example of a "pseudo-science."[15] He says, "By making their interpretations and prophecies sufficiently vague they [astrologers] were able to explain away anything that might have been a refutation of the theory had the theory and the prophecies been more precise. In order to escape falsification they destroyed the testability of the theory."[16] Those generalizations catch something of the spirit of the astrological enterprise. But taken at all literally, as they must be if they are to provide a demarcation criterion, they are impossible to support. The history of astrology during the centuries when it was intellectually reputable records many predictions that categorically failed.[17] Not even astrology's most convinced and vehement exponents doubted the recurrence of such failures. Astrology cannot be barred from the sciences because of the form in which its predictions were cast.

14. Popper, *Conjectures and Refutations*, p. 34.

15. The index to *Conjectures and Refutations* has eight entries under the heading "astrology as a typical pseudo science."

16. Popper, *Conjectures and Refutations*, p. 37.

17. For examples, see Thorndike, *A History of Magic and Experimental Science*, 8 vols. (1923–58), 5:225 ff.; 6:71, 101, 114.

Nor can it be barred because of the way its practitioners explained failure. Astrologers pointed out, for example, that, unlike general predictions about, say, an individual's propensities or a natural calamity, the forecast of an individual's future was an immensely complex task, demanding the utmost skill, and extremely sensitive to minor errors in relevant data. The configuration of the stars and eight planets was constantly changing; the astronomical tables used to compute the configuration at an individual's birth were notoriously imperfect; few men knew the instant of their birth with the requisite precision.[18] No wonder, then, that forecasts often failed. Only after astrology itself became implausible did these arguments come to seem question begging.[19] Similar arguments are regularly used today when explaining, for example, failures in medicine or meteorology. In times of trouble they are also deployed in the exact sciences, fields like physics, chemistry, and astronomy.[20] There was nothing unscientific about the astrologer's explanation of failure.

Nevertheless, astrology was not a science. Instead it was a craft, one of the practical arts, with close resemblances to engineering, meteorology, and medicine as these fields were practised until little more than a century ago. The parallels to an older medicine and to contemporary psychoanalysis are, I think, particularly close. In each of these fields shared theory was adequate only to establish the plausibility of the discipline and to provide a rationale for the various craft rules which governed practice. These rules had proved their use in the past, but no practitioner supposed they were sufficient to prevent recurrent failure. A more articulated theory and more powerful rules were desired, but it would have been absurd to abandon a plausible and badly needed discipline with a tradition of limited success simply because these desiderata were not yet at hand. In their absence, however, neither the astrologer nor the

18. For reiterated explanations of failure, see ibid. 1:11, 514–15; 4:368; 5:279.

19. A perceptive account of some reasons for astrology's loss of plausibility is included in Stahlman, "Astrology in Colonial America: An Extended Query," *William and Mary Quarterly* 13 (1956): 551–63. For an explanation of the previous appeal of astrology, see Thorndike, "The True Place of Astrology in the History of Science," *Isis* 46 (1955): 273–78.

20. Cf. my *Structure of Scientific Revolutions*, pp. 66–76.

doctor could do research. Though they had rules to apply, they had no puzzles to solve and therefore no science to practice.[21]

Compare the situations of the astronomer and the astrologer. If an astronomer's prediction failed and his calculations checked, he could hope to set the situation right. Perhaps the data were at fault: old observations could be reexamined and new measurements made, tasks which posed a host of calculational and instrumental puzzles. Or perhaps theory needed adjustment, either by the manipulation of epicycles, eccentrics, equants, and the like, or by more fundamental reforms of astronomical technique. For more than a millennium these were the theoretical and mathematical puzzles around which, together with their instrumental counterparts, the astronomical research tradition was constituted. The astrologer, by contrast, had no such puzzles. The occurrence of failures could be explained, but particular failures did not give rise to research puzzles, for no man, however skilled, could make use of them in a constructive attempt to revise the astrological tradition. There were too many possible sources of difficulty, most of them beyond the astrologer's knowledge, control, or responsibility. Individual failures were correspondingly uninformative, and they did not reflect on the competence of the prognosticator in the eyes of his professional compeers.[22] Though astronomy and astrology

21. This formulation suggests that Sir Karl's criterion of demarcation might be saved by a minor restatement entirely in keeping with his apparent intent. For a field to be a science its conclusions must be *logically derivable* from *shared premises*. On this view astrology is to be barred not because its forecasts were not testable but because only the most general and least testable ones could be derived from accepted theory. Since any field that did satisfy this condition *might* support a puzzle-solving tradition, the suggestion is clearly helpful. It comes close to supplying a sufficient condition for a field's being a science. But in this form, at least, it is not even quite a sufficient condition, and it is surely not a necessary one. It would, for example, admit surveying and navigation as sciences, and it would bar taxonomy, historical geology, and the theory of evolution. The conclusions of a science may be both precise and binding without being fully derivable by logic from accepted premises. See my *Structure of Scientific Revolutions*, pp. 35–51, and also the discussion below.

22. This is not to suggest that astrologers did not criticize each other. [illegible] the contrary, like the practitioners of philosophy and of some social [illegible]nces, they belonged to a variety of different schools, and the interschool [illegible]e was sometimes bitter. But these debates ordinarily revolved around [illegible] *implausibility* of the particular theory employed by one or another

were regularly practiced by the same people, including Ptolemy, Kepler, and Tycho Brahe, there was never an astrological equivalent of the puzzle-solving astronomical tradition. And without puzzles, able first to challenge and then to exemplify the ingenuity of the individual practitioner, astrology could not have become a science even if the stars had, in fact, controlled human destiny.

In short, though astrologers made testable predictions and recognized that these predictions sometimes failed, they did not and could not engage in the sorts of activities that normally characterize all recognized sciences. Sir Karl is right to exclude astrology from the sciences, but his overconcentration on revolutionary changes of scientific theory prevents his seeing the surest reason for doing so.

That fact, in turn, may explain another oddity of Sir Karl's historiography. Though he repeatedly underlines the role of tests in the replacement of scientific theories, he is also constrained to recognize that many theories, for example the Ptolemaic, were replaced before they had in fact been tested.[23] On some occasions, at least, tests are not requisite to the revolutions through which science advances. But that is not true of puzzles. Though the theories Sir Karl cites had not been put to the test before their displacement, none of these was replaced before it had ceased adequately to support a puzzle-solving tradition. The state of astronomy was a scandal in the early sixteenth century. Most astronomers nevertheless felt that normal adjustments of a basically Ptolemaic model would set the situation right. In this sense the theory had not failed a test. But a few astronomers, Copernicus among them, felt that the difficulties must lie in the Ptolemaic approach itself rather than in the particular versions of Ptolemaic theory so far developed, and the results of that conviction are already recorded. The situation is typical.[24] With or without tests, a puzzle-solving tradition can prepare the way for its own displacement. To rely on testing as the mark of a science is to miss what scientists mostly do and, with it, the most characteristic feature of their enterprise.

With the background supplied by the preceding remarks we can quickly discover the occasion and consequences of another of Sir

school. Failures of individual predictions played very little role. Compare Thorndike, *A History of Magic and Experimental Science*, 5:233.

23. See Popper, *Conjectures and Refutations*, p. 246.

24. See my *Structure of Scientific Revolutions*, pp. 77–87.

Karl's favorite locutions. The preface to *Conjectures and Refutations* opens with the sentence: "The essays and lectures of which this book is composed, are variations upon one very simple theme —the thesis that *we can learn from our mistakes*." The emphasis is Sir Karl's; the thesis recurs in his writing from an early date;[25] taken in isolation, it inevitably commands assent. Everyone can and does learn from his mistakes; isolating and correcting them is an essential technique in teaching children. Sir Karl's rhetoric has roots in everyday experience. Nevertheless, in the contexts for which he invokes this familiar imperative, its application seems decisively askew. I am not sure a mistake has been made, at least not a mistake to learn from.

One need not confront the deeper philosophical problems presented by mistakes to see what is presently at issue. It is a mistake to add three plus three and get five, or to conclude from "All men are mortal" that "All mortals are men." For different reasons, it is a mistake to say, "He is my sister," or to report the presence of a strong electric field when test charges fail to indicate it. Presumably there are still other sorts of mistakes, but all the normal ones are likely to share the following characteristics. A mistake is made, or is committed, at a specifiable time and place by a particular individual. That individual has failed to obey some established rule of logic, or of language, or of the relations between one of these and experience. Or he may instead have failed to recognize the consequences of a particular choice among the alternatives which the rules allow him. The individual can learn from his mistake only because the group whose practice embodies these rules can isolate the individual's failure in applying them. In short, the sorts of mistakes to which Sir Karl's imperative most obviously applies are an individual's failure of understanding or of recognition within an activity governed by preestablished rules. In the sciences such mistakes occur most frequently and perhaps exclusively within the practice of normal puzzle-solving research.

That is not, however, where Sir Karl seeks them, for his concept

25. The quotation is from Popper, *Conjectures and Refutations*, p. vii, in a preface dated 1962. Earlier Sir Karl had equated "learning from our mistakes" with "learning by trial and error" (ibid., p. 216), and the trial-and-error formulation dates from 1937 at least (ibid., p. 312) and is in spirit older than that. Much of what is said below about Sir Karl's notion of "mistake" applies equally to his concept of "error."

of science obscures even the existence of normal research. Instead, he looks to the extraordinary or revolutionary episodes in scientific development. The mistakes to which he points are not usually acts at all but rather out-of-date scientific theories: Ptolemaic astronomy, the phlogiston theory, or Newtonian dynamics. And "learning from our mistakes" is, correspondingly, what occurs when a scientific community rejects one of these theories and replaces it with another.[26] If this does not immediately seem an odd usage, that is mainly because it appeals to the residual inductivist in us all. Believing that valid theories are the product of correct inductions from facts, the inductivist must also hold that a false theory is the result of a mistake in induction. In principle, at least, he is prepared to answer the questions: what mistake was made, what rule broken, when and by whom, in arriving at, say, the Ptolemaic system? To the man for whom those are sensible questions and to him alone, Sir Karl's locution presents no problems.

But neither Sir Karl nor I is an inductivist. We do not believe that there are rules for inducing correct theories from facts, or even that theories, correct or incorrect, are induced at all. Instead we view them as imaginative posits, invented in one piece for application to nature. And though we point out that such posits can and usually do at last encounter puzzles they cannot solve, we also recognize that those troublesome confrontations rarely occur for some time after a theory has been both invented and accepted. In our view, then, no mistake was made in arriving at the Ptolemaic system, and it is therefore difficult for me to understand what Sir Karl has in mind when he calls that system, or any other out-of-date theory, a mistake. At most one may wish to say that a theory

26. Ibid., pp. 215 and 220. In these pages Sir Karl outlines and illustrates his thesis that science grows through revolutions. He does not, in the process, ever juxtapose the term "mistake" with the name of an out-of-date scientific theory, presumably because his sound historic instinct inhibits so gross an anachronism. Yet the anachronism is fundamental to Sir Karl's rhetoric, which does repeatedly provide clues to more substantial differences between us. Unless out-of-date theories are mistakes, there is no way to reconcile, say, the opening paragraph of Sir Karl's preface (ibid., p. vii: "learn from our mistakes"; "our often mistaken attempts to solve our problems"; "tests which may help us in the discovery of our mistakes") with the view (ibid., p. 215) that "the growth of scientific knowledge . . . [consists in] the repeated overthrow of scientific theories and their replacement by better or more satisfactory ones."

which was not previously a mistake has become one or that a scientist has made the mistake of clinging to a theory for too long. And even these locutions, of which at least the first is extremely awkward, do not return us to the sense of mistake with which we are most familiar. Those mistakes are the normal ones which a Ptolemaic (or a Copernican) astronomer makes within his system, perhaps in observation, calculation, or the analysis of data. They are, that is, the sort of mistake which can be isolated and then at once corrected, leaving the original system intact. In Sir Karl's sense, on the other hand, a mistake infects an entire system and can be corrected only by replacing the system as a whole. No locutions and no similarities can disguise these fundamental differences, nor can it hide the fact that before infection set in the system had the full integrity of what we now call sound knowledge.

Quite possibly Sir Karl's sense of "mistake" can be salvaged, but a successful salvage operation must deprive it of certain still current implications. Like the term "testing," "mistake" has been borrowed from normal science, where its use is reasonably clear, and applied to revolutionary episodes, where its application is at best problematic. That transfer creates, or at least reinforces, the prevalent impression that whole theories can be judged by the same sort of criteria employed when judging the individual research applications of a theory. The discovery of applicable criteria then becomes a primary desideratum for many people. That Sir Karl should be among them is strange, for the search runs counter to the most original and fruitful thrust in his philosophy of science. But I can understand his methodological writings since the *Logik der Forschung* in no other way. I shall now suggest that he has, despite explicit disclaimers, consistently sought evaluation procedures which can be applied to theories with the apodictic assurance characteristic of the techniques by which one identifies mistakes in arithmetic, logic, or measurement. I fear that he is pursuing a will-o'-the-wisp born from the same conjunction of normal and extraordinary science which made tests seem so fundamental a feature of the sciences.

In his *Logik der Forschung*, Sir Karl underlined the asymmetry of a generalization and its negation in their relation to empirical evidence. A scientific theory cannot be shown to apply successfully to all its possible instances, but it can be shown to be unsuccessful in particular applications. Emphasis upon that logical truism and its implications seems to me a forward step from which there must be no retreat. The same asymmetry plays a fundamental role in my *Structure of Scientific Revolutions*, where a theory's failure to provide rules that identify solvable puzzles is viewed as the source of professional crises which often result in the theory's being replaced. My point is very close to Sir Karl's, and I may well have taken it from what I had heard of his work.

But Sir Karl describes as "falsification" or "refutation" what happens when a theory fails in an attempted application, and these are the first of a series of related locutions that again strike me as extremely odd. Both "falsification" and "refutation" are antonyms of "proof." They are drawn principally from logic and from formal mathematics; the chains of argument to which they apply end with a "Q.E.D."; invoking these terms implies the ability to compel assent from any member of the relevant professional community. No member of this audience, however, still needs to be told that, where a whole theory or often even a scientific law is at stake, arguments are seldom so apodictic. All experiments can be challenged, either for their relevance or their accuracy. All theories can be modified by a variety of ad hoc adjustments without ceasing to be, in their main lines, the same theories. It is important, furthermore, that this should be so, for it is often by challenging observations or adjusting theories that scientific knowledge grows. Challenges and adjustments are a standard part of normal research in empirical science, and adjustments, at least, play a dominant role in informal mathematics as well. Dr. Lakatos's brilliant analysis of the permissible rejoinders to mathematical refutations provides the most telling arguments I know against a naive falsificationist position.[27]

Sir Karl is not, of course, a naive falsificationist. He knows all that has just been said and has emphasized it from the beginning of his career. Very early in his *Logic of Scientific Discovery*, for example, he writes: "In point of fact, no conclusive disproof of a theory can ever be produced; for it is always possible to say that the experimental results are not reliable or that the discrepancies which are asserted to exist between the experimental results and the the-

27. I. Lakatos, "Proofs and Refutations," *British Journal for the Philosophy of Science* 14 (1963–64): 1–25, 120–39, 221–43, 296–342.

ory are only apparent and that they will disappear with the advance of our understanding."[28] Statements like these display one more parallel between Sir Karl's view of science and my own, but what we make of them could scarcely be more different. For my view they are fundamental, both as evidence and as source. For Sir Karl's, in contrast, they are an essential qualification which threatens the integrity of his basic position. Having barred conclusive disproof, he has provided no substitute for it, and the relation he does employ remains that of logical falsification. Though he is not a naive falsificationist, Sir Karl may, I suggest, legitimately be treated as one.

If his concern were exclusively with demarcation, the problems posed by the unavailability of conclusive disproofs would be less severe and perhaps eliminable. Demarcation might, that is, be achieved by an exclusively syntactic criterion.[29] Sir Karl's view would then be, and perhaps is, that a theory is scientific if and only if *observation statements*—particularly the negations of singular existential statements—can be logically deduced from it, perhaps in conjunction with stated background knowledge. The difficulties (to which I shall shortly turn) in deciding whether the outcome of a particular laboratory operation justifies asserting a particular observation statement would then be irrelevant. Perhaps, though the basis for doing so is less apparent, the equally grave difficulties in deciding whether an observation statement deduced from an approximate (e.g., mathematically manageable) version of the theory should be considered consequences of the theory itself could be eliminated in the same way. Problems like these would belong not to the syntactics but to the pragmatics or semantics of the language in which the theory was cast, and they would therefore have no role in determining its status as a science. To be scientific a theory need be falsifiable only by an observation statement not by actual observation. The relation between statements, unlike that between a

28. Popper, *Logic of Scientific Discovery*, p. 50.

29. Though my point is somewhat different, I owe my recognition of the need to confront this issue to C. G. Hempel's strictures on those who misinterpret Sir Karl by attributing to him a belief in absolute rather than relative falsification. See Hempel, *Aspects of Scientific Explanation* (1965), p. 45. I am also indebted to Professor Hempel for a close and perceptive critique of this paper in draft.

statement and an observation, could be the conclusive disproof familiar from logic and mathematics.

For reasons suggested above (note 21) and elaborated immediately below, I doubt that scientific theories can without decisive change be cast in a form which permits the purely syntactic judgments which this version of Sir Karl's criterion requires. But even if they could, these reconstructed theories would provide a basis only for his demarcation criterion, not for the logic of knowledge so closely associated with it. The latter has, however, been Sir Karl's most persistent concern, and his notion of it is quite precise. "The logic of knowledge . . . ," he writes, "consists solely in investigating the methods employed in those systematic tests to which every new idea must be subjected if it is to be seriously entertained."[30] From this investigation, he continues, result methodological rules or conventions like the following: "Once a hypothesis has been proposed and tested, and has proved its mettle, it may not be allowed to drop out without 'good reason.' A 'good reason' may be, for instance . . . the falsification of one of the consequences of the hypothesis."[31]

Rules like these, and with them the entire logical enterprise described above, are no longer simply syntactic in their import. They require that both the epistemological investigator and the research scientist be able to relate sentences derived from a theory not to other sentences but to actual observations and experiments. This is the context in which Sir Karl's term "falsification" must function, and Sir Karl is entirely silent about how it can do so. What is falsification if it is not conclusive disproof? Under what circumstances does the *logic* of knowledge require a scientist to abandon a previously accepted theory when confronted, not with statements about experiments, but with experiments themselves? Pending clarification of these questions, I am not sure that what Sir Karl has given us is a logic of knowledge at all. In my conclusion I shall suggest that, though equally valuable, it is something else entirely. Rather than a logic, Sir Karl has provided an ideology; rather than methodological rules, he has supplied procedural maxims.

That conclusion must, however, be postponed until after a last deeper look at the source of the difficulties with Sir Karl's notion

30. Popper, *Logic of Scientific Discovery*, p. 31.
31. Ibid., pp. 53–54.

of falsification. It presupposes, as I have already suggested, that a theory is cast, or can without distortion be recast, in a form that permits scientists to classify each conceivable event as either a confirming instance, a falsifying instance, or irrelevant to the theory. That is obviously required if a general law is to be falsifiable: to test the generalization $(x)\phi(x)$ by applying it to the constant a, we must be able to tell whether or not a lies within the range of the variable x and whether or not $\phi(a)$. The same presupposition is even more apparent in Sir Karl's recently elaborated measure of verisimilitude. It requires that we first produce the class of all logical consequences of the theory and then choose from among these, with the aid of background knowledge, the classes of all true and of all false consequences.[32] At least, we must do this if the criterion of verisimilitude is to result in a *method* of theory choice. None of these tasks can, however, be accomplished unless the theory is fully articulated logically and unless the terms through which it attaches to nature are sufficiently defined to determine their applicability in each possible case. In practice, however, no scientific theory satisfies these rigorous demands, and many people have argued that a theory would cease to be useful in research if it did so.[33] I have myself elsewhere introduced the term "paradigm" to underscore the dependence of scientific research upon concrete examples that bridge what would otherwise be gaps in the specification of the content and application of scientific theories. The relevant arguments cannot be repeated here. But a brief example, though it will temporarily alter my mode of discourse, may be even more useful.

My example takes the form of a constructed epitome of some elementary scientific knowledge. That knowledge concerns swans, and to isolate its presently relevant characteristics I shall ask three questions about it: (*a*) How much can one know about swans without introducing explicit generalizations like "All swans are

32. Popper, *Conjectures and Refutations*, pp. 233–35. Notice also, at the foot of the last of these pages, that Sir Karl's comparison of the relative verisimilitude of two theories depends upon there being "no revolutionary changes in our background knowledge," an assumption which he nowhere argues and which is hard to reconcile with his conception of scientific change by revolutions.

33. Braithwaite, *Scientific Explanation* (1953), pp. 50–87, esp. p. 76, and my *Structure of Scientific Revolutions*, pp. 97–101.

white"? (*b*) Under what circumstances and with what consequences are such generalizations worth adding to what was known without them? (*c*) Under what circumstances are generalizations rejected once they have been made? In raising these questions my object is to suggest that, though logic is a powerful and ultimately an essential tool of scientific enquiry, one can have sound knowledge in forms to which logic can scarcely be applied. Simultaneously, I shall suggest that logical articulation is not a value for its own sake, but is to be undertaken only when and to the extent that circumstances demand it.

Imagine that you have been shown and can remember ten birds which have authoritatively been identified as swans; that you have a similar acquaintance with ducks, geese, pigeons, doves, gulls, and the like; and that you are informed that each of these types constitutes a natural family. A natural family you already know as an observed cluster of like objects, sufficiently important and sufficiently discrete to command a generic name. More precisely, though here I introduce more simplification than the concept requires, a natural family is a class whose members resemble each other more closely than they resemble the members of other natural families.[34] The experience of generations has to date confirmed that all observed objects fall into one or another natural family. It has, that is, shown that the entire population of the world can always be divided (though not once and for all) into perceptually discontinuous categories. In the perceptual spaces between these categories there are believed to be no objects at all.

What you have learned about swans from exposure to paradigms is very much like what children first learn about dogs and cats, tables and chairs, mothers and fathers. Its precise scope and content are, of course, impossible to specify, but it is sound knowledge nonetheless. Derived from observation, it can be infirmed by further observation, and it meanwhile provides a basis for rational action. Seeing a bird much like the swans you already know, you may reasonably presume that it will require the same food as the

34. Note that the resemblance between members of a natural family is here a learned relationship and one which can be unlearned. Contemplate the old saw, "To an occidental, all chinamen look alike." That example also highlights the most drastic of the simplifications introduced at this point. A fuller discussion would have to allow for hierarchies of natural families with resemblance relations between families at the higher levels.

others and will breed with them. Provided swans are a natural family, no bird which closely resembles them on sight should display radically different characteristics on closer acquaintance. Of course you may have been misinformed about the natural integrity of the swan family. But that can be discovered from experience, for example, by the discovery of a number of animals (note that more than one is required) whose characteristics bridge the gap between swans and, say, geese by barely perceptible intervals.[35] Until that does occur, however, you will know a great deal about swans though you will not be altogether sure what you know or what a swan is.

Suppose now that all the swans you have actually observed are white. Should you embrace the generalization, "All swans are white"? Doing so will change what you know very little; that change will be of use only in the unlikely event that you meet a nonwhite bird that otherwise resembles a swan; by making the change you increase the risk that the swan family will prove not to be a natural family after all. Under those circumstances you are likely to refrain from generalizing unless there are special reasons for doing so. Perhaps, for example, you must describe swans to men who cannot be directly exposed to paradigms. Without superhuman caution both on your part and on that of your readers, your description will acquire the force of a generalization; this is often the problem of the taxonomist. Or perhaps you have discovered some grey birds that look otherwise like swans but eat different food and have an unfortunate disposition. You may then generalize to avoid a behavioral mistake. Or you may have a more theoretical reason for thinking the generalization worthwhile. For example, you may have observed that the members of other natural families share coloration. Specifying this fact in a form which permits the application of powerful logical techniques to what you know may enable you to learn more about animal color in general or about animal breeding.

35. This experience would not necessitate the abandonment of either the category "swans" or the category "geese," but it would necessitate the introduction of an *arbitrary* boundary between them. The families "swans" and "geese" would no longer be natural families, and you could conclude nothing about the character of a new swanlike bird that was not also true of geese. Empty perceptual space is essential if family membership is to have cognitive content.

Now, having made the generalization, what will you do if you encounter a black bird that looks otherwise like a swan? Almost the same things, I suggest, as if you had not previously committed yourself to the generalization at all. You will examine the bird with care, externally and perhaps internally as well, to find other characteristics that distinguish this specimen from your paradigms. That examination will be particularly long and thorough if you have theoretical reasons for believing that color characterizes natural families or if you are deeply ego involved with the generalization. Very likely the examination will disclose other differentiae, and you will announce the discovery of a new natural family. Or you may fail to find such differentiae and may then announce that a black swan has been found. Observation cannot, however, force you to that falsifying conclusion, and you would occasionally be the loser if it could do so. Theoretical considerations may suggest that color alone is sufficient to demarcate a natural family: the bird is not a swan because it is black. Or you may simply postpone the issue pending the discovery and examination of other specimens. Only if you have previously committed yourself to a full definition of "swan," one which will specify its applicability to every conceivable object, can you be logically *forced* to rescind your generalization.[36] And why should you have offered such a definition? It could serve no cognitive function and would expose you to tremendous risks.[37] Risks, of course, are often worth taking, but to say more than one knows solely for the sake of risk is foolhardy.

I suggest that scientific knowledge, though logically more articulate and far more complex, is of this sort. The books and teachers

36. Further evidence for the unnaturalness of any such definitions is provided by the following question. Should "whiteness" be included as a defining characteristic of swans? If so, the generalization "All swans are white" is immune to experience. But if "whiteness" is excluded from the definition, then some other characteristic must be included for which "whiteness" might have substituted. Decisions about which characteristics are to be parts of a definition and which are to be available for the statement of general laws are often arbitrary and, in practice, are seldom made. Knowledge is not usually articulated in that way.

37. This incompleteness of definitions is often called "open texture" or "vagueness of meaning," but those phrases seem decisively askew. Perhaps the definitions are incomplete, but nothing is wrong with the meanings. That is the way meanings behave!

from whom it is acquired present concrete examples together with a multitude of theoretical generalizations. Both are essential carriers of knowledge, and it is therefore Pickwickian to seek a methodological criterion that supposes the scientist can specify in advance whether each imaginable instance fits or would falsify his theory. The criteria at his disposal, explicit and implicit, are sufficient to answer that question only for the cases that clearly do fit or that are clearly irrelevant. These are the cases he expects, the ones for which his knowledge was designed. Confronted with the unexpected, he must always do more research in order further to articulate his theory in the area that has just become problematic. He may then reject it in favor of another and for good reason. But no exclusively logical criteria can entirely dictate the conclusion he must draw.

Almost everything said so far rings changes on a single theme. The criteria with which scientists determine the validity of an articulation or an application of existing theory are not by themselves sufficient to determine the choice between competing theories. Sir Karl has erred by transferring selected characteristics of everyday research to the occasional revolutionary episodes in which scientific advance is most obvious and by thereafter ignoring the everyday enterprise entirely. In particular, he has sought to solve the problem of theory choice during revolutions by logical criteria that are applicable in full only when a theory can already be presupposed. That is the largest part of my thesis in this paper, and it could be the entire thesis if I were content to leave altogether open the questions that have been raised. How do the scientists make the choice between competing theories? How are we to understand the way in which science does progress?

Let me at once be clear that having opened that Pandora's box, I shall close it quickly. There is too much about these questions that I do not understand and must not pretend to. But I believe I see the directions in which answers to them must be sought, and I shall conclude with an attempt briefly to mark the trail. Near its end we shall once more encounter a set of Sir Karl's characteristic locutions.

I must first ask what it is that still requires explanation. Not that scientists discover the truth about nature, nor that they approach ever closer to the truth. Unless, as one of my critics suggests,[38] we simply define the approach to truth as the result of what scientists do, we cannot recognize progress toward that goal. Rather we must explain why science—our surest example of sound knowledge—progresses as it does, and we must first find out how, in fact, it does progress.

Surprisingly little is yet known about the answer to that descriptive question. A vast amount of thoughtful empirical investigation is still required. With the passage of time, scientific theories taken as a group are obviously more and more articulated. In the process, they are matched to nature at an increasing number of points and with increasing precision. Or again, the number of subject matters to which the puzzle-solving approach can be applied clearly grows with time. There is a continuing proliferation of scientific specialties, partly by an extension of the boundaries of science and partly by the subdivision of existing fields.

Those generalizations are, however, only a beginning. We know, for example, almost nothing about what a group of scientists will sacrifice in order to achieve the gains that a new theory invariably offers. My own impression, though it is no more than that, is that a scientific community will seldom or never embrace a new theory unless it solves all or almost all the quantitative, numerical puzzles that have been treated by its predecessor.[39] They will, on the other hand, occasionally sacrifice explanatory power, however reluctantly, sometimes leaving previously resolved questions open and sometimes declaring them altogether unscientific.[40] Turning to another area, we know little about historical changes in the unity of the sciences. Despite occasional spectacular successes, communication across the boundaries between scientific specialties becomes worse and worse. Does the number of incompatible viewpoints employed by the increasing number of communities of specialists grow with time? Unity of the sciences is clearly a value for scientists, but for what will they give it up? Or again, though the bulk of scientific knowledge clearly increases with time, what are we to

38. D. Hawkins, review of *Structure of Scientific Revolutions* in *American Journal of Physics* 31 (1963): 554–55.

39. Cf. Kuhn, "The Function of Measurement in Modern Physical Science, *Isis* 52 (1961): 161–93 (pp. 178–224, below).

40. Cf. Kuhn, *Structure of Scientific Revolutions,* pp. 102–8.

say about ignorance? The problems solved during the last thirty years did not exist as open questions a century ago. In any age, the scientific knowledge already at hand virtually exhausts what there is to know, leaving visible puzzles only at the horizon of existing knowledge. Is it not possible, or perhaps even likely, that contemporary scientists know less of what there is to know about their world than the scientists of the eighteenth century knew of theirs? Scientific theories, it must be remembered, attach to nature only here and there. Are the interstices between those points of attachment perhaps now larger and more numerous than ever before?

Until we can answer more questions like these, we shall not know quite what scientific progress is and cannot therefore quite hope to explain it. On the other hand, answers to those questions will very nearly provide the explanation sought. The two come almost together. Already it should be clear that the explanation must, in the final analysis, be psychological or sociological. It must, that is, be a description of a value system, an ideology, together with an analysis of the institutions through which that system is transmitted and enforced. Knowing what scientists value, we may hope to understand what problems they will undertake and what choices they will make in particular circumstances of conflict. I doubt that there is another sort of answer to be found.

What form that answer will take is, of course, another matter. At this point, too, my sense that I control my subject matter ends. But again, some sample generalizations will illustrate the sorts of answers which must be sought. For a scientist, the solution of a difficult conceptual or instrumental puzzle is a principal goal. His success in that endeavour is rewarded through recognition by other members of his professional group and by them alone. The practical merit of his solution is at best a secondary value, and the approval of men outside the specialist group is a negative value or none at all. These values, which do much to dictate the form of normal science, are also significant at times when a choice must be made between theories. A man trained as a puzzle solver will wish
as many as possible of the prior puzzle solutions ob-
s group, and he will also wish to maximize the number
hat can be solved. But even these values frequently con-
here are others that make the problem of choice still
ult. It is just in this connection that a study of what sci-

entists will give up would be most significant. Simplicity, precision, and congruence with the theories used in other specialties are all significant value for the scientists, but they do not all dictate the same choice nor will they all be applied in the same way. That being the case, it is also important that group unanimity be a paramount value, causing the group to minimize the occasions for conflict and to reunite quickly about a single set of rules for puzzle solving even at the price of subdividing the specialty or excluding a formerly productive member.[41]

I do not suggest that these are the right answers to the problem of scientific progress, but only that they are the types of answers that must be sought. Can I hope that Sir Karl will join me in this view of the task still to be done? For some time I have assumed he would not, as a set of phrases that recurs in his work seems to bar the position to him. Again and again he has rejected "the psychology of knowledge" or the "subjective" and insisted that his concern was instead with the "objective" or "the logic of knowledge."[42] The title of his most fundamental contribution to our field is *The* Logic *of Scientific Discovery*, and it is there that he most positively asserts that his concern is with the logical spurs to knowledge rather than with the psychological drives of individuals. Until very recently I have supposed that this view of the problem must bar the sort of solution I have advocated.

But now I am less certain, for there is another aspect of Sir Karl's work, not quite compatible with what precedes. When he rejects "the psychology of knowledge," Sir Karl's explicit concern is only to deny the methodological relevance to an *individual's* source of inspiration or of an individual's sense of certainty. With that much I cannot disagree. It is, however, a long step from the rejection of the psychological idiosyncrasies of an individual to the rejection of the common elements induced by nurture and training in the psychological make-up of the licensed membership of a *scientific group*. One need not be dismissed with the other. And this, too, Sir Karl seems sometimes to recognize. Though he insists he is writing about the logic of knowledge, an essential role in his

41. Ibid., pp. 161–69.

42. Popper, *Logic of Scientific Discovery*, pp. 22 and 31–32, 46; *Conjectures and Refutations*, p. 52.

methodology is played by passages which I can only read as attempts to inculcate moral imperatives in the membership of the scientific group.

"Assume," Sir Karl writes, "that we have deliberately made it our task to live in this unknown world of ours; to adjust ourselves to it as well as we can; . . . . and to explain it, *if* possible (we need not assume that it is) and as far as possible, with help of laws and explanatory theories. *If we have made this our task, then there is no more rational procedure than the method of . . . conjecture and refutation:* of boldly proposing theories; of trying our best to show that these are erroneous; and of accepting them tentatively if our critical efforts are unsuccessful."[43] We shall not, I suggest, understand the success of science without understanding the full force of rhetorically induced and professionally shared imperatives like these. Institutionalized and articulated further (and also somewhat differently) such maxims and values may explain the outcome of choices that could not have been dictated by logic and experiment alone. The fact that passages like these occupy a prominent place in Sir Karl's writing is therefore further evidence of the resemblance of our views. That he does not, I think, ever see them for the social-psychological imperatives that they are is further evidence of the gestalt switch that still divides us deeply.

43. Popper, *Conjectures and Refutations*, p. 51 (italics in original).

12 Second Thoughts on Paradigms

Reprinted by permission from *The Structure of Scientific Theories*, ed. Frederick Suppe (Urbana: University of Illinois Press, 1974), pp. 459–82.

It has now been several years since a book of mine, *The Structure of Scientific Revolutions*, was published. Reactions to it have been varied and occasionally strident, but the book continues to be widely read and much discussed. By and large I take great satisfaction from the interest it has aroused, including much of the criticism. One aspect of the response does, however, from time to time dismay me. Monitoring conversations, particularly among the book's enthusiasts, I have sometimes found it hard to believe that all parties to the discussion had been engaged with the same volume. Part of the reason for its success is, I regretfully conclude, that it can be too nearly all things to all people.

For that excessive plasticity, no aspect of the book is so much responsible as its introduction of the term "paradigm,"[1] a word that figures more often than any other, excepting the grammatical

1. Other problems and sources of misunderstanding are discussed in my essay, "Logic of Discovery or Psychology of Research," in *Criticism and the Growth of Knowledge*, ed. I. Lakatos and A. Musgrave (Cambridge: Cambridge University Press, 1970); see pp. 266–92 above. That book, which also includes an extended "Response to Critics," constitutes the fourth volume of the proceedings of the International Colloquium in the Philosophy of Science held at Bedford College, London, during July, 1965. A briefer but more balanced discussion of critical reactions to *The Structure of Scientific Revolutions* (Chicago: University of Chicago Press, 1962) has been prepared for the Japanese translation of that book. An English version has been in-

particles, in its pages. Challenged to explain the absence of an index, I regularly point out that its most frequently consulted entry would be: "paradigm, 1–172, passim." Critics, whether sympathetic or not, have been unanimous in underscoring the large number of different senses in which the term is used.[2] One commentator, who thought the matter worth systematic scrutiny, prepared a partial subject index and found at least twenty-two different usages, ranging from "a concrete scientific achievement" (p. 11) to a "characteristic set of beliefs and preconceptions" (p. 17), the latter including instrumental, theoretical, and metaphysical commitments together (pp. 39–42).[3] Though neither the compiler of that index nor I think the situation so desperate as those divergences suggest, clarification is obviously called for. Nor will clarification by itself suffice. Whatever their number, the usages of "paradigm" in the book divide into two sets which require both different names and separate discussion. One sense of "paradigm" is global, embracing all the shared commitments of a scientific group; the other isolates a particularly important sort of commitment and is thus a subset of the first. In what follows I shall try initially to disentangle them and then to scrutinize the one that I believe most urgently needs philosophical attention. However imperfectly I understood paradigms when I wrote the book, I still think them worth much attention.

In the book the term "paradigm" enters in close proximity, both physical and logical, to the phrase "scientific community" (pp. 10–11). A paradigm is what the members of a scientific community, and they alone, share. Conversely, it is their possession of a common paradigm that constitutes a scientific community of a group of otherwise disparate men. As empirical generalizations, both those statements can be defended. But in the book they func-

cluded in subsequent American printings. Parts of these papers carry on where this one leaves off and thus clarify the relations of the ideas developed here to such notions as incommensurability and revolutions.

2. The most thoughtful and thorough negative account of this problem is Dudley Shapere's "The Structure of Scientific Revolutions," *Philosophical Review* 73 (1964): 383–94.

3. Margaret Masterman, "The Nature of a Paradigm," in *Criticism and the Growth of Knowledge*, ed. I. Lakatos and A. Musgrave. Parenthetical page references in the text are to my *Structure of Scientific Revolutions*.

tion at least partly as definitions, and the result is a circularity with at least a few vicious consequences.[4] If the term "paradigm" is to be successfully explicated, scientific communities must first be recognized as having an independent existence.

In fact, the identification and study of scientific communities has recently emerged as a significant research subject among sociologists. Preliminary results, many of them still unpublished, suggest that the requisite empirical techniques are nontrivial, but some are already in hand, and others are sure to be developed.[5] Most practicing scientists respond at once to questions about their community affiliations, taking it for granted that responsibility for the various current specialties and research techniques is distributed among groups of at least roughly determinate membership. I shall therefore assume that more systematic means for their identification will be forthcoming, and content myself here with a brief articulation of an intuitive notion of community, one widely shared by scientists, sociologists, and a number of historians of science.

4. The most damaging of these consequences grows out of my use of the term "paradigm" when distinguishing an earlier from a later period in the development of an individual science. During what is called, in *Structure of Scientific Revolutions*, the "preparadigm period," the practitioners of a science are split into a number of competing schools, each claiming competence for the same subject matter but approaching it in quite different ways. This developmental stage is followed by a relatively rapid transition, usually in the aftermath of some notable scientific achievement, to a so-called postparadigm period characterized by the disappearance of all or most schools, a change which permits far more powerful professional behavior to the members of the remaining community. I still think that pattern both typical and important, but it can be discussed without reference to the first achievement of a paradigm. Whatever paradigms may be, they are possessed by any scientific community, including the schools of the so-called preparadigm period. My failure to see that point clearly has helped make a paradigm seem a quasi-mystical entity or property which, like charisma, transforms those infected by it. There is a transformation, but it is not induced by the acquisition of a paradigm.

5. W. O. Hagstrom, *The Scientific Community* (New York: Basic Books, 1965), chaps. 4 and 5; D. J. Price and D. de B. Beaver, "Collaboration in an Invisible College," *American Psychologist* 21 (1966): 1011–18; Diana Crane, "Social Structure in a Group of Scientists: A Test of the 'Invisible College' Hypothesis," *American Sociological Review* 34 (1969): 335–52; N. C. Mullins, "Social Networks among Biological Scientists," (Ph.D. thesis, Harvard University, 1966), and "The Development of a Scientific Specialty," *Minerva* 10 (1972): 51–82.

A scientific community consists, in this view, of the practitioners of a scientific specialty. Bound together by common elements in their education and apprenticeship, they see themselves and are seen by others as the men responsible for the pursuit of a set of shared goals, including the training of their successors. Such communities are characterized by the relative fullness of communication within the group and by the relative unanimity of the group's judgment in professional matters. To a remarkable extent the members of a given community will have absorbed the same literature and drawn similar lessons from it.[6] Because the attention of different communities is focused on different matters, professional communication across group lines is likely to be arduous, often gives rise to misunderstanding, and may, if pursued, isolate significant disagreement.

Clearly, communities in this sense exist at numerous levels. Perhaps all natural scientists form a community. (We ought not, I think, allow the storm surrounding C. P. Snow to obscure those points about which he has said the obvious.) At an only slightly lower level, the main scientific professional groups provide examples of communities: physicists, chemists, astronomers, zoologists, and the like. For these major communities group membership is readily established, except at the fringes. Subject of highest degree, membership in professional societies, and journals read are ordinarily more than sufficient. Similar techniques will also isolate the major subgroups: organic chemists and perhaps protein chemists among them, solid state and high energy physicists, radio astronomers, and so on. It is only at the next lower level that empirical difficulties emerge. How, prior to its public acclaim, would an outsider have isolated the phage group? For this, one must have recourse to attendance at summer institutes and special conferences, to preprint distribution lists, and above all to formal and informal communication networks, including the linkages among citations.[7]

6. For the historian, to whom interview and questionnaire techniques are ordinarily unavailable, shared source materials often provide the most significant clues to community structure. That is one of the reasons why widely read works like Newton's *Principia* are, in *Structure of Scientific Revolutions*, so often referred to as paradigms. I should now describe them as particularly important sources of the elements in a community's disciplinary matrix.

7. E. Garfield, *The Use of Citation Data in Writing the History of Science* (Philadelphia: Institute for Scientific Information, 1964); M. M. Kessler,

I take it that the job can and will be done, and that it will typically yield communities of perhaps a hundred members, sometimes significantly fewer. Individual scientists, particularly the ablest, will belong to several such groups, either simultaneously or in succession. Though it is not yet clear just how far empirical analysis can take us, there is excellent reason to suppose that the scientific enterprise is distributed among and carried forward by communities of this sort.

Let me now suppose that we have, by whatever techniques, identified one such community. What shared elements account for the relatively unproblematic character of professional communication and for the relative unanimity of professional judgment? To this question *The Structure of Scientific Revolutions* licences the answer "a paradigm" or "a set of paradigms." That is one of the two main senses in which the term occurs in the book. For it I might now adopt the notation "paradigm_1," but less confusion will result if I instead replace it with the phrase "disciplinary matrix"—"disciplinary" because it is the common possession of the practitioners of a professional discipline and "matrix" because it is composed of ordered elements of various sorts, each requiring further specification. Constituents of the disciplinary matrix include most or all of the objects of group commitment described in the book as paradigms, parts of paradigms, or paradigmatic.[8] I shall not at this time even attempt an exhaustive list but will instead briefly identify three of these which, because they are central to the cognitive operation of the group, should particularly concern philosophers of science. Let me refer to them as symbolic generalizations, models, and exemplars.

The first two are already familiar objects of philosophical attention. Symbolic generalizations, in particular, are those expressions, deployed without question by the group, which can readily be cast in some logical form like $(x)(y)(z)\phi(x, y, z)$. They are the formal, or the readily formalizable, components of the disciplinary matrix. Models, about which I shall have nothing further to say in this paper, are what provide the group with preferred analogies

"Comparison of the Results of Bibliographic Coupling and Analytic Subject Indexing," *American Documentation* 16 (1965): 223–33; D. J. Price, "Networks of Scientific Papers," *Science* 149 (1965): 510–15.

8. See *Structure of Scientific Revolutions*, pp. 38–42.

or, when deeply held, with an ontology. At one extreme they are heuristic: the electric circuit may fruitfully be regarded as a steady-state hydrodynamic system, or a gas behaves like a collection of microscopic billiard balls in random motion. At the other, they are the objects of metaphysical commitment: the heat of a body *is* the kinetic energy of its constituent particles, or, more obviously metaphysical, all perceptible phenomena are due to the motion and interaction of qualitatively neutral atoms in the void.[9] Exemplars, finally, are concrete problem solutions, accepted by the group as, in a quite usual sense, paradigmatic. Many of you will already have guessed that the term "exemplar" provides a new name for the second, and more fundamental, sense of "paradigm" in the book.

To understand how a scientific community functions as a producer and validator of sound knowledge, we must ultimately, I think, understand the operation of at least these three components of the disciplinary matrix. Alterations in any one can result in changes of scientific behavior affecting both the locus of a group's research and its standards of verification. Here I shall not attempt to defend a thesis quite so general. My primary concern is now with exemplars. To make room for them, however, I must first say something about symbolic generalizations.

In the sciences, particularly in physics, generalizations are often found already in symbolic form: $f = ma$, $I = V/R$, or $\nabla^2\psi + 8\pi^2 m/h^2(E - V)\psi = 0$. Others are ordinarily expressed in words: "action equals reaction," "chemical composition is in fixed proportions by weight," or "all cells come from cells." No one will question that the members of a scientific community do routinely deploy expressions like these in their work, that they ordinarily do so without felt need for special justification, and that they are seldom challenged at such points by other members of their group. That behavior is important, for without a shared commitment to a set of symbolic generalizations, logic and mathematics could not routinely be applied in the community's work. The example of taxonomy suggests that a science can exist with few, perhaps with no, such

9. It is not usual to include, say, atoms, fields, or forces acting at a distance under the rubric of models, but I presently see no harm in the broadened usage. Obviously the degree of a community's commitment varies as one goes from heuristic to metaphysical models, but the nature of the models' cognitive functions seems to remain the same.

generalizations. I shall later suggest how this could be the case. But I see no reason to doubt the widespread impression that the power of a science increases with the number of symbolic generalizations its practitioners have at their disposal.

Note, however, how small a measure of agreement we have yet attributed to the members of our community. When I say they share a commitment to, say, the symbolic generalization $f = ma$, I mean only that they will raise no difficulties for the man who inscribes the four symbols f, $=$, m, and a in succession on a line, who manipulates the resulting expression by logic and mathematics, and who exhibits a still symbolic result. For us at this point in the discussion, though not for the scientists who use them, these symbols and the expressions formed by compounding them are uninterpreted, still empty of empirical meaning or application. A shared commitment to a set of generalizations justifies logical and mathematical manipulation and induces commitment to the result. It need not, however, imply agreement about the manner in which the symbols, individually and collectively, are to be correlated with the results of experiment and observation. To this extent the shared symbolic generalizations function as yet like expressions in a pure mathematical system.

The analogy between a scientific theory and a pure mathematical system has been widely exploited in twentieth-century philosophy of science and has been responsible for some extremely interesting results. But it is only an analogy and can therefore be misleading. I believe that in several respects we have been victimized by it. One of them has immediate relevance to my argument.

When an expression like $f = ma$ appears in a pure mathematical system, it is, so to speak, there once and for all. If, that is, it enters into the solution of a mathematical problem posed within the system, it always enters in the form $f = ma$ or in a form reducible to that one by the substitutivity of identities or by some other syntactic substitution rule. In the sciences symbolic generalizations ordinarily behave very differently. They are not so much generalizations as generalization-sketches, schematic forms whose detailed symbolic expression varies from one application to the next. For the problem of free fall, $f = ma$ becomes $mg = md^2s/dt^2$. For the simple pendulum, it becomes $mg\mathrm{Sin}\theta = -m\mathrm{d}^2\mathrm{s}/\mathrm{dt}^2$. For coupled harmonic oscillators it becomes two equations, the first of which may be written $m_1\mathrm{d}^2\mathrm{s}_1/\mathrm{dt}^2 + \mathrm{k}_1\mathrm{s}_1 = \mathrm{k}_2(\mathrm{d} + \mathrm{s}_2 - \mathrm{s}_1)$. More interesting mechanical problems, for example, the motion of a gyroscope, would display still greater disparity between $f = ma$ and the actual symbolic generalization to which logic and mathematics are applied; but the point should already be clear. Though uninterpreted symbolic expressions are the common possession of the members of a scientific community, and though it is such expressions which provide the group with an entry point for logic and mathematics, it is not to the shared generalization that these tools are applied but to one or another special version of it. In a sense, each such class requires a new formalism.[10]

An interesting conclusion follows, one with likely relevance to the status of theoretical terms. Those philosophers who exhibit scientific theories as uninterpreted formal systems often remark that empirical reference enters such theories from the bottom up, moving from an empirically meaningful basic vocabulary into the theoretical terms. Despite the well-known difficulties that cluster about the notion of a basic vocabulary, I cannot doubt the importance of that route in the transformation of an uninterpreted symbol into the sign for a particular physical concept. But it is not the only route. Formalisms in science also attach to nature at the top, without intervening deduction which eliminates theoretical terms. Before he can begin the logical and mathematical manipulations which eventuate with the prediction of meter readings, the scientist must inscribe the particular form of $f = ma$ that applies to, say, the vibrating string or the particular form of the Schrödinger equation which applies to, say, the helium atom in a magnetic field. Whatever procedure he employs in doing so, it cannot be purely syntactic. Empirical content must enter formalized theories from the top as well as the bottom.

One cannot, I think, escape this conclusion by suggesting that

10. This difficulty cannot be evaded by stating the laws of Newtonian mechanics in, say, a Lagrangian or Hamiltonian form. On the contrary, the latter formulations are explictly law sketches rather than laws, as Newton's formulation of mechanics is not. Starting with Hamilton's equations or Lagrange's, one must still write down a particular Hamiltonian or Lagrangian for the particular problem at hand. Notice, however, that a crucial advantage of these formulations is that they make it far easier to identify the particular formalism suitable to a particular problem. Contrasted with Newton's formulation, they thus illustrate a typical direction of normal scientific development.

the Schrödinger equation or $f = ma$ be construed as an abbreviation for a conjunction of the numerous particular symbolic forms which these expressions take for application to particular physical problems. In the first place, scientists would still require criteria to tell them which particular symbolic version should be applied to which problem, and these criteria, like the correlation rules that are said to transport meaning from a basic vocabulary to theoretical terms, would be a vehicle for empirical content. Besides, no conjunction of particular symbolic forms would exhaust what the members of a scientific community can properly be said to know about how to apply symbolic generalizations. Confronted with a new problem, they can often agree on the particular symbolic expression appropriate to it, even though none of them has seen that particular expression before.

Any account of the cognitive apparatus of a scientific community may reasonably be asked to tell us something about the way in which the group's members, in advance of *directly* relevant empirical evidence, identify the special formalism appropriate to a particular problem, especially to a new problem. That clearly is one of the functions which scientific knowledge does serve. It does not, of course, always do so correctly; there is room, indeed need, for empirical checks on a special formalism proposed for a new problem. The deductive steps and the comparison of their end products with experiment remain prerequisites of science. But special formalisms are regularly accepted as plausible or rejected as implausible in advance of experiment. With remarkable frequency, furthermore, the community's judgments prove to be correct. Designing a special formalism, a new version of the formalization, cannot therefore be quite like inventing a new theory. Among other things, the former can be taught as theory invention cannot. That is what the problems at the ends of chapters in science texts are principally for. What can it be that students learn while solving them?

To that question most of the remainder of this paper is devoted, but I shall approach it indirectly, asking at first a more usual one: How do scientists attach symbolic expressions to nature? That is, in fact, two questions in one, for it may be asked either about a special symbolic generalization designed for a particular experimental situation or about a singular symbolic consequence of that generalization deduced for comparison with experiment. For present purposes, however, we may treat these two questions as one. In scientific practice, also, they are ordinarily answered together.

Since the abandonment of hope for a sense-datum language, the usual answer to this question has been in terms of correspondence rules. These have ordinarily been taken to be either operational definitions of scientific terms or else a set of necessary and sufficient conditions for the terms' applicability.[11] I do not myself doubt that the examination of a given scientific community would disclose a number of such rules shared by its members. Probably a few others could legitimately be induced from close observation of their behavior. But, for reasons I have given elsewhere and shall advert to briefly below, I do doubt that the correspondence rules discovered in this way would be nearly sufficient in number or force to account for the actual correlations between formalism and experiment made regularly and unproblematically by members of the

11. Since this paper was read, I have realized that eliding the two questions mentioned in the preceding paragraph introduces a possible source of confusion at this point and below. In normal philosophical usage, correspondence rules connect words only to other words, not to nature. Thus theoretical terms acquire meaning via the correspondence rules that attach them to a previously meaningful basic vocabulary. Only the latter attach directly to nature. Part of my argument is directed to this standard view and should therefore create no problems. The distinction between a theoretical and a basic vocabulary will not do in its present form because many theoretical terms can be shown to attach to nature in the same way, whatever it may be, as basic terms. But I am in addition concerned to inquire how "direct attachment" may work, whether of a theoretical or basic vocabulary. In the process I attack the often implicit assumption that anyone who knows how to use a basic term correctly has access, conscious or unconscious, to a set of criteria which define that term or provide necessary and sufficient conditions governing its application. For that mode of attachment by criteria I am also here using the term "correspondence rule," and that does violate normal usage. My excuse for the extension is my belief that explicit reliance on correspondence rules and implicit reliance on criteria introduce the same procedure and misdirect attention in the same ways. Both make the deployment of language seem more a matter of convention than it is. As a result they disguise the extent to which a man who acquires either an everyday or a scientific language simultaneously learns things about nature which are not themselves embodied in verbal generalizations.

group.[12] If the philosopher wants an adequate body of correspondence rules, he will have to supply most of them for himself.[13]

Almost surely that is a job he can do. Examining the collected examples of past community practice, the philosopher may reasonably expect to construct a set of correspondence rules adequate, in conjunction with known symbolic generalizations, to account for them all. Very likely he would be able to construct several alternate sets. Nevertheless, he ought to be extraordinarily wary about describing any one of them as a reconstruction of the rules held by the community under study. Though each of his sets of rules would be equivalent with respect to the community's past practice, they need not be equivalent when applied to the very next problem faced by the discipline. In that sense they would be reconstructions of slightly different theories, none of which need be the one held by the group. The philosopher might well, by behaving as a scientist, have improved the group's theory, but he would not, as a philosopher, have analyzed it.

Suppose, for example, that the philosopher is concerned with

12. See *Structure of Scientific Revolutions*, pp. 43–51.

13. It is, I think, remarkable how little attention philosophers of science have paid to the language-nature link. Surely, the epistemic force of the formalists' enterprise depends upon the possibility of making it unproblematic. One reason for this neglect is, I suspect, a failure to notice how much has been lost, from an epistemological standpoint, in the transition from a sense-datum language to a basic vocabulary. While the former seemed viable, definitions and correspondence rules required no special attention. "Green patch there" scarcely needed further operational specification; "benzene boils at 80° centigrade" is, however, a very different sort of statement. In addition, as I shall suggest below, formalists have frequently conflated the task of *improving* the clarity and structure of the formal elements of a scientific theory with the quite different job of *analyzing* scientific knowledge, and only the latter raises the problems of present concern. Hamilton produced a better formulation of Newtonian mechanics than had Newton, and the philosopher may hope to effect further improvements by further formalization. But he may not take for granted that he emerges with the same theory with which he began nor that the formal elements of either version of the theory are coextensive with the theory itself. For a typical example of the assumption that a perfected formalism is ipso facto an account of the knowledge deployed by the community which uses the formalism to be improved, see Patrick Suppes, "The Desirability of Formalization in Science," *Journal of Philosophy* 65 (1968): 651–64.

Ohm's law, $I = V/R$, and that he knows that the members of the group he studies measure voltage with an electrometer and current with a galvanometer. Seeking a correspondence rule for resistance, he may choose the quotient of voltage divided by current, in which case Ohm's law becomes a tautology. Or he may instead choose to correlate the value of resistance with the results of measurements made on a Wheatstone Bridge, in which case Ohm's law provides information about nature. For past practice the two reconstructions may be equivalent, but they will not dictate the same future behavior. Imagine, in particular, that an especially adept experimentalist in the community applies higher voltages than any realized before and discovers that the voltage-to-current ratio changes gradually at high voltage. According to the second, the Wheatstone Bridge, reconstruction, he has discovered that there are deviations from Ohm's law at high voltage. On the first reconstruction, however, Ohm's law is a tautology and deviations from it are unimaginable. The experimentalist has discovered, not a deviation from the law, but rather that resistance changes with voltage. The two reconstructions lead to different localizations of the difficulty and to different patterns of follow-up research.[14]

14. A less artificial example would require the simultaneous manipulation of several symbolic generalizations and would thus demand more space than is presently available. But historical examples that display the differential effects of generalizations held as laws and as definitions are not hard to find (see the discussion of Dalton and the Proust-Berthollet controversy in *Structure of Scientific Revolutions*, pp. 129–34), nor is the present example without historical foundation. Ohm did measure resistance by dividing current into voltage. His law thus provided a part of a definition of resistance. One of the reasons it proved so notably difficult to accept (neglect of Ohm is one of the most famous examples of resistance to innovation offered by the history of science) is that it was incompatible with the concept of resistance accepted prior to Ohm's work. Just because it demanded redefinition of electrical concepts, the assimilation of Ohm's law produced a revolution in electrical theory. (For parts of this story, see T. M. Brown, "The Electric Current in Early Nineteenth-Century Electricity," *Historical Studies in the Physical Sciences* 1 (1969): 61–103, and M. L. Schagrin, "Resistance to Ohm's Law," *American Journal of Physics* 31 (1963): 536–47.) I suspect that, quite generally, scientific revolutions can be distinguished from normal scientific developments in that the former require, as the latter do not, the modification of generalizations which had previously been regarded as quasi-analytic. Did Einstein discover the relativity of simultaneity or did he destroy a previously tautologous implication of that term?

Nothing in the preceding discussion proves that there is no set of correspondence rules adequate to explain the behavior of the community under study. A negative of that sort scarcely can be proven. But the discussion may lead us to take a bit more seriously some aspects of scientific training and behavior that philosophers have often managed to look right through. Very few correspondence rules are to be found in science texts or science teaching. How can the members of a scientific community have acquired a sufficient set? It is also noteworthy that if asked by a philosopher to provide such rules, scientists regularly deny their relevance and thereafter sometimes grow uncommonly inarticulate. When they cooperate at all, the rules they produce may vary from one member of the community to another, and all may be defective. One begins to wonder whether more than a few such rules are deployed in community practice, whether there is not some alternate way in which scientists correlate their symbolic expressions with nature.

A phenomenon familiar both to students of science and to historians of science provides a clue. Having been both, I shall speak from experience. Students of physics regularly report that they have read through a chapter of their text, understood it perfectly, but nonetheless had difficulty solving the problems at the end of the chapter. Almost invariably their difficulty is in setting up the appropriate equations, in relating the words and examples given in the text to the particular problems they are asked to solve. Ordinarily, also, those difficulties dissolve in the same way. The student discovers a way to see his problem as like a problem he has already encountered. Once that likeness or analogy has been seen, only manipulative difficulties remain.

The same pattern shows clearly in the history of science. Scientists model one problem solution on another, often with only a minimal recourse to symbolic generalizations. Galileo found that a ball rolling down an incline acquires just enough velocity to return it to the same vertical height on a second incline of any slope, and he learned to see that experimental situation as like the pendulum with a point-mass for a bob. Huyghens then solved the problem of the center of oscillation of a physical pendulum by imagining that the extended body of the latter was composed of Galilean point-pendula, the bonds between which could be instantaneously released at any point in the swing. After the bonds were released, the individual point-pendula would swing freely, but their collective center of gravity, like that of Galileo's pendulum, would rise only to the height from which the center of gravity of the extended pendulum had begun to fall. Finally, Daniel Bernoulli, still with no aid from Newton's laws, discovered how to make the flow of water from an orifice in a storage tank resemble Huyghens's pendulum. Determine the descent of the center of gravity of the water in tank and jet during an infinitesimal interval of time. Next imagine that each particle of water afterwards moves separately upward to the maximum height obtainable with the velocity it possessed at the end of the interval of descent. The ascent of the center of gravity of the separate particles must then equal the descent of the center of gravity of the water in tank and jet. From that view of the problem, the long-sought speed of efflux followed at once.[15]

Lacking time to multiply examples, I suggest that an acquired ability to see resemblances between apparently disparate problems plays in the sciences a significant part of the role usually attributed to correspondence rules. Once a new problem is seen to be analogous to a problem previously solved, both an appropriate formalism and a new way of attaching its symbolic consequences to nature follow. Having seen the resemblance, one simply uses the attachments that have proved effective before. That ability to recognize group-licensed resemblances is, I think, the main thing students acquire by doing problems, whether with pencil and paper or in a well-designed laboratory. In the course of their training a vast number of such exercises are set for them, and students entering the same specialty regularly do very nearly the same ones, for example, the inclined plane, the conical pendulum, Kepler ellipses, and so on. These concrete problems with their solutions are what I previously referred to as exemplars, a community's standard examples. They constitute the third main sort of cognitive component of the disciplinary matrix, and they illustrate the second main func-

15. For the example, see René Dugas, *A History of Mechanics*, trans. J. R. Maddox (Neuchâtel: Éditiones du Griffon and New York: Central Book Co., 1955), pp. 135–36, 186–93, and Daniel Bernoulli, *Hydrodynamica, sive de viribus et motibus fluidorum, commentarii opus academicum* (Strasbourg: J. R. Dulseckeri, 1738), sec. 3. For the extent to which mechanics progressed during the first half of the eighteenth century by modeling one problem solution on another, see Clifford Truesdell, "Reactions of Late Baroque Mechanics to Success, Conjecture, Error, and Failure in Newton's *Principia*," *Texas Quarterly* 10 (1967): 238–58.

tion of the term "paradigm" in *The Structure of Scientific Revolutions*.[16] Acquiring an arsenal of exemplars, just as much as learning symbolic generalizations, is integral to the process by which a student gains access to the cognitive achievements of his disciplinary group.[17] Without exemplars he would never learn much of what the group knows about such fundamental concepts as force and field, element and compound, or nucleus and cell.

I shall shortly attempt, by means of a simple example, to explicate the notion of a learned similarity relationship, an acquired perception of analogy. Let me first, however, sharpen the problem at which that explication will be aimed. It is a truism that anything is similar to, and also different from, anything else. It depends, we usually say, on the criteria. To the man who speaks of similarity or of analogy, we therefore at once pose the question: similar with respect to what? In this case, however, that is just the question that must not be asked, for an answer would at once provide us with correspondence rules. Acquiring exemplars would teach the student nothing that such rules, in the form of criteria of resemblance, could not equally well have supplied. Doing problems would then be mere practice in the application of rules, and there would be no need for talk of similarity.

Doing problems, however, I have already argued, is not like that. Much more nearly it resembles the child's puzzle in which one is asked to find the animal shapes or faces hidden in the drawing of shrubbery or clouds. The child seeks forms that are like those of the animals or faces he knows. Once they are found, they do not again retreat into the background, for the child's way of seeing the

16. It is, of course, the sense of "paradigm" as standard example that led originally to my choice of that term. Unfortunately, most readers of *The Structure of Scientific Revolutions* have missed what was for me its central function and use "paradigm" in a sense close to that for which I now suggest "disciplinary matrix." I see little chance of recapturing "paradigm" for its original use, the only one that is philologically at all appropriate.

17. Note that exemplars (and also models) are far more effective determinants of community substructure than are symbolic generalizations. Many scientific communities share, for example, the Schrödinger equation, and their members encounter that formula correspondingly early in their scientific education. But, as that training continues, say toward solid state physics on the one hand and field theory on the other, the exemplars they encounter diverge. Thereafter it is only the uninterpreted, not the interpreted, Schrödinger equation they can unequivocally be said to share.

picture has been changed. In the same way, the science student, confronted with a problem, seeks to see it as like one or more of the exemplary problems he has encountered before. Where rules exist to guide him, he, of course, deploys them. But his basic criterion is a perception of similarity that is both logically and psychologically prior to any of the numerous criteria by which that same identification of similarity might have been made. After the similarity has been seen, one may ask for criteria, and it is then often worth doing so. But one need not. The mental or visual set acquired while learning to see two problems as similar can be applied directly. Under appropriate circumstances, I now want to argue, there is a means of processing data into similarity sets which does not depend on a prior answer to the question, similar with respect to what?

My argument begins with a brief digression on the term "data." Philologically it derives from "the given." Philosophically, for reasons deeply engrained in the history of epistemology, it isolates the minimal stable elements provided by our senses. Though we no longer hope for a sense-datum language, phrases like "green there," "triangle here," or "hot down there" continue to connote our paradigms for a datum, the given in experience. In several respects, they should play this role. We have no access to elements of experience more minimal than these. Whenever we consciously process data, whether to identify an object, to discover a law, or to invent a theory, we necessarily manipulate sensations of this sort or compounds of them. Nevertheless, from another point of view, sensations and their elements are not the given. Viewed theoretically rather than experientially, that title belongs rather to stimuli. Though we have access to them only indirectly, via scientific theory, it is stimuli, not sensations, that impinge on us as organisms. A vast amount of neural processing takes place between our receipt of a stimulus and the sensory response which is our datum.

None of this would be worth saying if Descartes had been right in positing a one-to-one correspondence between stimuli and sensations. But we know that nothing of the sort exists. The perception of a given color can be evoked by an infinite number of differently combined wavelengths. Conversely, a given stimulus can evoke a variety of sensations, the image of a duck in one recipient, the image of a rabbit in another. Nor are responses like these entirely innate. One can learn to discriminate colors or patterns which were indistinguishable prior to training. To an extent still unknown, the production of data from stimuli is a learned procedure. After the learning process, the same stimulus evokes a different datum. I conclude that, though data are the minimal elements of our individual experience, they need be shared responses to a given stimulus only within the membership of a relatively homogeneous community, educational, scientific, or linguistic.[18]

Return now to my main argument, but not to scientific examples. Inevitably the latter prove excessively complex. Instead I ask that you imagine a small child on a walk with his father in a zoological garden. The child has previously learned to recognize birds and to discriminate robin redbreasts. During the afternoon now at hand, he will learn for the first time to identify swans, geese, and ducks. Anyone who has taught a child under such circumstances knows that the primary pedagogic tool is ostension. Phrases like "all swans are white" may play a role, but they need not. I shall for the moment omit them from consideration, my object being to isolate a different mode of learning in its purest form. Johnny's education then proceeds as follows. Father points to a bird, saying, "Look, Johnny, there's a swan." A short time later Johnny himself points to a bird, saying, "Daddy, another swan." He has not yet, however, learned what swans are and must be corrected: "No, Johnny, that's a goose." Johnny's next identification of a swan proves to be correct, but his next "goose" is, in fact, a duck, and he is again set straight. After a few more such encounters, however, each with its appropriate correction or reinforcement, Johnny's ability to identify these waterfowl is as great as his father's. Instruction has been quickly completed.

I ask now what has happened to Johnny, and I urge the plausibility of the following answer. During the afternoon, part of the neural mechanism by which he processes visual stimuli has been

18. In *The Structure of Scientific Revolutions*, particularly chap. 10, I repeatedly insist that members of different scientific communities live in different worlds and that scientific revolutions change the world in which a scientist works. I would now want to say that members of different communities are presented with different data by the same stimuli. Notice, however, that that change does not make phrases like "a different world" inappropriate. The given world, whether everyday or scientific, is not a world of stimuli.

reprogrammed, and the data he receives from stimuli which would all earlier have evoked "bird" have changed. When he began his walk, the neural program highlighted the differences between individual swans as much as those between swans and geese. By the end of the walk, features like the length and curvature of the swan's neck have been highlighted and others have been suppressed so that swan data match each other and differ from goose and duck data as they had not before. Birds that had previously all looked alike (and also different) are now grouped in discrete clusters in perceptual space.

A process of this sort can readily be modeled on a computer; I am in the early stages of such an experiment myself. A stimulus, in the form of a string of *n* ordered digits, is fed to the machine. There it is transformed to a datum by the application of a preselected transformation to each of the *n* digits, a different transformation being applied to each position in the string. Every datum thus obtained is a string of *n* numbers, a position in what I shall call an *n*-dimensional quality space. In this space the distance between two data, measured with a euclidean or a suitable noneuclidean metric, represents their similarity. Which stimuli transform to similar or nearby data depends, of course, on the choice of transformation functions. Different sets of functions produce different clusters of data, different patterns of similarity and difference, in perceptual space. But the transformation functions need not be manmade. If the machine is given stimuli which can be grouped in clusters and if it is informed which stimuli must be placed in the same and which in different clusters, it can design an appropriate set of transformation functions for itself. Note that both conditions are essential. Not all stimuli can be transformed to form data clusters. Even when they can, the machine, like the child, must be told at first which ones belong together and which apart. Johnny did not discover for himself that there were swans, geese, and ducks. Rather he was taught it.

If we now represent Johnny's perceptual space in a two-dimensional diagram, the process he has undergone is rather like the transition from figure 1 to figure 2.[19] In the first, ducks, geese, and swans are mixed together. In the second, they have clustered in

19. For the drawings I am indebted to both the pen and patience of Sarah Kuhn.

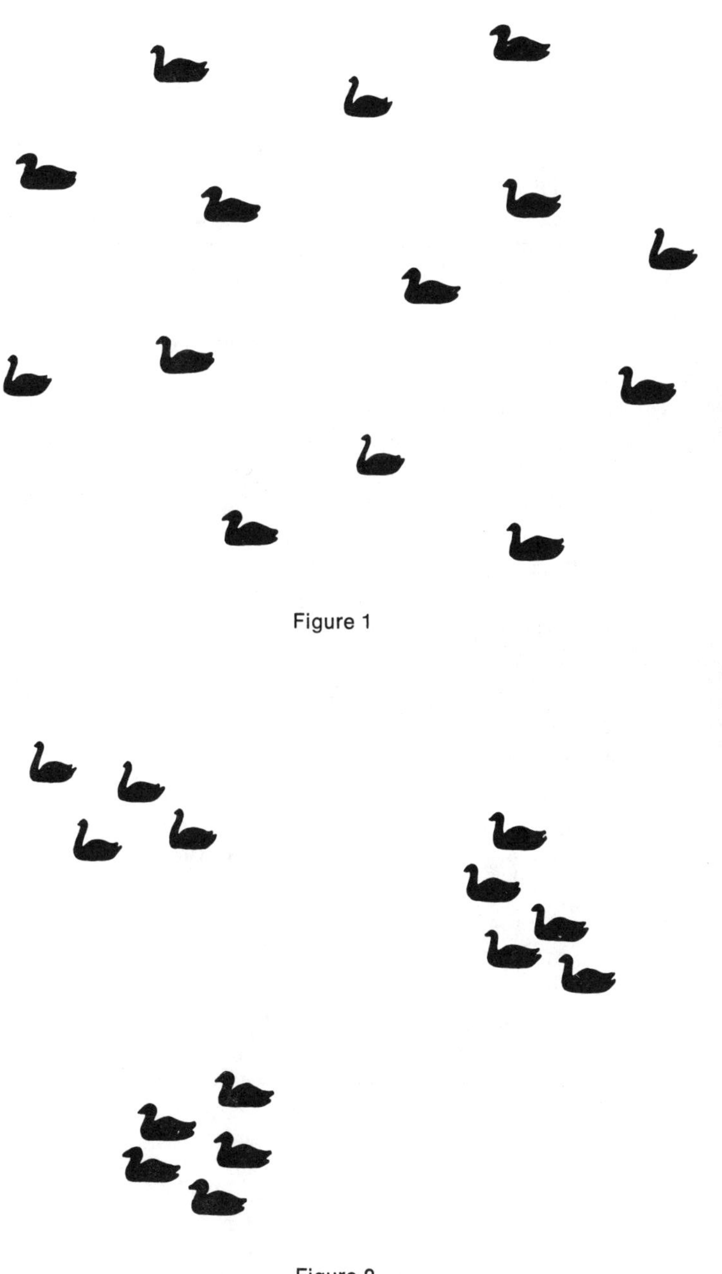

Figure 1

Figure 2

discrete sets with appreciable distances between them.[20] Since Johnny's father has, in effect, told him that ducks, geese, and swans are members of discrete natural families, Johnny has every right to expect that all future ducks, geese, and swans will fall naturally into or at the edge of one of these families, and that he will encounter no datum that falls in the region midway between them. That expectation may be violated, perhaps during a visit to Australia. But it will serve him well while he remains a member of the community that has discovered from experience the utility and viability of these particular perceptual discriminations and has transmitted the ability to make them from one generation to the next.

By being programmed to recognize what his prospective community already knows, Johnny has acquired consequential information. He has learned that geese, ducks, and swans form discrete natural families and that nature offers no swan-geese or goose-ducks. Some quality constellations go together; others are not found at all. If the qualities in his clusters include aggressiveness, his afternoon in the park may have had behavioral as well as everyday zoological functions. Geese, unlike swans and ducks, hiss and bite. What Johnny has learned is thus worth knowing. But does he know what the terms "goose," "duck," and "swan" mean? In any useful sense, yes, for he can apply these labels unequivocally and without effort, drawing behavioral conclusions from their application, either directly, or via general statements. On the other hand, he has learned all this without acquiring, or at least without needing to acquire, even one criterion for identifying swans, geese, or ducks. He can point to a swan and tell you there must be water nearby, but he may well be unable to tell you what a swan is.

Johnny, in short, has learned to apply symbolic labels to nature without anything like definitions or correspondence rules. In their absence he employs a learned but nonetheless primitive perception of similarity and difference. While acquiring the perception, he has learned something about nature. This knowledge can thereafter be embedded, not in generalizations or rules, but in the similarity relationship itself. I do not, let me emphasize, at all suppose

20. It will become apparent below that everything which is special about this method of processing stimuli depends upon the possibility of grouping data in clusters with empty space between them. In the absence of empty space, there is no alternative to the processing strategy that, designed for a world of all possible data, relies upon definitions and rules.

Johnny's technique is the only one by which knowledge is acquired and stored. Nor do I think it likely that very much human knowledge is acquired and stored with so little recourse to verbal generalizations. But I do urge the recognition of the integrity of a cognitive process like the one just outlined. In combination with more familiar processes, like symbolic generalization and modeling, it is, I think, essential to an adequate reconstruction of scientific knowledge.

Need I now say that the swans, geese, and ducks which Johnny encountered during his walk with father were what I have been calling exemplars? Presented to Johnny with their labels attached, they were solutions to a problem that the members of his prospective community had already resolved. Assimilating them is part of the socialization procedure by which Johnny is made part of that community and, in the process, learns about the world which the community inhabits. Johnny is, of course, no scientist, nor is what he has learned yet science. But he may well become a scientist, and the technique employed on his walk will still be viable. That he does, in fact, use it will be most obvious if he becomes a taxonomist. The herbaria, without which no botanist can function, are storehouses for professional exemplars, and their history is coextensive with that of the discipline they support. But the same technique, if in a less pure form, is essential to the more abstract sciences as well. I have already argued that assimilating solutions to such problems as the inclined plane and the conical pendulum is part of learning what Newtonian physics is. Only after a number of such problems have been assimilated, can a student or a professional proceed to identify other Newtonian problems for himself. That assimilation of examples is, furthermore, part of what enables him to isolate the forces, masses, and constraints within a new problem and to write down a formalism suitable for its solution. Despite its excessive simplicity, Johnny's case should suggest why I continue to insist that shared examples have essential cognitive functions prior to a specification of criteria with respect to which they are exemplary.

I conclude my argument by returning to a crucial question discussed earlier in connection with symbolic generalizations. Suppose scientists do assimilate and store knowledge in shared examples, need the philosopher concern himself with the process? May he not instead study the examples and derive correspondence rules which, together with the formal elements of the theory, would make the examples superfluous? To that question I have already suggested the following answer. The philosopher is at liberty to substitute rules for examples and, at least in principle, he can expect to succeed in doing so. In the process, however, he will alter the nature of the knowledge possessed by the community from which his examples were drawn. What he will be doing, in effect, is to substitute one means of data processing for another. Unless he is extraordinarily careful he will weaken the community's cognition by doing so. Even with care, he will change the nature of the community's future responses to some experimental stimuli.

Johnny's education, though not in science, provides a new sort of evidence for these claims. To identify swans, geese, and ducks by correspondence rules rather than by perceived similarity is to draw closed nonintersecting curves around each of the clusters in figure 2. What results is a simple Venn diagram, displaying three nonoverlapping classes. All swans lie in one, all geese in another, and so on. Where, however, should the curves be drawn? There are infinite possibilities. One of them is illustrated in figure 3, where boundaries are drawn very close to the bird figures in the three clusters. Given such boundaries, Johnny now can say what the criteria are for membership in the class of swans, geese, or ducks. On the other hand, he may be troubled by the very next waterfowl he sees. The outlined shape in the diagram is obviously a swan by the perceived distance criterion, but it is neither swan, goose, nor duck by the newly introduced correspondence rules for class membership.

Boundaries ought not, therefore, be drawn too near the edges of a cluster of exemplars. Let us therefore go to the other extreme, figure 4, and draw boundaries which exhaust most of the relevant parts of Johnny's perceptual space. With this choice, no bird that appears near one of the existing clusters will present a problem, but in avoiding that difficulty we have created another. Johnny used to know that there are no swan-geese. The new reconstruction of his knowledge deprives him of that information. Instead it supplies something he is extremely unlikely to need, the name that applies to a bird datum deep in the unoccupied space between swans and geese. To replace what has been lost we may imagine adding to John's cognitive apparatus a density function that describes the likelihood of his encountering a swan at various positions within

Figure 3

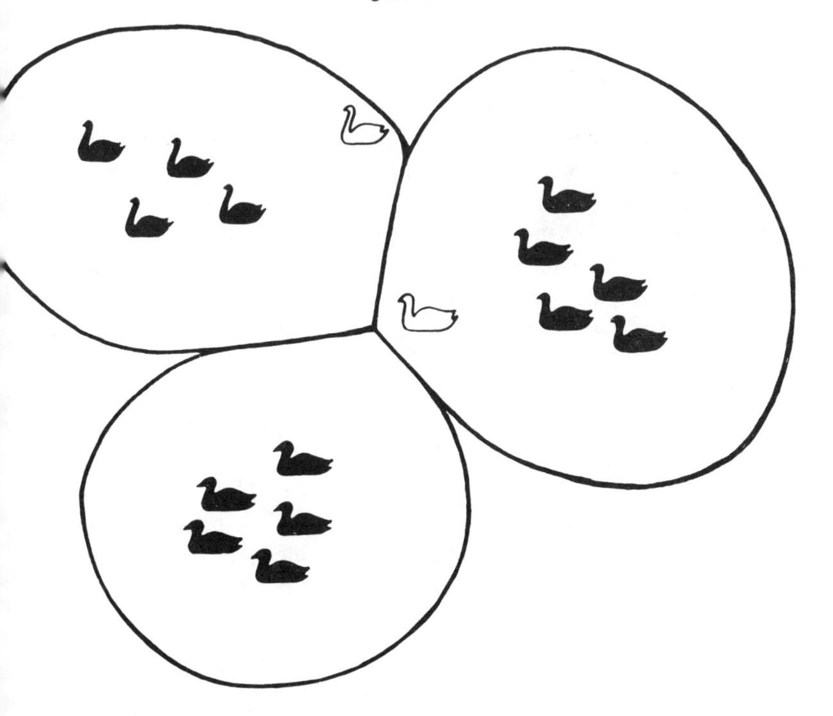

Figure 4

the swan boundary, together with similar functions for geese and ducks. But the original similarity criterion supplied those already. In effect we would just have returned to the data-processing mechanism we had meant to replace.

Clearly, neither of the extreme techniques for drawing class boundaries will do. The compromise indicated in figure 5 is an obvious improvement. Any bird which appears near one of the existing clusters belongs to it. Any bird which appears midway between clusters has no name, but there is unlikely ever to be such a datum. With class boundaries like these, Johnny should be able to operate successfully for some time. Yet he has gained nothing by substituting class boundaries for his original similarity criterion, and there has been some loss. If the strategic suitability of these boundaries is to be maintained, their location may need to be changed each time Johnny encounters another swan.

Figure 6 shows what I have in mind. Johnny has encountered one more swan. It lies, as it should, entirely within the old class boundary. There has been no problem of identification. But there may be one next time unless new boundaries, here shown as dotted lines, are drawn to take account of the altered shape of the swan cluster. Without the outward adjustment of the swan boundary, the very next bird encountered, though unproblematically a swan by the resemblance criterion, may fall on or even outside the old boundary. Without the simultaneous retraction of the duck boundary, the empty space, which Johnny's more experienced seniors have assured him can be preserved, would have become excessively narrow. If that is so—if, that is, each new experience can demand some adjustment of the class boundaries, one may well ask whether Johnny was wise to allow philosophers to draw any such boundaries for him. The primitive resemblance criterion he had previously acquired would have handled all these cases unproblematically and without continual adjustment. There is, I feel sure, such a thing as meaning change or change in the range of application of a term. But only the notion that meaning or applicability depends on predetermined boundaries could make us want to deploy any such phraseology here.[21]

21. By the same token one should here withhold phrases like "vagueness of meaning" or "open texture of concepts." Both imply an imperfection, something lacking that may later be supplied. That sense of imperfection is,

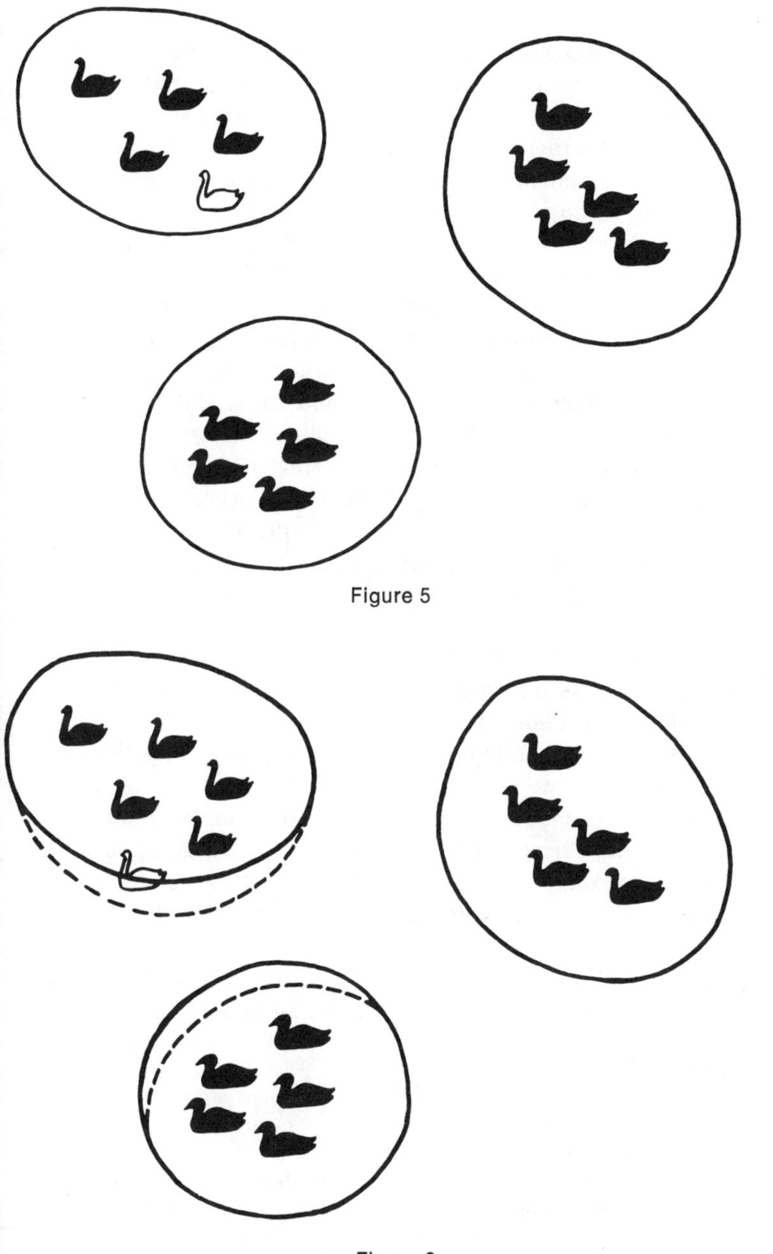

Figure 5

Figure 6

I am not, let me now emphasize, suggesting that there are never good reasons to draw boundaries or adopt correspondence rules. If Johnny had been presented with a series of birds that bridged the empty space between swans and geese, he would have been forced to resolve the resulting quandary with a line that divided the swan-goose continuum by definition. Or, if there were independent reasons for supposing that color is a stable criterion for the identification of waterfowl, Johnny might wisely have committed himself to the generalization, "all swans are white."[22] That strategy might save valuable data-processing time. In any case, the generalization would provide an entry point for logical manipulation. There are appropriate occasions for switching to the well-known strategy that relies upon boundaries and rules. But it is not the only available strategy for either stimuli- or data-processing. An alternative does exist, one based upon what I have been calling a learned perception of similarity. Observation, whether of language learning, scientific education, or scientific practice, suggests that it is, in fact, widely used. By ignoring it in epistemological discussion, we may do much violence to our understanding of the nature of knowledge.

Return, finally, to the term "paradigm." It entered *The Structure of Scientific Revolutions* because I, the book's historian-author, could not, when examining the membership of a scientific community, retrieve enough shared rules to account for the group's unproblematic conduct of research. Shared examples of successful practice could, I next concluded, provide what the group lacked in rules. Those examples were its paradigms, and as such essential to its continued research. Unfortunately, having gotten that far, I al-

however, created solely by a standard that demands our possessing necessary and sufficient conditions for the applicability of a word or phrase in a world of all possible data. In a world in which some data never appear, such a criterion is superfluous.

22. Note that Johnny's commitment to "all swans are white" may be a commitment either to a law about swans or to a (partial) definition of swans. He may, that is, receive the generalization either as analytic or synthetic. As suggested in note 14, above, the difference can prove consequential, particularly if Johnny next encounters a black waterfowl that, in other respects, closely resembles a swan. Laws drawn directly from observation are corrigible piecemeal as definitions generally are not.

lowed the term's applications to expand, embracing all shared group commitments, all components of what I now wish to call the disciplinary matrix. Inevitably, the result was confusion, and it obscured the original reasons for introducing a special term. But those reasons still stand. Shared examples can serve cognitive functions commonly attributed to shared rules. When they do, knowledge develops differently from the way it does when governed by rules. This paper has, above all, been an effort to isolate, clarify, and drive home those essential points. If they can be seen, we shall be able to dispense with the term "paradigm," though not with the concept that led to its introduction.

13

Objectivity, Value Judgment, and Theory Choice

Previously unpublished Machette Lecture, delivered at Furman University, 30 November 1973.

In the penultimate chapter of a controversial book first published fifteen years ago, I considered the ways scientists are brought to abandon one time-honored theory or paradigm in favor of another. Such decision problems, I wrote, "cannot be resolved by proof." To discuss their mechanism is, therefore, to talk "about techniques of persuasion, or about argument and counterargument in a situation in which there can be no proof." Under these circumstances, I continued, "lifelong resistance [to a new theory] . . . is not a violation of scientific standards. . . . Though the historian can always find men—Priestley, for instance—who were unreasonable to resist for as long as they did, he will not find a point at which resistance becomes illogical or unscientific."[1] Statements of that sort obviously raise the question of why, in the absence of binding criteria for scientific choice, both the number of solved scientific problems and the precision of individual problem solutions should increase so markedly with the passage of time. Confronting that issue, I sketched in my closing chapter a number of characteristics that scientists share by virtue of the training which licenses their membership in one or another community of specialists. In the absence of criteria able to dictate the choice of each individual, I argued,

1. *The Structure of Scientific Revolutions*, 2d ed. (Chicago, 1970), pp. 148, 151–52, 159. All the passages from which these fragments are taken appeared in the same form in the first edition, published in 1962.

we do well to trust the collective judgment of scientists trained in this way. "What better criterion could there be," I asked rhetorically, "than the decision of the scientific group?"[2]

A number of philosophers have greeted remarks like these in a way that continues to surprise me. My views, it is said, make of theory choice "a matter for mob psychology."[3] Kuhn believes, I am told, that "the decision of a scientific group to adopt a new paradigm cannot be based on good reasons of any kind, factual or otherwise."[4] The debates surrounding such choices must, my critics claim, be for me "mere persuasive displays without deliberative substance."[5] Reports of this sort manifest total misunderstanding, and I have occasionally said as much in papers directed primarily to other ends. But those passing protestations have had negligible effect, and the misunderstandings continue to be important. I conclude that it is past time for me to describe, at greater length and with greater precision, what has been on my mind when I have uttered statements like the ones with which I just began. If I have been reluctant to do so in the past, that is largely because I have preferred to devote attention to areas in which my views diverge more sharply from those currently received than they do with respect to theory choice.

What, I ask to begin with, are the characteristics of a good scientific theory? Among a number of quite usual answers I select five, not because they are exhaustive, but because they are individually important and collectively sufficiently varied to indicate what is at stake. First, a theory should be accurate: within its domain, that is, consequences deducible from a theory should be in demonstrated agreement with the results of existing experiments and observations. Second, a theory should be consistent, not only inter-

2. Ibid., p. 170.

3. Imre Lakatos, "Falsification and the Methodology of Scientific Research Programmes," in I. Lakatos and A. Musgrave, eds., *Criticism and the Growth of Knowledge* (Cambridge, 1970), pp. 91–195. The quoted phrase, which appears on p. 178, is italicized in the original.

4. Dudley Shapere, "Meaning and Scientific Change," in R. G. Colodny, ed., *Mind and Cosmos: Essays in Contemporary Science and Philosophy*, University of Pittsburgh Series in the Philosophy of Science, vol. 3 (Pittsburgh, 1966), pp. 41–85. The quotation will be found on p. 67.

5. Israel Scheffler, *Science and Subjectivity* (Indianapolis, 1967), p. 81.

nally or with itself, but also with other currently accepted theories applicable to related aspects of nature. Third, it should have broad scope: in particular, a theory's consequences should extend far beyond the particular observations, laws, or subtheories it was initially designed to explain. Fourth, and closely related, it should be simple, bringing order to phenomena that in its absence would be individually isolated and, as a set, confused. Fifth—a somewhat less standard item, but one of special importance to actual scientific decisions—a theory should be fruitful of new research findings: it should, that is, disclose new phenomena or previously unnoted relationships among those already known.[6] These five characteristics—accuracy, consistency, scope, simplicity, and fruitfulness—are all standard criteria for evaluating the adequacy of a theory. If they had not been, I would have devoted far more space to them in my book, for I agree entirely with the traditional view that they play a vital role when scientists must choose between an established theory and an upstart competitor. Together with others of much the same sort, they provide *the* shared basis for theory choice.

Nevertheless, two sorts of difficulties are regularly encountered by the men who must use these criteria in choosing, say, between Ptolemy's astronomical theory and Copernicus's, between the oxygen and phlogiston theories of combustion, or between Newtonian mechanics and the quantum theory. Individually the criteria are imprecise: individuals may legitimately differ about their application to concrete cases. In addition, when deployed together, they repeatedly prove to conflict with one another; accuracy may, for example, dictate the choice of one theory, scope the choice of its competitor. Since these difficulties, especially the first, are also relatively familiar, I shall devote little time to their elaboration. Though my argument does demand that I illustrate them briefly, my views will begin to depart from those long current only after I have done so.

Begin with accuracy, which for present purposes I take to include not only quantitative agreement but qualitative as well. Ulti-

6. The last criterion, fruitfulness, deserves more emphasis than it has yet received. A scientist choosing between two theories ordinarily knows that his decision will have a bearing on his subsequent research career. Of course he is especially attracted by a theory that promises the concrete successes for which scientists are ordinarily rewarded.

mately it proves the most nearly decisive of all the criteria, partly because it is less equivocal than the others but especially because predictive and explanatory powers, which depend on it, are characteristics that scientists are particularly unwilling to give up. Unfortunately, however, theories cannot always be discriminated in terms of accuracy. Copernicus's system, for example, was not more accurate than Ptolemy's until drastically revised by Kepler more than sixty years after Copernicus's death. If Kepler or someone else had not found other reasons to choose heliocentric astronomy, those improvements in accuracy would never have been made, and Copernicus's work might have been forgotten. More typically, of course, accuracy does permit discriminations, but not the sort that lead regularly to unequivocal choice. The oxygen theory, for example, was universally acknowledged to account for observed weight relations in chemical reactions, something the phlogiston theory had previously scarcely attempted to do. But the phlogiston theory, unlike its rival, could account for the metals' being much more alike than the ores from which they were formed. One theory thus matched experience better in one area, the other in another. To choose between them on the basis of accuracy, a scientist would need to decide the area in which accuracy was more significant. About that matter chemists could and did differ without violating any of the criteria outlined above, or any others yet to be suggested.

However important it may be, therefore, accuracy by itself is seldom or never a sufficient criterion for theory choice. Other criteria must function as well, but they do not eliminate problems. To illustrate I select just two—consistency and simplicity—asking how they functioned in the choice between the heliocentric and geocentric systems. As astronomical theories both Ptolemy's and Copernicus's were internally consistent, but their relation to related theories in other fields was very different. The stationary central earth was an essential ingredient of received physical theory, a tight-knit body of doctrine which explained, among other things, how stones fall, how water pumps function, and why the clouds move slowly across the skies. Heliocentric astronomy, which required the earth's motion, was inconsistent with the existing scientific explanation of these and other terrestrial phenomena. The consistency criterion, by itself, therefore, spoke unequivocally for the geocentric tradition.

Simplicity, however, favored Copernicus, but only when evaluated in a quite special way. If, on the one hand, the two systems were compared in terms of the actual computational labor required to predict the position of a planet at a particular time, then they proved substantially equivalent. Such computations were what astronomers did, and Copernicus's system offered them no labor-saving techniques; in that sense it was not simpler than Ptolemy's. If, on the other hand, one asked about the amount of mathematical apparatus required to explain, not the detailed quantitative motions of the planets, but merely their gross qualitative features—limited elongation, retrograde motion, and the like—then, as every schoolchild knows, Copernicus required only one circle per planet, Ptolemy two. In that sense the Copernican theory was the simpler, a fact vitally important to the choices made by both Kepler and Galileo and thus essential to the ultimate triumph of Copernicanism. But that sense of simplicity was not the only one available, nor even the one most natural to professional astronomers, men whose task was the actual computation of planetary position.

Because time is short and I have multiplied examples elsewhere, I shall here simply assert that these difficulties in applying standard criteria of choice are typical and that they arise no less forcefully in twentieth-century situations than in the earlier and better-known examples I have just sketched. When scientists must choose between competing theories, two men fully committed to the same list of criteria for choice may nevertheless reach different conclusions. Perhaps they interpret simplicity differently or have different convictions about the range of fields within which the consistency criterion must be met. Or perhaps they agree about these matters but differ about the relative weights to be accorded to these or to other criteria when several are deployed together. With respect to divergences of this sort, no set of choice criteria yet proposed is of any use. One can explain, as the historian characteristically does, why particular men made particular choices at particular times. But for that purpose one must go beyond the list of shared criteria to characteristics of the individuals who make the choice. One must, that is, deal with characteristics which vary from one scientist to another without thereby in the least jeopardizing their adherence to the canons that make science scientific. Though such canons do exist and should be discoverable (doubtless the criteria of choice with which I began are among them), they are not by themselves sufficient to determine the decisions of individual scientists. For that purpose the shared canons must be fleshed out in ways that differ from one individual to another.

Some of the differences I have in mind result from the individual's previous experience as a scientist. In what part of the field was he at work when confronted by the need to choose? How long had he worked there; how successful had he been; and how much of his work depended on concepts and techniques challenged by the new theory? Other factors relevant to choice lie outside the sciences. Kepler's early election of Copernicanism was due in part to his immersion in the Neoplatonic and Hermetic movements of his day; German Romanticism predisposed those it affected toward both recognition and acceptance of energy conservation; nineteenth-century British social thought had a similar influence on the availability and acceptability of Darwin's concept of the struggle for existence. Still other significant differences are functions of personality. Some scientists place more premium than others on originality and are correspondingly more willing to take risks; some scientists prefer comprehensive, unified theories to precise and detailed problem solutions of apparently narrower scope. Differentiating factors like these are described by my critics as subjective and are contrasted with the shared or objective criteria from which I began. Though I shall later question that use of terms, let me for the moment accept it. My point is, then, that every individual choice between competing theories depends on a mixture of objective and subjective factors, or of shared and individual criteria. Since the latter have not ordinarily figured in the philosophy of science, my emphasis upon them has made my belief in the former hard for my critics to see.

What I have said so far is primarily simply descriptive of what goes on in the sciences at times of theory choice. As description, furthermore, it has not been challenged by my critics, who reject instead my claim that these facts of scientific life have philosophic import. Taking up that issue, I shall begin to isolate some, though I think not vast, differences of opinion. Let me begin by asking how philosophers of science can for so long have neglected the subjective elements which, they freely grant, enter regularly into the actual theory choices made by individual scientists? Why have these elements seemed to them an index only of human weakness, not at all of the nature of scientific knowledge?

One answer to that question is, of course, that few philosophers, if any, have claimed to possess either a complete or an entirely well-articulated list of criteria. For some time, therefore, they could reasonably expect that further research would eliminate residual imperfections and produce an algorithm able to dictate rational, unanimous choice. Pending that achievement, scientists would have no alternative but to supply subjectively what the best current list of objective criteria still lacked. That some of them might still do so even with a perfected list at hand would then be an index only of the inevitable imperfection of human nature.

That sort of answer may still prove to be correct, but I think no philosopher still expects that it will. The search for algorithmic decision procedures has continued for some time and produced both powerful and illuminating results. But those results all presuppose that individual criteria of choice can be unambiguously stated and also that, if more than one proves relevant, an appropriate weight function is at hand for their joint application. Unfortunately, where the choice at issue is between scientific theories, little progress has been made toward the first of these desiderata and none toward the second. Most philosophers of science would, therefore, I think, now regard the sort of algorithm which has traditionally been sought as a not quite attainable ideal. I entirely agree and shall henceforth take that much for granted.

Even an ideal, however, if it is to remain credible, requires some demonstrated relevance to the situations in which it is supposed to apply. Claiming that such demonstration requires no recourse to subjective factors, my critics seem to appeal, implicitly or explicitly, to the well-known distinction between the contexts of discovery and of justification.[7] They concede, that is, that the subjective factors I invoke play a significant role in the discovery or invention of new theories, but they also insist that that inevitably intuitive process lies outside of the bounds of philosophy of science and is irrelevant to the question of scientific objectivity. Objectivity enters science, they continue, through the processes by which the-

7. The least equivocal example of this position is probably the one developed in Scheffler, *Science and Subjectivity*, chap. 4.

ories are tested, justified, or judged. Those processes do not, or at least need not, involve subjective factors at all. They can be governed by a set of (objective) criteria shared by the entire group competent to judge.

I have already argued that that position does not fit observations of scientific life and shall now assume that that much has been conceded. What is now at issue is a different point: whether or not this invocation of the distinction between contexts of discovery and of justification provides even a plausible and useful idealization. I think it does not and can best make my point by suggesting first a likely source of its apparent cogency. I suspect that my critics have been misled by science pedagogy or what I have elsewhere called textbook science. In science teaching, theories are presented together with exemplary applications, and those applications may be viewed as evidence. But that is not their primary pedagogic function (science students are distressingly willing to receive the word from professors and texts). Doubtless *some* of them were *part* of the evidence at the time actual decisions were being made, but they represent only a fraction of the considerations relevant to the decision process. The context of pedagogy differs almost as much from the context of justification as it does from that of discovery.

Full documentation of that point would require longer argument than is appropriate here, but two aspects of the way in which philosophers ordinarily demonstrate the relevance of choice criteria are worth noting. Like the science textbooks on which they are often modelled, books and articles on the philosophy of science refer again and again to the famous crucial experiments: Foucault's pendulum, which demonstrates the motion of the earth; Cavendish's demonstration of gravitational attraction; or Fizeau's measurement of the relative speed of sound in water and air. These experiments are paradigms of good reason for scientific choice; they illustrate the most effective of all the sorts of argument which could be available to a scientist uncertain which of two theories to follow; they are vehicles for the transmission of criteria of choice. But they also have another characteristic in common. By the time they were performed no scientist still needed to be convinced of the validity of the theory their outcome is now used to demonstrate. Those decisions had long since been made on the basis of significantly more equivocal evidence. The exemplary crucial experiments to which philosophers again and again refer would have been historically relevant to theory choice only if they had yielded unexpected results. Their use as illustrations provides needed economy to science pedagogy, but they scarcely illuminate the character of the choices that scientists are called upon to make.

Standard philosophical illustrations of scientific choice have another troublesome characteristic. The only arguments discussed are, as I have previously indicated, the ones favorable to the theory that, in fact, ultimately triumphed. Oxygen, we read, could explain weight relations, phlogiston could not; but nothing is said about the phlogiston theory's power or about the oxygen theory's limitations. Comparisons of Ptolemy's theory with Copernicus's proceed in the same way. Perhaps these examples should not be given since they contrast a developed theory with one still in its infancy. But philosophers regularly use them nonetheless. If the only result of their doing so were to simplify the decision situation, one could not object. Even historians do not claim to deal with the full factual complexity of the situations they describe. But these simplifications emasculate by making choice totally unproblematic. They eliminate, that is, one essential element of the decision situations that scientists must resolve if their field is to move ahead. In those situations there are always at least some good reasons for each possible choice. Considerations relevant to the context of discovery are then relevant to justification as well; scientists who share the concerns and sensibilities of the individual who discovers a new theory are ipso facto likely to appear disproportionately frequently among that theory's first supporters. That is why it has been difficult to construct algorithms for theory choice, and also why such difficulties have seemed so thoroughly worth resolving. Choices that present problems are the ones philosophers of science need to understand. Philosophically interesting decision procedures must function where, in their absence, the decision might still be in doubt.

That much I have said before, if only briefly. Recently, however, I have recognized another, subtler source for the apparent plausibility of my critics' position. To present it, I shall briefly describe a hypothetical dialogue with one of them. Both of us agree that each scientist chooses between competing theories by deploying some Bayesian algorithm which permits him to compute a value for $p(T,E)$, i.e., for the probability of a theory T on the evidence E available both to him and to the other members of his professional group at a particular period of time. "Evidence," furthermore, we both interpret broadly to include such considerations as simplicity and fruitfulness. My critic asserts, however, that there is only one such value of p, that corresponding to objective choice, and he believes that all rational members of the group must arrive at it. I assert, on the other hand, for reasons previously given, that the factors he calls objective are insufficient to determine in full any algorithm at all. For the sake of the discussion I have conceded that each individual has an algorithm and that all their algorithms have much in common. Nevertheless, I continue to hold that the algorithms of individuals are all ultimately different by virtue of the subjective considerations with which each must complete the objective criteria before any computations can be done. If my hypothetical critic is liberal, he may now grant that these subjective differences do play a role in determining the hypothetical algorithm on which each individual relies during the early stages of the competition between rival theories. But he is also likely to claim that, as evidence increases with the passage of time, the algorithms of different individuals converge to the algorithm of objective choice with which his presentation began. For him the increasing unanimity of individual choices is evidence for their increasing objectivity and thus for the elimination of subjective elements from the decision process.

So much for the dialogue, which I have, of course, contrived to disclose the non sequitur underlying an apparently plausible position. What converges as the evidence changes over time need only be the values of p that individuals compute from their individual algorithms. Conceivably those algorithms themselves also become more alike with time, but the ultimate unanimity of theory choice provides no evidence whatsoever that they do so. If subjective factors are required to account for the decisions that initially divide the profession, they may still be present later when the profession agrees. Though I shall not here argue the point, consideration of the occasions on which a scientific community divides suggests that they actually do so.

My argument has so far been directed to two points. It first provided evidence that the choices scientists make between competing theories depend not only on shared criteria—those my critics call objective—but also on idiosyncratic factors dependent on individual biography and personality. The latter are, in my critics' vocabulary, subjective, and the second part of my argument has attempted to bar some likely ways of denying their philosophic import. Let me now shift to a more positive approach, returning briefly to the list of shared criteria—accuracy, simplicity, and the like—with which I began. The considerable effectiveness of such criteria does not, I now wish to suggest, depend on their being sufficiently articulated to dictate the choice of each individual who subscribes to them. Indeed, if they were articulated to that extent, a behavior mechanism fundamental to scientific advance would cease to function. What the tradition sees as eliminable imperfections in its rules of choice I take to be in part responses to the essential nature of science.

As so often, I begin with the obvious. Criteria that influence decisions without specifying what those decisions must be are familiar in many aspects of human life. Ordinarily, however, they are called, not criteria or rules, but maxims, norms, or values. Consider maxims first. The individual who invokes them when choice is urgent usually finds them frustratingly vague and often also in conflict one with another. Contrast "He who hesitates is lost" with "Look before you leap," or compare "Many hands make light work" with "Too many cooks spoil the broth." Individually maxims dictate different choices, collectively none at all. Yet no one suggests that supplying children with contradictory tags like these is irrelevant to their education. Opposing maxims alter the nature of the decision to be made, highlight the essential issues it presents, and point to those remaining aspects of the decision for which each individual must take responsibility himself. Once invoked, maxims like these alter the nature of the decision process and can thus change its outcome.

Values and norms provide even clearer examples of effective guidance in the presence of conflict and equivocation. Improving the quality of life is a value, and a car in every garage once followed from it as a norm. But quality of life has other aspects, and the old norm has become problematic. Or again, freedom of speech is a value, but so is preservation of life and property. In application, the two often conflict, so that judicial soul-searching, which still continues, has been required to prohibit such behavior as inciting to riot or shouting fire in a crowded theater. Difficulties like these are an appropriate source for frustration, but they rarely result in charges that values have no function or in calls for their abandonment. That response is barred to most of us by an acute consciousness that there are societies with other values and that these value differences result in other ways of life, other decisions about what may and what may not be done.

I am suggesting, of course, that the criteria of choice with which I began function not as rules, which determine choice, but as values, which influence it. Two men deeply committed to the same values may nevertheless, in particular situations, make different choices as, in fact, they do. But that difference in outcome ought not to suggest that the values scientists share are less than critically important either to their decisions or to the development of the enterprise in which they participate. Values like accuracy, consistency, and scope may prove ambiguous in application, both individually and collectively; they may, that is, be an insufficient basis for a *shared* algorithm of choice. But they do specify a great deal: what each scientist must consider in reaching a decision, what he may and may not consider relevant, and what he can legitimately be required to report as the basis for the choice he has made. Change the list, for example by adding social utility as a criterion, and some particular choices will be different, more like those one expects from an engineer. Subtract accuracy of fit to nature from the list, and the enterprise that results may not resemble science at all, but perhaps philosophy instead. Different creative disciplines are characterized, among other things, by different sets of shared values. If philosophy and engineering lie too close to the sciences, think of literature or the plastic arts. Milton's failure to set *Paradise Lost* in a Copernican universe does not indicate that he agreed with Ptolemy but that he had things other than science to do.

Recognizing that criteria of choice can function as values when incomplete as rules has, I think, a number of striking advantages. First, as I have already argued at length, it accounts in detail for aspects of scientific behavior which the tradition has seen as anomalous or even irrational. More important, it allows the standard criteria to function fully in the earliest stages of theory choice, the period when they are most needed but when, on the traditional view, they function badly or not at all. Copernicus was responding to them during the years required to convert heliocentric astronomy from a global conceptual scheme to mathematical machinery for predicting planetary position. Such predictions were what astronomers valued; in their absence, Copernicus would scarcely have been heard, something which had happened to the idea of a moving earth before. That his own version convinced very few is less important than his acknowledgment of the basis on which judgments would have to be reached if heliocentricism were to survive. Though idiosyncrasy must be invoked to explain why Kepler and Galileo were early converts to Copernicus's system, the gaps filled by their efforts to perfect it were specified by shared values alone.

That point has a corollary which may be more important still. Most newly suggested theories do not survive. Usually the difficulties that evoked them are accounted for by more traditional means. Even when this does not occur, much work, both theoretical and experimental, is ordinarily required before the new theory can display sufficient accuracy and scope to generate widespread conviction. In short, before the group accepts it, a new theory has been tested over time by the research of a number of men, some working within it, others within its traditional rival. Such a mode of development, however, *requires* a decision process which permits rational men to disagree, and such disagreement would be barred by the shared algorithm which philosophers have generally sought. If it were at hand, all conforming scientists would make the same decision at the same time. With standards for acceptance set too low, they would move from one attractive global viewpoint to another, never giving traditional theory an opportunity to supply equivalent attractions. With standards set higher, no one satisfying the criterion of rationality would be inclined to try out the new theory, to articulate it in ways which showed its fruitfulness or displayed its accuracy and scope. I doubt that science would survive the change. What from one viewpoint may seem the looseness and imperfection of choice criteria conceived as rules may, when the same criteria are seen as values, appear an indispensable means of spreading the risk which the introduction or support of novelty always entails.

Even those who have followed me this far will want to know how a value-based enterprise of the sort I have described can develop as a science does, repeatedly producing powerful new techniques for prediction and control. To that question, unfortunately, I have no answer at all, but that is only another way of saying that I make no claim to have solved the problem of induction. If science did progress by virtue of some shared and binding algorithm of choice, I would be equally at a loss to explain its success. The lacuna is one I feel acutely, but its presence does not differentiate my position from the tradition.

It is, after all, no accident that my list of the values guiding scientific choice is, as nearly as makes any difference, identical with the tradition's list of rules dictating choice. Given any concrete situation to which the philosopher's rules could be applied, my values would function like his rules, producing the same choice. Any justification of induction, any explanation of why the rules worked, would apply equally to my values. Now consider a situation in which choice by shared rules proves impossible, not because the rules are wrong but because they are, as rules, intrinsically incomplete. Individuals must then still choose and be guided by the rules (now values) when they do so. For that purpose, however, each must first flesh out the rules, and each will do so in a somewhat different way even though the decision dictated by the variously completed rules may prove unanimous. If I now assume, in addition, that the group is large enough so that individual differences distribute on some normal curve, then any argument that justifies the philosopher's choice by rule should be immediately adaptable to my choice by value. A group too small, or a distribution excessively skewed by external historical pressures, would, of course, prevent the argument's transfer.[8] But those are just the circumstances under which scientific progress is itself problematic. The transfer is not then to be expected.

8. If the group is small, it is more likely that random fluctuations will result in its members' sharing an atypical set of values and therefore making choices different from those that would be made by a larger and more representative group. External environment—intellectual, ideological, or economic—must systematically affect the value system of much larger groups, and the consequences can include difficulties in introducing the scientific enterprise to societies with inimical values or perhaps even the end of that enterprise within societies where it had once flourished. In this area, however, great caution is required. Changes in the environment where science is practiced can also have fruitful effects on research. Historians often resort, for example, to differences between national environments to explain why particular innovations were initiated and at first disproportionately pursued in particular countries, e.g., Darwinism in Britain, energy conservation in Germany. At present we know substantially nothing about the minimum requisites of the social milieux within which a sciencelike enterprise might flourish.

I shall be glad if these references to a normal distribution of individual differences and to the problem of induction make my position appear very close to more traditional views. With respect to theory choice, I have never thought my departures large and have been correspondingly startled by such charges as "mob psychology," quoted at the start. It is worth noting, however, that the positions are not quite identical, and for that purpose an analogy may be helpful. Many properties of liquids and gases can be accounted for on the kinetic theory by supposing that all molecules travel at the same speed. Among such properties are the regularities known as Boyle's and Charles's law. Other characteristics, most obviously evaporation, cannot be explained in so simple a way. To deal with them one must assume that molecular speeds differ, that they are distributed at random, governed by the laws of chance. What I have been suggesting here is that theory choice, too, can be explained only in part by a theory which attributes the same properties to all the scientists who must do the choosing. Essential aspects of the process generally known as verification will be understood only by recourse to the features with respect to which men may differ while still remaining scientists. The tradition takes it for granted that such features are vital to the process of discovery, which it at once and for that reason rules out of philosophical bounds. That they may have significant functions also in the philosophically central problem of justifying theory choice is what philosophers of science have to date categorically denied.

What remains to be said can be grouped in a somewhat miscellaneous epilogue. For the sake of clarity and to avoid writing a book, I have throughout this paper utilized some traditional concepts and locutions about the viability of which I have elsewhere expressed serious doubts. For those who know the work in which I have done so, I close by indicating three aspects of what I have said which would better represent my views if cast in other terms, simultaneously indicating the main directions in which such recasting should proceed. The areas I have in mind are: value invariance, subjectivity, and partial communication. If my views of scientific development are novel—a matter about which there is legitimate room for doubt—it is in areas such as these, rather than theory choice, that my main departures from tradition should be sought.

Throughout this paper I have implicitly assumed that, whatever their initial source, the criteria or values deployed in theory choice are fixed once and for all, unaffected by their participation in transitions from one theory to another. Roughly speaking, but only very roughly, I take that to be the case. If the list of relevant values is kept short (I have mentioned five, not all independent) and if their specification is left vague, then such values as accuracy, scope, and fruitfulness are permanent attributes of science. But little knowledge of history is required to suggest that both the application of these values and, more obviously, the relative weights attached to them have varied markedly with time and also with the field of application. Furthermore, many of these variations in value have been associated with particular changes in scientific theory. Though the experience of scientists provides no philosophical justification for the values they deploy (such justification would solve the problem of induction), those values are in part learned from that experience, and they evolve with it.

The whole subject needs more study (historians have usually taken scientific values, though not scientific methods, for granted), but a few remarks will illustrate the sort of variations I have in mind. Accuracy, as a value, has with time increasingly denoted quantitative or numerical agreement, sometimes at the expense of qualitative. Before early modern times, however, accuracy in that sense was a criterion only for astronomy, the science of the celestial region. Elsewhere it was neither expected nor sought. During the seventeenth century, however, the criterion of numerical agreement was extended to mechanics, during the late eighteenth and early nineteenth centuries to chemistry and such other subjects as electricity and heat, and in this century to many parts of biology. Or think of utility, an item of value not on my initial list. It too has figured significantly in scientific development, but far more strongly and steadily for chemists than for, say, mathematicians and physicists. Or consider scope. It is still an important scientific value, but important scientific advances have repeatedly been achieved at its expense, and the weight attributed to it at times of choice has diminished correspondingly.

What may seem particularly troublesome about changes like these is, of course, that they ordinarily occur in the aftermath of a theory change. One of the objections to Lavoisier's new chemistry was the roadblocks with which it confronted the achievement of what had previously been one of chemistry's traditional goals: the explanation of qualities, such as color and texture, as well as of their changes. With the acceptance of Lavoisier's theory such explanations ceased for some time to be a value for chemists; the ability to explain qualitative variation was no longer a criterion relevant to the evaluation of chemical theory. Clearly, if such value changes had occurred as rapidly or been as complete as the theory changes to which they related, then theory choice would be value choice, and neither could provide justification for the other. But, historically, value change is ordinarily a belated and largely unconscious concomitant of theory choice, and the former's magnitude is regularly smaller than the latter's. For the functions I have here ascribed to values, such relative stability provides a sufficient basis. The existence of a feedback loop through which theory change affects the values which led to that change does not make the decision process circular in any damaging sense.

About a second respect in which my resort to tradition may be misleading, I must be far more tentative. It demands the skills of an ordinary language philosopher, which I do not possess. Still, no very acute ear for language is required to generate discomfort with the ways in which the terms "objectivity" and, more especially, "subjectivity" have functioned in this paper. Let me briefly suggest the respects in which I believe language has gone astray. "Subjective" is a term with several established uses: in one of these it is opposed to "objective," in another to "judgmental." When my critics describe the idiosyncratic features to which I appeal as subjective, they resort, erroneously I think, to the second of these senses. When they complain that I deprive science of objectivity, they conflate that second sense of subjective with the first.

A standard application of the term "subjective" is to matters of taste, and my critics appear to suppose that that is what I have made of theory choice. But they are missing a distinction standard since Kant when they do so. Like sensation reports, which are also subjective in the sense now at issue, matters of taste are undiscussable. Suppose that, leaving a movie theater with a friend after seeing a western, I exclaim: "How I liked that terrible potboiler!" My friend, if he disliked the film, may tell me I have low tastes, a matter about which, in these circumstances, I would readily agree. But, short of saying that I lied, he cannot disagree with my report that I liked the film or try to persuade me that what I said about my reaction was wrong. What is discussable in my remark is not my characterization of my internal state, my exemplification of taste, but rather my *judgment* that the film was a potboiler. Should my friend disagree on that point, we may argue most of the night, each comparing the film with good or great ones we have seen, each revealing, implicitly or explicitly, something about how he *judges* cinematic merit, about his aesthetic. Though one of us may, before retiring, have persuaded the other, he need not have done so to demonstrate that our difference is one of judgment, not taste.

Evaluations or choices of theory have, I think, exactly this character. Not that scientists never say merely, I like such and such a theory, or I do not. After 1926 Einstein said little more than that about his opposition to the quantum theory. But scientists may always be asked to explain their choices, to exhibit the bases for their judgments. Such judgments are eminently discussable, and the man who refuses to discuss his own cannot expect to be taken seriously. Though there are, very occasionally, leaders of scientific taste, their existence tends to prove the rule. Einstein was one of the few, and his increasing isolation from the scientific community in later life shows how very limited a role taste alone can play in theory choice. Bohr, unlike Einstein, did discuss the bases for his judgment, and he carried the day. If my critics introduce the term "subjective" in a sense that opposes it to judgmental—thus suggesting that I make theory choice undiscussable, a matter of taste—they have seriously mistaken my position.

Turn now to the sense in which "subjectivity" is opposed to "objectivity," and note first that it raises issues quite separate from those just discussed. Whether my taste is low or refined, my report that I liked the film is objective unless I have lied. To my judgment that the film was a potboiler, however, the objective-subjective distinction does not apply at all, at least not obviously and directly. When my critics say I deprive theory choice of objectivity, they must, therefore, have recourse to some very different sense of subjective, presumably the one in which bias and personal likes or dislikes function instead of, or in the face of, the actual facts. But that sense of subjective does not fit the process I have been describing any better than the first. Where factors dependent on individual biography or personality must be introduced to make values applicable, no standards of factuality or actuality are being set aside. Conceivably my discussion of theory choice indicates some limitations of objectivity, but not by isolating elements properly called subjective. Nor am I even quite content with the notion that what I have been displaying are limitations. Objectivity ought to be analyzable in terms of criteria like accuracy and consistency. If these criteria do not supply all the guidance that we have customarily expected of them, then it may be the meaning rather than the limits of objectivity that my argument shows.

Turn, in conclusion, to a third respect, or set of respects, in which this paper needs to be recast. I have assumed throughout that the discussions surrounding theory choice are unproblematic, that the facts appealed to in such discussions are independent of theory, and that the discussions' outcome is appropriately called a choice. Elsewhere I have challenged all three of these assumptions, arguing that communication between proponents of different theories is inevitably partial, that what each takes to be facts depends in part on the theory he espouses, and that an individual's transfer of allegiance from theory to theory is often better described as conversion than as choice. Though all these theses are problematic as well as controversial, my commitment to them is undiminished. I shall not now defend them, but must at least attempt to indicate how what I have said here can be adjusted to conform with these more central aspects of my view of scientific development.

For that purpose I resort to an analogy I have developed in other places. Proponents of different theories are, I have claimed, like native speakers of different languages. Communication between them goes on by translation, and it raises all translation's familiar difficulties. That analogy is, of course, incomplete, for the vocabulary of the two theories may be identical, and most words function in the same ways in both. But some words in the basic as well as in the theoretical vocabularies of the two theories—words like "star" and "planet," "mixture" and "compound," or "force" and "matter"—do function differently. Those differences are unexpected and will be discovered and localized, if at all, only by repeated experience of communication breakdown. Without pursuing the matter further, I simply assert the existence of significant limits to what the proponents of different theories can communicate to one another. The same limits make it difficult or, more likely, impossible for an individual to hold both theories in mind together and compare them point by point with each other and with nature.

That sort of comparison is, however, the process on which the appropriateness of any word like "choice" depends.

Nevertheless, despite the incompleteness of their communication, proponents of different theories can exhibit to each other, not always easily, the concrete technical results achievable by those who practice within each theory. Little or no translation is required to apply at least some value criteria to those results. (Accuracy and fruitfulness are most immediately applicable, perhaps followed by scope. Consistency and simplicity are far more problematic.) However incomprehensible the new theory may be to the proponents of tradition, the exhibit of impressive concrete results will persuade at least a few of them that they must discover how such results are achieved. For that purpose they must learn to translate, perhaps by treating already published papers as a Rosetta stone or, often more effective, by visiting the innovator, talking with him, watching him and his students at work. Those exposures may not result in the adoption of the theory; some advocates of the tradition may return home and attempt to adjust the old theory to produce equivalent results. But others, if the new theory is to survive, will find that at some point in the language-learning process they have ceased to translate and begun instead to speak the language like a native. No process quite like choice has occurred, but they are practicing the new theory nonetheless. Furthermore, the factors that have led them to risk the conversion they have undergone are just the ones this paper has underscored in discussing a somewhat different process, one which, following the philosophical tradition, it has labelled theory choice.

14 Comment on the Relations of Science and Art

Reprinted by permission from *Comparative Studies in Society and History* 11 (1969): 403–12.

For reasons which will appear, the problem of the avant-garde, as presented by Professors Ackerman and Kubler, has caught my interest in unexpected and, I hope, fruitful ways. Nevertheless, both on grounds of competence and because of the nature of my assignment, my present remarks are directed primarily to Professor Hafner's rapprochement of science and art. As a former physicist now mainly engaged with the history of that science, I remember well my own discovery of the close and persistent parallels between the two enterprises I had been taught to regard as polar. A belated product of that discovery is the book on Scientific Revolutions to which my fellow contributors have referred. Discussing either developmental patterns or the nature of creative innovation in the sciences, it treats such topics as the role of competing schools and of incommensurable traditions, of changing standards of value, and of altered modes of perception. Topics like these have long been basic for the art historian but are minimally represented in writings on the history of science. Not surprisingly, therefore, the book that makes them central to science is also concerned to deny, at least by strong implication, that art can readily be distinguished from science by application of the classic dichotomies between, for example, the world of value and the world of fact, the subjective and the objective, or the intuitive and the inductive. Gombrich's work, which tends in many of the same directions, has been a source of great encouragement to me, and so is Hafner's essay. Under these circumstances, I must concur in its major conclusion: "The more carefully we try to distinguish artist from scientist, the more difficult our task becomes." Certainly that statement describes my own experience.

Unlike Hafner, however, I find the experience disquieting and the conclusion unwelcome. Surely it is only when we take particular care, deploying our subtlest analytic apparatus, that the distinction between artist and scientist or between their products seems to evade us. The casual observer, however well educated, has no such difficulties except when, as in some of Hafner's examples, carefully selected objects are removed from their normal context and placed in one which systematically misleads. If *careful* analysis makes art and science seem so implausibly alike, that may be due less to their intrinsic similarity than to the failure of the tools we use for close scrutiny. Lacking space to reiterate arguments developed at length elsewhere, I shall simply assert my conviction that the problem of discrimination is at present very real, that the fault is with our tools, and that an alternate set is urgently needed. Close analysis must again be enabled to display the obvious: that science and art are very different enterprises or at least have become so during the last century and a half. About how that is to be achieved I remain unclear (the closing chapter of the book mentioned above illustrates the difficulties), but Hafner's paper has provided some long-sought clues. His parallels between science and art are drawn principally from three areas: the products of the scientist and artist, the activities from which these products result, and, finally, the response of the public to them. I shall comment on all three, though not in quite systematic order, hoping to find points of entry to the still elusive problem of discrimination, a problem which he and I share but toward which our attitudes are very different.

With respect to the parallelism of products, one difficulty has already been noted. The examples of scientific and artistic work juxtaposed in Hafner's fascinating examples are drawn from a very restricted range of the available material. Virtually all the scientific illustrations he refers to, for example, are photomicrographs of organic and inorganic substances. That such striking parallels can be exhibited at all does, of course, raise important problems of influence to which neither he nor I is prepared to speak. But enterprises need not be similar in order to influence one another; the case for intrinsic similarity would profit from a less systematically selected group of examples.

A more illuminating difficulty arises from the artificial context in which the parallel illustrations are displayed. Both are shown as works of art against the same ground, a fact which considerably obscures the difference between the senses in which they can be labelled "products" of their respective enterprises. However atypical and however imperfect, the paintings are end-products of artistic activity. They are the sorts of object which the painter aims to produce, and his reputation is a function of their appeal. The scientific illustrations, on the other hand, are at best by-products of scientific activity. Usually they are made and sometimes they are analyzed by technicians rather than by the scientist for whose research they provide data. Once the research result is published, the original pictures may even be destroyed. In Hafner's striking parallels, an end-product of art is juxtaposed with a tool of science. During the latter's transition from laboratory to exhibition, ends and means have been transposed.

A closely related difficulty appears if one examines the apparently parallel use of mathematical concepts and standards in art and science. Undoubtedly, as Hafner emphasizes, considerations of symmetry, of simplicity and elegance in symbolic expression, and of other forms of the mathematical aesthetic play important roles in both disciplines. But in the arts, the aesthetic is itself the goal of the work. In the sciences it is, at best, again a tool: a criterion of choice between theories which are in other respects comparable, or a guide to the imagination seeking a key to the solution of an intractable technical puzzle. Only if it unlocks the puzzle, only if the scientist's aesthetic turns out to coincide with nature's, does it play a role in the development of science. In the sciences the aesthetic is seldom an end in itself and never the primary one.

One example may underscore the point. It is sometimes suggested that ancient and medieval astronomers were bound by the aesthetic perfection of the circle and that the new spatial perceptions of the Renaissance were therefore required before the ellipse could play a role in science. The point cannot be altogether wrong. But no change of aesthetic could have made the ellipse significant to astronomy before the late sixteenth century. Whatever its beauty, the figure had no use in astronomical theories based on a central earth. Only after Copernicus had placed the sun at the center could the ellipse contribute to solving an astronomical problem, and Kepler, who so used it, was among the very first of the mathematically proficient converts to Copernicanism. There was no lag between the possibility and its realization. Undoubtedly Kepler's Pythagorean vision of the mathematical harmonies in nature was instrumental in his discovery that elliptical orbits fit nature. But it was only instrumental: the right tool at the right time for the resolution of a pressing technical puzzle, the description of the observed motion of Mars.

People like Hafner and me, to whom the similarities of science and art came as a revelation, have been concerned to stress that the artist, too, like the scientist, faces persistent technical problems which must be resolved in the pursuit of his craft. Even more we emphasize that the scientist, like the artist, is guided by aesthetic considerations and governed by established modes of perception. Those parallels still need to be both underlined and developed. We have only begun to discover the benefits of seeing science and art as one. But an exclusive emphasis upon these parallels obscures a vital difference. Whatever the term "aesthetic" may mean, the artist's goal is the production of aesthetic objects; technical puzzles are what he must resolve in order to produce such objects. For the scientist, on the other hand, the solved technical puzzle is the goal, and the aesthetic is a tool for its attainment. Whether in the realm of products or of activities, what are ends for the artist are means for the scientist, and vice versa. That transposition, furthermore, may point to another of even greater importance—between the public and private, the explicit and the inarticulate components of vocational identity. Members of a scientific community share, both in their own eyes and in the public's, a set of problem solutions, but their aesthetic responses and research styles, often painfully eliminated from their published work, are to a considerable degree private and varied. For the arts I am not competent to generalize, but is there not a sense in which the members of an artistic school share and are identified rather with a style and aesthetic, one which is prior to shared problem solutions as a determinant of the cohesion of their group?

Look next at another of Hafner's parallels, public reaction. Widespread public estrangement is a characteristic contemporary response to both science and art. Often the reaction is expressed in similar terms. But there are revealing differences as well. Those who today spurn the science of their age do not suggest that their five-year-old child could do as well. Nor do they proclaim that what today results from the activity most admired by scientists is not really science at all but rather fraud. For the sciences it is hard to imagine a clear equivalent of the cartoon with which Hafner's essay begins. These differences can be phrased more generally. Public rejection of science, derived in part simply from anxiety, is ordinarily a rejection of the enterprise as a whole: "I don't like science." Public rejection of art, on the other hand, is a rejection of one movement in favor of another: "Modern art is not really art at all; give me pictures with subjects I can recognize."

These divergences of response point to a more fundamental difference in the public's relation to art and to science. Both enterprises depend ultimately upon the public for support. Directly or through selected institutions, the public is a consumer both of art and of the technological products of science. But only for art, not for science, is there a public audience. Even the *Scientific American* is, I believe, read predominantly by scientists and engineers. Scientists compose the audience for science, and, for the man in a particular specialty, the relevant audience is even smaller, consisting entirely of that specialty's other practitioners. Only they look critically at his work, and only their judgment affects the further development of his career. Scientists who attempt to find a wider audience for professional work are condemned by their peers. Artists, of course, also judge each other's work. Often, as Ackerman points out, a small group of fellow practitioners provides the innovator's only support against the assembled condemnation of the entire public and most fellow artists. But many people scrutinize the innovator's work, and his career depends on that scrutiny as well as on the response of critics, galleries, and museums, none of which has any parallel in the life of science. Whether the artist values or rejects such institutions, he is vitally affected by their existence as the very vehemence of his rejection sometimes attests. Art is an intrinsically other-directed enterprise in ways and to an extent which science is not.

These divergences, both in audience and in the identity of ends and means, have to this point been educed merely as isolated symptoms of a more central and consequential constellation of differences between science and art. Ultimately it should be possible to identify these deeper divergences and show that the symptoms follow directly from them. Currently I am unprepared to attempt anything of the sort, partly because I know too little of art as activity. But I can suggest how the symptoms so far examined are interrelated and how they tie to still other symptoms of difference. Seeing them as parts of a pattern may enable us to glimpse what a future treatment of our problem should articulate and make explicit.

For this purpose recall a difference between scientists and artists to which both Ackerman and I have already referred, their sharply divergent responses to their discipline's past. Though contemporaries address them with an altered sensibility, the past products of artistic activity are still vital parts of the artistic scene. Picasso's success has not relegated Rembrandt's paintings to the storage vaults of art museums. Masterpieces from the near and distant past still play a vital role in the formation of public taste and in the initiation of many artists to their craft. This role is, furthermore, strangely unaffected by the fact that neither the artist nor his audience would accept these same masterpieces as legitimate products of contemporary activity. In no area is the contrast between art and science clearer. Science textbooks are studded with the names and sometimes with portraits of old heroes, but only historians read old scientific works. In science new breakthrough do initiate the removal of suddenly outdated books and journals from their active position in a science library to the desuetude of a general depository. Few scientists are ever seen in science museums, of which the function is, in any case, to memorialize or recruit, not to inculcate craftsmanship or enlighten public taste. Unlike art, science destroys its past.

As Ackerman emphasizes, however, it has ordinarily been through the products of past traditions, not through contemporary innovation, that the tenuous communion between artists and their public audience is mediated. That is the function of museums and similar institutions which, as institutions will, generally lag innovation by a generation or more. Ackerman even suggests that the elimination of this lag—the acceptance of innovation for its own sake prior to its assay by other artists—is subversive of the artistic enterprise itself. On this view, which I find both plausible and appealing, the development of art has been shaped in some essential respects by the existence of an audience whose members do not create art and whose tastes were formed by institutions resistant to innovation. One reason, I suggest, why there is no such public audience for science (and why it proves so difficult to create one) is that mediating institutions like the museum have no function in the professional life of the scientist. The products through which he might maintain communion with the public, though sometimes only a generation old, are, for him, dead and gone.

There is a second aspect to the problem of audience, but another part of the pattern of symptom relations must be examined first. Why is the museum essential to the artist, functionless for the scientist? The answer, I think, relates to the previously discussed difference in their goals, but I lack one vital ingredient of the argument. What I need to know and have so far been unable to discover is what the artist says to himself as he admires an old masterpiece for its aesthetic achievement, simultaneously recognizing that to paint in the same way himself would violate basic tenets of an artist's credo. I can only recognize and value, but not internalize or understand, an attitude which accepts the works of, say, Rembrandt as living art but rejects as forgeries works that can at this time be readily distinguished from Rembrandt's (or his school's) only by scientific test. (The transfer of the word "forgery" to this context is interesting because slightly strained.) In the sciences there is no such problem, and forgery, excepting the literal sort, is correspondingly unimaginable. Asked why his work is like that of, say, Einstein and Schrödinger rather than Galileo and Newton, the scientist replies that Galileo and Newton, whatever their genius, were wrong, made a mistake. My problem, then, is to know what takes the place of "right" and "wrong," "correct" and "incorrect," in an ideology which declares a tradition dead but its products living. Resolving that question seems to me prerequisite to a deeper understanding of the difference between art and science. Recognizing its existence, however, may permit some progress.

Like most puzzles, those that scientists aim to solve are seen as having only one solution or one best solution. Finding it is the scientist's goal; once it is found all earlier attempts lose their felt relevance to research. For the scientist they become excess baggage, a needless burden which must be set aside in the interests of his discipline. With them into discard go most traces of the private and idiosyncratic factors, the merely historical and aesthetic, which led the discoverer to his solution. (Compare the place of honor given artists' preliminary sketches with the fate of the equivalent drafts by scientists. The former guide the viewer to a fuller appreciation; the latter, when compared with subsequent more finished versions, illuminate only their author's intellectual biography, not the solution of his puzzle.) That is why neither out-of-date theories nor even the original formulations of current theory are of much concern to practitioners. Put differently, it is why science, as a puzzle-solving enterprise, has no place for museums. The artist, of course, also has puzzles to solve, whether of perspective, coloration, brush technique, or framing edge. Their solution is not, however, the aim of his work but rather a means to its attainment. His goal, which I have already confessed my inability adequately to characterize, is the aesthetic object, a more global product to which the law of the excluded middle does not apply. Having seen Matisse's *Odalisque*, one may regard Ingres' with new eyes but one does not stop looking. Both can therefore be museum pieces as two solutions to a scientist's puzzle cannot.

The different position of puzzle solutions in the ends-means spectrum also provides a second, perhaps more fundamental, solution to the problem of a public audience for art and for science. Both disciplines present puzzles to their practitioners, and in both cases the solutions to these puzzles are technical and esoteric. As such they are of intense interest to other practitioners, artists and scientists, respectively, but of almost no concern to a general audience. Members of that larger group cannot usually recognize for themselves either a puzzle or a solution whether in art or in science. What interests them is rather the more global products of the enterprises, works of art, on the one hand, and theories of nature, on the other. But unlike works of art for the artist, theories for the scientist are principally tools. He is trained, as I have argued at length elsewhere, to take them for granted and to use them, not to change or to produce them. Except in very special cases, which do in fact evoke public response, what would most interest the public in science is for the scientist a decidedly secondary concern.

The value placed on past products, the identity of ends and means, and the existence of a public audience can thus all be seen as parts of a single pattern of related differences between art and science. Probably that pattern would emerge more clearly from an analysis which penetrated to greater depth, but I have as yet little notion of the concepts best deployed to that end. What I can do, however, as a preface to a few concluding remarks, is extend the pattern to embrace some additional symptoms of difference, in this case symptoms drawn from an examination of the ways in which art and science develop in time. Elsewhere, as Ackerman points out, I have been concerned to emphasize the similarity of the evolutionary lines of the two disciplines. In both the historian can discover periods during which practice conforms to a tradition based upon one or another stable constellation of values, techniques, and models. In both he is also able to isolate periods of relatively rapid change in which one tradition and one set of values and models gives way to another. That much, however, can probably be said about the development of any human enterprise. With respect to gross developmental pattern my originality, if any, was only the insistence that what has long been recognized about the development of, say, the arts or philosophy applies to science as well. Recognizing that fundamental resemblance can therefore be no more than a first step. Having made it, one must also be prepared to discover a number of revealing differences in developmental fine structure. Several of them prove quite easy to find.

For example, just because the success of one artistic tradition does not render another wrong or mistaken, art can support, far more readily than science, a number of simultaneous incompatible traditions or schools. For the same reason, when traditions do change, the accompanying controversies are usually resolved far more rapidly in science than in art. In the latter, Ackerman suggests, controversy over innovation is not usually settled until some new school arises to draw the fire of irate critics; even then, I presume, the end of controversy often means only the acceptance of the new tradition not the end of the old. In the sciences, on the other hand, victory or defeat is not so long postponed, and the side which loses is then banished. Its remaining adherents, if any, are considered to have left the field. Or again, though resistance to innovation is a characteristic common to both art and science, posthumous recognition recurs with regularity only in the arts. Most scientists whose contributions are ever recognized at all live long enough to experience the rewards of their achievements. In the exceptional cases, like that of Mendel, the contribution for which the scientist receives belated recognition is one that had to be independently rediscovered by others. Mendel's case is typical of posthumous recognition for scientific achievement in that his brilliant papers had no effect on the subsequent development of his field. The parallel to art fails, because, from Mendel's death to the rediscovery of his work, there was no Mendelian school, one which worked in isolation for a time but was at last embraced within the main scientific tradition.

These differences are drawn from the group behavior of artists and scientists, but they may also show in the development of individual careers. Artists can and sometimes do voluntarily undertake dramatic changes in style on one or more occasions during their lives. Or again, most artists begin by painting in the style of their masters, only later discovering the idiom for which they are ultimately known. Similar changes occur, though far more rarely, in the career of an individual scientist, but they are not voluntary. (The exception, itself illuminating, is provided by men who abandon one scientific field entirely in favor of another, e.g., change from physics to biology.) Instead, they are forced upon him either by acute internal difficulties within the tradition he had at first embraced or by the particular success within his special field of an innovation produced by someone else. And even when they are reluctantly undertaken, for to change style within a scientific field is to confess that one's earlier products and that of one's masters are wrong.

A perceptive remark of Ackerman's points the way, I think, to the center of this constellation of developmental differences. In the evolution of art, he suggests, there is nothing quite like the internal crises which a scientific tradition encounters when the puzzles it aims to solve cease to respond as they should. I agree and would add only that some such difference is inevitable between an enterprise which aims at puzzle solving and one which does not. (Note that, with respect to many of the differences under discussion, the development of mathematics resembles that of art more closely than of science, and that crises in mathematics are correspondingly rare. Few mathematical puzzles are recognized before the moment of their solution. In any case, failure to solve such a puzzle, unless it lies at the foundation of mathematics, never casts doubt on the presuppositions of the field, but only on the skill of its practitioners. In the sciences, on the other hand, any puzzle raises foundation problems if it strenuously resists solutions.) Ackerman's observation ought to be true, and, when seen as part of a pattern, it proves to be extremely consequential.

The function of crisis in the sciences is to signal the need for innovation, to direct the attention of scientists toward the area from which fruitful innovation may arise, and to evoke clues to the nature of that innovation. Just because the discipline possesses this built-in signal system, innovation itself need not be a prime value for scientists, and innovation for its own sake can be condemned. Science has its elite and may have its rear guard, its producers of Kitsch. But there is no scientific avant-garde, and the existence of one would threaten science. In scientific development, innovation must remain a response, often reluctant, to concrete challenges posed by concrete puzzles. Ackerman suggests that, to the arts also, the contemporary response to the avant-garde poses a threat, and he may be right. But that must not disguise the historic function which the existence of an avant-garde makes manifest. Both individually and in groups, artists do seek new things to express and new ways to express them. They do make innovation a primary value, and they had begun to do so before the avant-garde gave that value an institutional expression. Since the Renaissance at least, this innovative component of the artist's ideology (it is not the only component nor easily compatible with all the others) has done for the development of art some part of what internal crises have done to promote revolution in science. To say with pride, as both artists and scientists do, that science is cumulative, art not, is to mistake the developmental pattern in both fields. Nevertheless, that often repeated generalization does express what may be the deepest of the differences we have been examining: the radically different value placed upon innovation for innovation's sake by scientists and artists.

I shall conclude by pleading personal or professional privilege, changing my topic abruptly, and commenting very briefly on Kubler's remarks about Ackerman's use of my book on Scientific Revolutions. The fault is surely mine, for the points to which Kubler refers are among the most obscure in the book, but it seems nonetheless worth pointing out that he mistakes both my views and their possible bearing on the problems under discussion. In the first place, I have never intended to limit the notions of paradigm and revolution "to major theories." On the contrary, I take the special importance of those concepts to be that they permit a fuller understanding of the oddly noncumulative character of events like the discovery of oxygen, of X rays, or of the planet Uranus. More important, paradigms are not to be entirely equated with theories. Most fundamentally, they are accepted concrete examples of scientific achievement, actual problem solutions which scientists study with care and upon which they model their own work. If the notion of paradigm can be useful to the art historian, it will be pictures not styles that serve as paradigms. That way of drawing the parallel could prove important, for I discover that the problems which drove me from talk of theories to talk of paradigms are very nearly identical with those which make Kubler disdain the notion of style. Both "style" and "theory" are terms used when describing a group of works which are recognizably similar. (They are "in the same style" or "applications of the same theory.") In both cases it proves difficult—I think ultimately impossible—to specify the nature of the shared elements that distinguish a given style or a given theory from another. My response to such difficulties has been to suggest that scientists can learn from paradigms or accepted models without any process like the abstraction of elements that could constitute a theory. Could something of the same sort be said of the manner in which artists learn by scrutinizing particular works of art?

Kubler makes one other, to me extremely important, generalization. "In effect," he says, "Kuhn's remarks are ethological, being addressed more to the behavior of a community than to the results they are getting." Here there is no misunderstanding. As a description, Kubler's remark catches nicely a number of my central concerns. Nevertheless, it disturbs me to find that such a description can be used, without even a discussion, to declare those concerns irrelevant to the issues currently being considered. What I have been trying to suggest, both in the book to which Kubler refers and in the preceding comments, is that many of the problems which have most vexed historians and philosophers of science and of art lose their air of paradox and become research subjects when they are viewed as ethological or sociological. That science and art are both products of human behavior is a truism, but not therefore inconsequential. The problems of both "style" and "theory" may, for example, be among the numerous prices we pay for ignoring the obvious.

Index

Compiled by Robert S. Bernstein